Audel™
Practical Electricity

Audel™
Practical Electricity

All New 5th Edition

Paul Rosenberg
Robert Middleton

Wiley Publishing, Inc.

Vice President and Executive Group Publisher: Richard Swadley
Vice President and Executive Publisher: Robert Ipsen
Vice President and Publisher: Joseph B. Wikert
Executive Editor: Carol A. Long
Editorial Manager: Kathryn A. Malm
Senior Production Manager: Fred Bernardi
Development Editors: Emilie Herman and Erica Weinstein
Production Editor: Vincent Kunkemueller
Text Design & Composition: TechBooks

Copyright © 2004 by Wiley Publishing, Inc. All rights reserved.

Published simultaneously in Canada

No part of this publication may be reproduced, stored in a retrieval system, or transmitted in any form or by any means, electronic, mechanical, photocopying, recording, scanning, or otherwise, except as permitted under Section 107 or 108 of the 1976 United States Copyright Act, without either the prior written permission of the Publisher, or authorization through payment of the appropriate per-copy fee to the Copyright Clearance Center, Inc., 222 Rosewood Drive, Danvers, MA 01923, (978) 750-8400, fax (978) 646-8600. Requests to the Publisher for permission should be addressed to the Legal Department, Wiley Publishing, Inc., 10475 Crosspoint Blvd., Indianapolis, IN 46256, (317) 572-3447, fax (317) 572-4447, E-mail: permcoordinator@wiley.com.

Limit of Liability/Disclaimer of Warranty: The publisher and the author make no representations or warranties with respect to the accuracy or completeness of the contents of this work and specifically disclaim all warranties, including without limitation warranties of fitness for a particular purpose. No warranty may be created or extended by sales or promotional materials. The advice and strategies contained herein may not be suitable for every situation. This work is sold with the understanding that the publisher is not engaged in rendering legal, accounting, or other professional services. If professional assistance is required, the services of a competent professional person should be sought. Neither the publisher nor the author shall be liable for damages arising herefrom. The fact that an organization or Web site is referred to in this work as a citation and/or a potential source of further information does not mean that the author or the publisher endorses the information the organization or Web site may provide or recommendations it may make. Further, readers should be aware that Internet Web sites listed in this work may have changed or disappeared between when this work was written and when it is read. For general information on our other products and services, please contact our Customer Care Department within the United States at (800) 762-2974, outside the United States at (317) 572-3993 or fax (317) 572-4002.

Trademarks: Wiley, the Wiley Publishing logo, and Audel are trademarks or registered trademarks of John Wiley & Sons, Inc., and/or its affiliates. All other trademarks are the property of their respective owners. Wiley Publishing, Inc., is not associated with any product or vendor mentioned in this book.

Wiley also publishes its books in a variety of electronic formats. Some content that appears in print may not be available in electronic books.

Library of Congress Cataloging-in-Publication Data:

Rosenberg, Paul.
    Practical electricity / Paul Rosenberg, Robert Middleton.—All new 5th ed.
       p. cm.
    At head of title: Audel.
    Rev. ed. of: Practical electricity / by Robert G. Middleton. c1988.
    Includes index.
    ISBN: 0-7645-4196-X (paper/website)
    1. Electric engineering. I. Middleton, Robert. Practical electricity.
    II. Title
    TK146.R539   2004
    621.319′24—dc22
                                                                            2004005525

Printed in the United States of America

10 9 8 7 6 5 4 3 2 1

Contents

Introduction		xiii
Chapter 1	**Magnetism and Electricity**	1
	Magnetic Poles	2
	Experiments with Magnets	3
	Formation of Permanent Magnets	7
	Aiding and Opposing Magnetic Fields	9
	Electromagnetism	11
	Volts, Amperes, and Ohms	17
	Electric and Magnetic Circuits	19
	Understanding Electric Circuits	25
	Kirchhoff's Voltage Law	30
	Electrical Power	30
	Quick-Check Instruments for Troubleshooting	32
	Summary	33
	Test Questions	34
Chapter 2	**Conductors and Insulators**	35
	Classes of Conductors	35
	Conducting Wire	36
	Circular Mil-Foot	37
	American Wire Gauge	39
	Stranded Wires	43
	Aluminum Wire	43
	Line Drop	44
	Temperature Coefficient of Resistance	44
	Earth (Ground) Conduction	45
	Conduction of Electricity by Air	45
	Conduction of Electricity by Liquids	50
	Insulators for Support of Wires	52
	Classes of Insulation	53
	Insulation Resistance	53
	Plastic-Insulated Sheathed Cables	55
	Summary	55
	Test Questions	56

v

vi Contents

Chapter 3	Electric Circuits	59
	Picture Diagrams and Schematic Diagrams	59
	Voltage Polarities in Series Circuits	60
	Voltage Measurements with Respect to Ground	62
	Resistance of a Battery	63
	Efficiency and Load Power	65
	Circuit Voltages in Opposition	66
	Principles of Parallel Circuits	69
	Shortcuts for Parallel Circuits	72
	Conductance Values	73
	Kirchhoff's Current Law for Parallel Circuits	78
	Practical Problems in Parallel Circuits	80
	Line Drop in Parallel Circuits	81
	Parallel Connection of Cells	84
	Summary	86
	Test Questions	87
Chapter 4	Series-Parallel Circuits	89
	Current Flow in a Series-Parallel Circuit	89
	Kirchhoff's Current Law	90
	Series-Parallel Connection of Cells	92
	Line Drop in Series-Parallel Circuits	93
	Use of a Wattmeter	94
	Circuit Reduction	94
	Power in a Series-Parallel Circuit	98
	Horsepower	99
	Three-Wire Distribution Circuit	100
	Summary	105
	Test Questions	106
Chapter 5	Electromagnetic Induction	107
	Principle of Electromagnetic Induction	107
	Laws of Electromagnetic Induction	109
	Self-Induction of a Coil	110
	Transformers	113

	Advantage of an Iron Core	119
	Choke Coils	119
	Reversal of Induced Secondary Voltage	122
	Switching Surges	122
	The Generator Principle	124
	Summary	126
	Test Questions	126
Chapter 6	**Principles of Alternating Currents**	**129**
	Frequency	129
	Instantaneous and Effective Voltages	129
	Ohm's Law in AC Circuits	133
	Power Laws in Resistive AC Circuits	136
	Combining AC Voltages	137
	Transformer Action in AC Circuits	142
	Core Loss and Core Lamination	148
	DC versus AC Resistance	150
	Summary	150
	Test Questions	151
Chapter 7	**Inductive and Capacitive AC Circuits**	**153**
	Inductive Circuit Action	153
	Power in an Inductive Circuit	157
	Resistance in an AC Circuit	159
	Impedance in AC Circuits	161
	Power in an Impedance	162
	Capacitive Reactance	163
	Capacitive Reactance and Resistance in Series	166
	Inductance, Capacitance, and Resistance in Series	172
	Inductance and Resistance in Parallel	175
	Capacitance and Resistance in Parallel	177
	Inductance, Capacitance, and Resistance in Parallel	179
	Summary	182
	Test Questions	183

Chapter 8	Electric Lighting	185
	Sources of Light	185
	Measurement of Amount of Illumination	186
	Tungsten Filaments	187
	Starting Surge of Current	188
	Aging Characteristics	190
	Lamp Bases and Bulbs	191
	Vapor-Discharge Lamps	194
	Metal Halide Lamps	201
	High-Pressure Sodium Lamps	202
	Low-Pressure Sodium Lamps	203
	Sun Lamps	204
	Ozone-Producing Lamps	205
	Flash Tubes	205
	Infrared Lamps	208
	Electroluminescent Panels	208
	Summary	209
	Test Questions	210
Chapter 9	Lighting Calculations	213
	Point-by-Point Method	213
	Lumen Method	217
	Lighting Equipment	224
	Level of Illumination	226
	Lamp Wattage	226
	Wire Capacity	227
	Load and Length of Run	229
	Panelboards	229
	Service and Feeders	229
	Watts per Square Foot	231
	Summary	232
	Test Questions	232
Chapter 10	Basic House Wiring	235
	Service Connections	236
	Residential Underground Service	240

Making Connections to the Entrance Panel	241
Circuitry	244
Individual Appliance Circuits	246
Regular Appliance Circuits	246
General-Purpose Circuits	246
New Work in Old Houses	250
Improving Circuits	251
Placing Outlets	251
Placing Switches	254
Roughing-In	254
Mounting Boxes	255
Old Work	256
Special Mounting Devices	258
Running Sheathed Cable	259
Metal Plates	261
Running Boards	261
Connecting Cable to Boxes	262
Fishing Cable in Old Work	263
Two-Person Technique	265
Bypassing One Floor of a Two-Story Home	266
Completing House Wiring: Splicing	268
Solderless Connectors	270
Installing Outlet Receptacles	271
Single-Pole Switches	273
Ceiling Box and Fixture Mounting	274
Summary	274
Test Questions	276

Chapter 11 Wiring with Armored Cable and Conduit   279

Cutting Armored Cable	279
Installing Armored Cable	280
Conduit Fittings	284
Wiring in Flexible Conduit	293

	Connectors	296
	Summary	297
	Test Questions	299
Chapter 12	**Home Circuiting, Multiple Switching, and Wiring Requirements**	**301**
	Lamp Control from Electrolier Switch	303
	Control of Lamps from More Than One Location with Three-Way and Four-Way Switches	307
	Stairway Lamp Control Lighting	307
	Color-Coding of Switch Wiring	307
	Functional Planning	310
	Planning of Equipment	311
	Lighting the Halls	312
	Lighting the Living Room	314
	Lighting the Dining Room	316
	Lighting the Kitchen	318
	Lighting the Bedroom	320
	Lighting the Bathroom	323
	Lighting the Attic	324
	Lighting the Basement	324
	Lighting the Garage	326
	Fluorescent Lights	327
	Ground Circuit Fault Interrupters	330
	Arc-Fault Circuit Interrupters	331
	Summary	331
	Test Questions	334
Chapter 13	**Electric Heating**	**337**
	General Installation Considerations	337
	Circuiting	341
	AC-Operated Relays	344
	Hot-Water Electric Heat	346
	Radiant Heater	346
	Electric Water Heaters	347
	Ground Circuit	352

	Heating Cables	354
	Summary	356
	Test Questions	357
Chapter 14	**Alarms and Intercoms**	359
	Alarm Devices	359
	Intercoms	365
	System Planning	365
	Planning Details	370
	Wireless Intercom Systems	371
	Music Distribution Systems	374
	Industrial Installations	375
	School Installations	375
	Bells and Chimes	376
	Photoelectric Control of Bell Circuits	378
	Summary	378
	Test Questions	379
Chapter 15	**Generating Stations and Substations**	381
	Generating Stations	381
	Single-Bus System	385
	Circuit Breakers	385
	Instrument Transformers	386
	Double-Bus System	387
	Paralleling Generators	397
	Distribution	398
	Substations	404
	Automatic Substations	404
	Semiautomatic Substations	408
	Summary	412
	Test Questions	413
Appendix		415
Glossary		417
Index		439

Introduction

Electricity is, almost unarguably, the most important basic technology in the world today. Almost every modern device, from cars to kitchen appliances to computers, is dependent upon it. Life, for most of us, would be almost unimaginable without electricity.

In fact, electricity cuts such a wide path through modern life that the teaching of electricity has developed into several different specialties. Typically, one learns electricity for computers, electricity for electronics, electricity for power wiring, or some other subcategory. And while this is somewhat understandable, it has often eliminated a real and basic coverage of electricity.

Practical Electricity, more than any other book available, covers electricity in broad terms. The first half of the book is not written for one particular specialty; it is written for all specialties. The fundamental forces of electricity are explained in terms that are understandable to almost everyone without eliminating anything of any importance. A computer technician can begin his or her training with the text, as could an electrician or a radio engineer.

Electrical circuits, test procedures, and electromagnetic induction are fully covered in Chapters 1 through 7.

In the second half of the book, Chapters 8 through 15 cover the most common applications of electricity, including house wiring, lighting, cables, electric heating, and generating. These apply not only to the electrical construction trade but also to anyone who lives in a wired dwelling.

One of the primary additions to this edition is a completely revised set of artwork. The many drawings contained in this text have been updated both for ease of use and for better application to modern technology and practices.

No doubt *Practical Electricity* will find broad use in technical schools, as well as for self-instruction. It should apply equally well to either. Complex mathematics is avoided, and when trigonometry must be used for certain circuit theory applications, it is carefully explained. Only a basic knowledge of arithmetic is required of the reader.

Finally, review questions are included with each chapter. This allows students to test themselves and gives instructors an extra learning tool.

Chapter 1

Magnetism and Electricity

Early experimenters dating back to the dawn of history discovered that certain hard black stones attracted small pieces of iron. Later, it was discovered that a *lodestone*, or leading stone, pointed north and south when freely suspended on a string, as shown in Figure 1-1. Lodestone is a magnetic ore that becomes magnetized if lightning happens to strike nearby. Today we use magnetized steel needles instead of lodestones in magnetic compasses. Figure 1-2 illustrates a typical pocket compass.

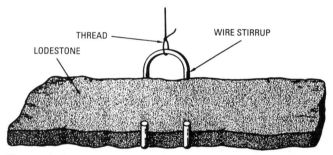

Figure 1-1 Lodestone is a magnetic ore.

Figure 1-2 A magnetic compass.

Magnetic Poles

Any magnet has a north and a south pole. We know that the earth is a huge, although weak, magnet. In Figure 1-1, the end of the lodestone that points toward the North Star is called its north-seeking pole; the opposite end of the lodestone is called its south-seeking pole. It is a basic law of magnetism that *like poles repel each other* and *unlike poles attract each other*. For example, a pair of north poles repel each other and a pair of south poles repel each other, but a north pole attracts a south pole.

Magnetic forces are invisible, but it is helpful to represent magnetic forces as imaginary lines. For example, we represent the earth's magnetism as shown in Figure 1-3. There are several important facts to be observed in this diagram. Since the north pole of a compass needle points toward the earth's geographical North Pole, we recognize that the earth's geographical North Pole has a magnetic south polarity. In other words, the north pole of a compass needle is attracted by magnetic south polarity.

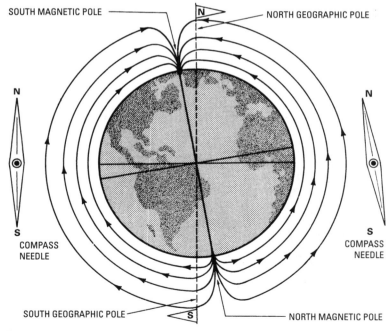

Figure 1-3 Earth's magnetic poles.

Another important fact shown in Figure 1-3 is the location of the earth's magnetic poles with respect to its geographic poles. The earth's magnetic poles are located some distance away from its geographic poles. Still another fact to be observed is that magnetic force lines have a direction, which can be indicated by arrows. Magnetic force lines are always directed out of the north pole of a magnet and directed into the south pole. Moreover, magnetic force lines are continuous; the lines always form closed paths. Thus, the earth's magnetic force lines in Figure 1-3 are continuous through the earth and around the outside of the earth.

The actual source of the earth's magnetism is still being debated by physicists. However, insofar as compass action is concerned, we may imagine that the earth contains a long lodestone along its axis. In turn, this imaginary lodestone will have its south pole near the earth's north geographic pole; the imaginary lodestone will have its north pole near the earth's south geographic pole.

Experiments with Magnets

If we bring the south pole of a magnet near the south pole of a suspended magnet, as shown in Figure 1-4, we know that the poles will repel each other. It can also be shown that magnetic attractive or repulsive forces vary inversely as the square of the distance between the poles. For example, if we double the distance between a pair of magnetic poles, the force between them will be decreased to one-fourth. It can also be shown that if the strength of the magnet in Figure 1-4 is doubled (as by holding a pair of similar magnets together with their south poles in the same direction), the force of repulsion is thereby doubled.

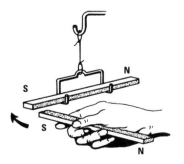

Figure 1-4 Showing repulsion between like magnetic poles.

The strength of a magnetic field is measured in *gauss* (G). For example, the strength of the earth's magnetic field is approximately

0.5 G. The gauss unit is a measurement of flux density—that is, it is a measure of the number of magnetic force lines that pass through a unit area. *One gauss is defined as one line of force per square centimeter.* In turn, one gauss is equal to 6.452 lines of force per square inch. For example, the strength of the earth's magnetic field is approximately 3.2 lines of force per square inch.

Note that there are 2.54 centimeters in 1 inch, or 0.3937 inches in 1 centimeter. Therefore, there are 6.452 square centimeters in 1 square inch, or 0.155 square inch in 1 square centimeter. Since one gauss is defined as one line of force per square centimeter, it follows that one gauss is also equal to 6.452 lines of force per square inch.

A unit of magnetic pole strength is measured in terms of force. That is, a *unit of magnetic force* is defined as one that exerts a force of one *dyne* on a similar magnetic pole at a distance of 1 centimeter. If we use a pair of like poles, this will be a repulsive force; if we use a pair of unlike poles, it will be an attractive force. There are 444,800 dynes in one pound; in other words, a dyne is equal to 1/444,800 of a pound. It is not necessary to remember these basic definitions and conversion factors. If you should need them at some future time, it is much more practical to look them up than to try to remember them.

Another important magnet experiment is shown in Figure 1-5. If we break a magnetized needle into two parts, each of the parts will become a complete magnet with north and south poles. No matter how many times we break a magnetized needle, we will not obtain a north pole by itself or a south pole by itself. This experiment leads us to another basic law of magnetism, which states that magnetic poles must always occur in opposite pairs. Many attempts have been made by scientists to find an isolated magnetic pole (called a *magnetic monopole*). All attempts to date have failed, though scientists are still trying.

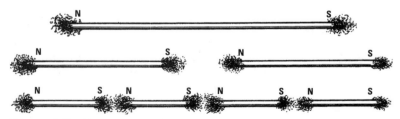

Figure 1-5 Showing the effects of breaking a magnet into several parts.

It has been found that iron and steel are the only substances that can be magnetized to any practical extent. However, certain *alloys,* such as *Alnico,* can be strongly magnetized. Substances such as hard steel and Alnico retain their magnetism after they have been magnetized and are called *permanent magnets.* Since a sewing needle is made from steel, it can be magnetized to form a permanent magnet. On the other hand, soft iron remains magnetized only as long as it is close to or in contact with a permanent magnet. The soft iron loses its magnetism as soon as it is removed from the vicinity of a permanent magnet. Therefore, soft iron is said to form a *temporary magnet.*

Permanent magnets for experimental work are commonly manufactured from hard steel or magnetic alloys in the form of *horseshoe magnets* and *bar magnets,* as shown in Figure 1-6. The space around the poles of a magnet is described as a *magnetic field* and is represented by magnetic lines of force. The space around a lodestone (Figure 1-1), around a compass needle (Figure 1-2), around the earth (Figure 1-3), and around a permanent magnet (Figure 1-4) are examples of magnetic fields. Since a magnetic field is invisible, we can demonstrate its presence only by its force of attraction for iron.

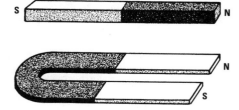

Figure 1-6 A bar magnet and a horseshoe magnet.

Consider the patterns formed by magnetic lines of force in various magnetic fields. One example has been shown in Figure 1-3. It can also be easily shown experimentally that when a bar magnet is held under a piece of cardboard and then iron filings are sprinkled on the cardboard, the filings will arrange themselves in curved-line patterns as shown in Figure 1-7. The pattern of iron filings formed provides a practical basis for our assumption of imaginary lines of force to describe a magnetic field. The total number of magnetic force lines surrounding a magnet, as shown in Figure 1-8, is called its *magnetic flux.*

A similar experiment with a horseshoe magnet is shown in Figure 1-9. The iron filings arrange themselves in curved lines that suggest the imaginary lines of force that we use to describe a magnetic field. Note that the magnetic field is strongest at the poles of the magnet in Figure 1-7. Since the field strength falls off as the square

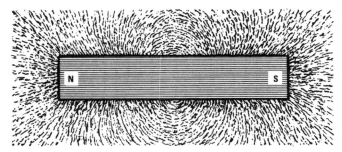

Figure 1-7 Pattern of iron filings around a bar magnet.

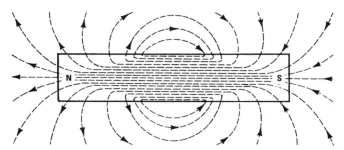

Figure 1-8 Field around a bar magnet represented by lines of force.

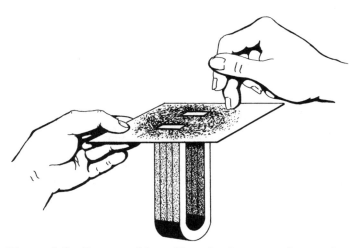

Figure 1-9 Pattern of iron filings in the space above a horseshoe magnet.

of the distance from a pole, a magnet exerts practically no force on a piece of iron at an appreciable distance. A magnet exerts its greatest force on a piece of iron when in direct contact.

Formation of Permanent Magnets

Another important and practical experiment is the magnetization of steel to form a permanent magnet. For example, if we wish to magnetize a steel needle, we may use any of the following methods:

- The needle can be stroked with one pole of a permanent magnet. The needle can be stroked several times to increase its magnetic strength, but each stroke must be made in the same direction.
- If the needle is held in a magnetic field (such as between the poles of a horseshoe magnet) and the needle is tapped sharply, it will become magnetized.
- We can heat a needle to dull red heat and then quickly cool the needle with cold water while holding it in a magnetic field, and the needle will become magnetized.

The formation of permanent magnets is explained in terms of molecular magnets. Each molecule in a steel bar is regarded as a tiny permanent magnet. As shown in Figure 1-10, the poles of these molecular magnets are distributed at random in an unmagnetized

(A) Unmagnetized.

(B) Partially magnetized.

(C) Magnetized.

Figure 1-10 Representation of molecular magnets in a steel bar.

steel bar. Therefore, the fields of the molecular magnets cancel out on the average, and the steel bar does not act as a magnet. On the other hand, when we stroke an unmagnetized steel bar with the pole of a permanent magnet, some of the molecular magnets respond by lining up end-to-end. In turn, the lined-up molecular magnets have a combined field that makes the steel bar a magnet. If the steel bar is stroked a number of times, more of the molecular magnets are lined up end-to-end, and a stronger permanent magnet is formed, as shown in Figure 1-11A.

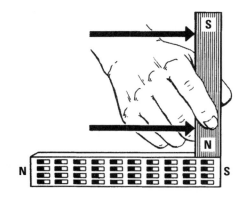

(A) Relative polarity produced in a steel bar.

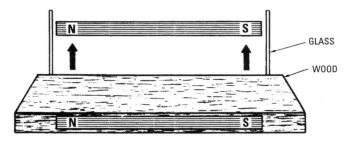

(B) One magnet floating in the field of another magnet.

Figure 1-11 Magnet characteristics.

Steel is much harder than iron; therefore, it is more difficult to line up the molecular magnets in a steel bar than in a soft-iron bar. To make a strong permanent magnet from a steel bar, we must stroke the bar many times with a strong permanent magnet. A soft-iron bar becomes fully magnetized as soon as it is touched by a permanent magnet but will return to its unmagnetized state as soon as it is removed from the field of a permanent magnet. Once the molecular magnets have been lined up in a hard steel bar, however, they will retain their positions and provide a permanent magnet.

Although there are very large numbers of molecules in an iron or steel bar, the number of molecules to be lined up is not infinite. Therefore, there is a limit to which the bar can be magnetized, no matter how strong a field we use. When all the molecular magnets are aligned in the same direction, the bar cannot be magnetized further, and the iron or steel is said to be *magnetically saturated*. The ability of a magnetic substance to retain its magnetism after the magnetizing force has been removed is called its *retentivity*. Thus, retentivity is very large in hard steel and almost absent in soft iron. Magnetic alloys such as Alnico V have a very high retentivity and are widely used in modern electrical and electronic equipment. The Alnico alloys contain iron, aluminum, nickel, copper, and cobalt in various proportions depending on the requirements.

A permanent magnet that weighs $1\frac{1}{2}$ lbs may have a strength of 900 G and will lift approximately 50 lbs of iron. This type of magnet is constructed in a horseshoe form and is less than 3 in. long. A 5-lb magnet may have strength of 2000 G and will lift approximately 100 lbs of iron. A 16-lb magnet $5\frac{1}{2}$ in. long may have a strength of 4800 G and will lift about 250 lbs of iron.

Bar magnets can be magnetized with sufficient strength so that one of the magnets will float in the field of the other magnet, as shown in Figure 1-11B. A similar demonstration of magnetic forces is provided by circular ceramic magnets. Each circular magnet is about $2\frac{1}{2}$ in. in diameter and has a hole in the center that is 1 in. in diameter. One surface of the disc is a north pole, and the opposite surface is a south pole. When placed on a nonmagnetic restraining pole, with like poles adjacent, the circular magnets float in the air, being held in suspension by repelling magnetic forces.

Aiding and Opposing Magnetic Fields

An experiment that demonstrates the repulsion of like magnetic poles was illustrated in Figure 1-4. The question is often asked how magnetic force lines act in aiding or opposing magnetic fields. Figure 1-12 shows the answer to this question. Note that when unlike poles are brought near each other, the lines of force in the air gap are in the *same* direction. Therefore, these are aiding fields, and the lines concentrate between the unlike poles. It is a basic law of magnetism that lines of force tend to shorten as much as possible; lines of force

10 Chapter 1

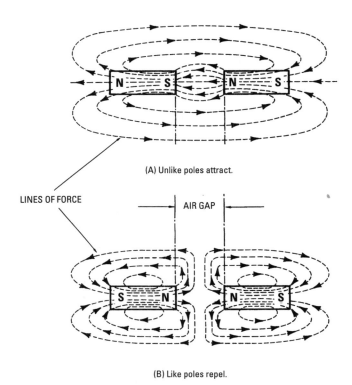

Figure 1-12 Lines of force between unlike and like poles.

have been compared to rubber bands in this respect. In turn, a force of attraction is exerted between the unlike poles in Figure 1-12A.

On the other hand, a pair of like poles have been brought near each other in Figure 1-12B. The lines of magnetic force are directed in opposition, and the lines from one pole oppose the lines from the other pole. In turn, none of the lines from one magnet enters the other magnet, and a force of repulsion is exerted between the magnets. We observe that the magnetic fields in Figure 1-12 are changed in shape, or are *distorted* with respect to the field shown in Figure 1-8. If the magnets in Figure 1-12 are brought more closely together, the fields become more distorted. Hence, we recognize that forces of attraction or repulsion between magnets are produced by distortion of their magnetic fields.

Electromagnetism

Electromagnetism is the production of magnetism by an electric current. An electric current is a flow of electrons; we can compare the flow of electrons in a wire to the flow of water in a pipe. Today we can read about electronics and electrons in newspapers, magazines, and many schoolbooks. However, the practical electrician needs to know more about electrons than a mechanic or machinist does. Therefore, let us see how electric current flows in a wire.

An atom is the smallest particle of any substance; thus, the smallest particle of copper is a copper atom. We often hear about *splitting the atom*. If a copper atom is split or broken down into smaller particles, we can almost say that it is built from extremely small particles of electricity. In other words, all substances, such as copper, iron, and wood, have the *same* building blocks, and these building blocks are particles of electricity. (This is technically not an entirely true statement, but it is close enough for our use here.) Copper and wood are different substances simply because these particles of electricity are arranged differently in their atoms. An atom can be compared to our solar system in which the planets revolve in orbits around the sun. For example, a copper atom has a *nucleus,* which consists of positive particles of electricity; electrons (negative particles of electricity) revolve in orbits around the nucleus.

Figure 1-13 shows three atoms in a metal wire. An electron in one atom can be transferred to the next atom under suitable conditions, and this movement of electrons from one end of the wire to the other end is called an electric current. Electric current is *electron flow.* To make electrons flow in a wire, an *electrical pressure* must be applied to the ends of the wire. This electrical pressure is a force called *electromotive force,* or *voltage.* For example, an ordinary dry cell is

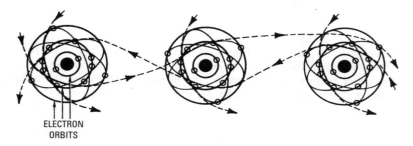

Figure 1-13 Atoms in a metal wire.

a source of electromotive force. A dry cell produces electromotive force by chemical action. If we connect a voltmeter across a dry cell, as shown in Figure 1-14, electrons flow through the voltmeter, which indicates the voltage of the cell.

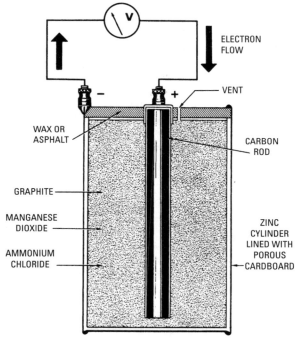

Figure 1-14 A dry cell produces electromotive force by chemical action.

We measure electromotive force (emf) in *volts* (V). If a dry cell is in good condition, it will have an emf of about 1.5 V. Note that a dry cell acts as a *charge separator.* In other words, the chemical action in the cell takes electrons away from the carbon rod and adds electrons to the zinc cylinder. Therefore, there is an electron pressure or emf at the zinc cylinder. When a voltmeter is connected across a dry cell (see Figure 1-14), this electron pressure forces electrons to flow in the connecting wire, as shown in Figure 1-13. We observe that electrons flow from the negative terminal of the dry cell, around the wire *circuit,* and back to the positive terminal of the dry cell.

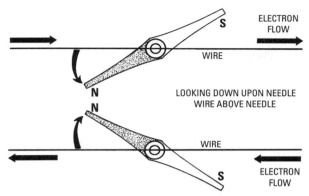

Figure 1-15 A compass needle is deflected in the vicinity of a current-carrying wire.

Next, we will find that an electric current produces a magnetic field. For example, if a compass needle is brought near a current-carrying wire, the compass needle turns, as shown in Figure 1-15. Since a compass needle is acted upon by a magnetic field, this experiment shows that the electric current is producing a magnetic field. This is the principle of electromagnetism. The magnetic lines of force surrounding a current-carrying wire can be demonstrated as shown in Figure 1-16. When iron filings are sprinkled over

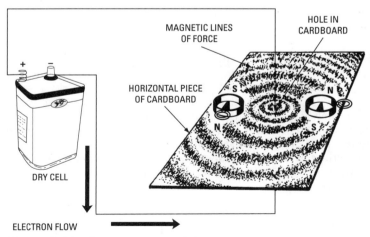

Figure 1-16 Demonstration of electromagnetism.

the cardboard, the filings arrange themselves in circles around the wire.

Although a current-carrying wire acts as a magnet, it is a form of *temporary magnet*. The magnetic field is not present until the wire is connected to the dry cell. Magnetic force lines are produced only while there is current in the wire. As soon as the circuit is opened (disconnected from the dry cell), the magnetic force lines disappear.

Let us observe the polarities of the compass needles in Figure 1-16. The magnetic lines of force are directed clockwise, looking down upon the cardboard. This experiment leads us to a basic rule of electricity called the *left-hand rule*. Figure 1-17 illustrates the left-hand rule; if a conductor is grasped with the left hand, with your thumb pointing in the direction of electron flow, then your fingers will point in the direction of the magnetic lines of force.

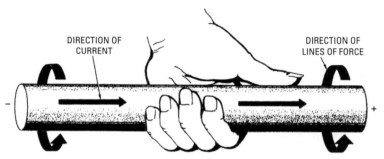

Figure 1-17 Left-hand rule used to determine direction of magnetic force lines around a current-carrying conductor.

Experiments show that the magnetic field around the wire in Figure 1-16 is weak, and we will now ask how the strength of an electromagnetic field can be increased. The magnetic field around a straight wire is comparatively weak because it is produced over a large volume of space. To reduce the space occupied by the magnetic field, a straight wire can be bent in the form of a loop, as shown in Figure 1-18. Now the magnetic flux lines are concentrated in the area enclosed by the loop. Therefore, the magnetic field strength is comparatively great inside the loop. This is an elementary form of *electromagnet*.

Next, to make an electromagnet with a much stronger magnetic field, we can wind a straight wire in the form of a helix with a number of turns, as shown in Figure 1-19. Since the field of one loop adds to the field of the next loop, the total field strength of

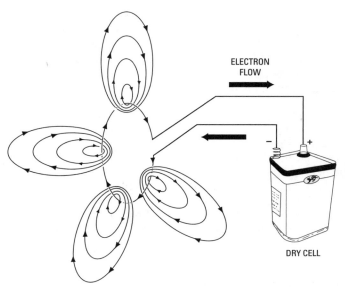

Figure 1-18 The magnetic field is concentrated by forming a conductor into a loop.

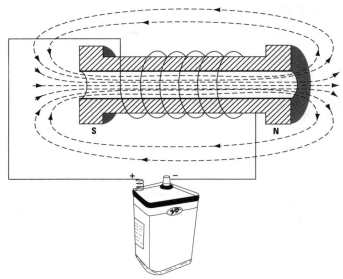

Figure 1-19 Magnetic field around an air-core solenoid.

the electromagnet is much greater than if a single turn were used. Note that if we use the same wire shown in Figure 1-16 to form the electromagnet in Figure 1-19, the current is the same in both circuits. That is, we have not changed the amount of current; we have merely concentrated the magnetic flux by winding the wire into a spiral. Electricians often call an electromagnet of this type a *solenoid*.

The name solenoid is applied to electromagnets that have an *air core*. For example, we might wind the coil in Figure 1-19 on a wooden spool. Since wood is not a magnetic substance, the electromagnet is essentially an air-core magnet. Note the polarity of the magnetic field in Figure 1-19 with respect to the direction of current flow. The left-hand rule applies to electromagnets, just as to straight wire. Thus, if we grasp an electromagnet as shown in Figure 1-20, with the fingers of the left hand in the direction of electron flow, then the thumb will point to the north pole of the electromagnet.

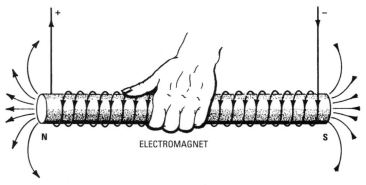

Figure 1-20 Method of finding the magnetic polarity of a coil by means of the left-hand rule.

Since iron is a magnetic substance, the strength of an electromagnet can be greatly increased by placing an iron core inside a solenoid. For example, if we place a soft-iron bar inside the wooden spool in Figure 1-19, we will find that the magnetic field strength becomes much greater. Let us see why this is so (with reference to Figure 1-10). The molecular magnets in the soft-iron bar, or *core*, are originally oriented in random directions. However, under the influence of the flux lines inside the electromagnet, these molecular magnets line up in the same direction. Therefore, the magnetic field of the molecular magnets is added to the magnetic field produced by the electric current, and the strength of the electromagnet is greatly increased.

Note that we have not changed the amount of current in the wire by placing an iron core in the solenoid. The magnetic field produced by the electric current remains unchanged. However, the electromagnet has a much greater field strength when an iron core is used because we then have two sources of magnetic field, which add up to produce the total field strength. If we open the circuit (shut off the current) of an iron-core electromagnet, both sources of magnetic field disappear. In other words, both solenoids and iron-core electromagnets are temporary magnets.

Volts, Amperes, and Ohms

To fully understand circuits such as the one shown in Figure 1-19, we must recognize another basic law of electricity, called *Ohm's law*. This is a simple law that states the relation between *voltage, current,* and *resistance*. We have become familiar with voltage, and we know that voltage (emf) is an electrical pressure. We also know that an electromotive force causes electric charges (electrons) to move through a wire. Electron flow is called an electric *current,* and a wire opposes (resists) electron flow. This opposition is called electrical *resistance*. We measure electromotive force (voltage) in *volts,* electric current in *amperes,* and resistance in *ohms*.

Figure 1-14 showed how the voltage of a dry cell is measured with a voltmeter. Current is measured with an *ammeter*. Resistance is measured with an *ohmmeter*. More electricians now use combination test instruments, such as the *multimeter* (a combination volt-ohm-milliammeter) shown in Figure 1-21. At this time, we are interested in the relation between voltage, current, and resistance. Ohm's law states that the current in a circuit is directly proportional to the applied voltage and inversely proportional to the circuit resistance. Thus, we write Ohm's law as follows:

$$\text{Electric current} = \frac{\text{electromotive force}}{\text{resistance}}$$

For example, Ohm's law states that if a wire has 1 ohm of resistance, an emf of 1 volt applied across the ends of the wire will cause 1 ampere of current to flow through the wire. Or, if we apply 2 volts across 1 ohm of resistance, 2 amperes of current will flow. Again,

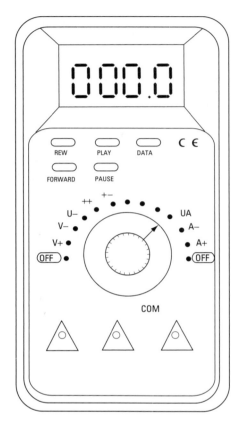

Figure 1-21 Electricians now use combination testing machines such as this digital multimeter.

if we apply 1 volt across 2 ohms of resistance, $1/2$ ampere of current will flow. We generally write Ohm's law with letters, as follows:

$$I = \frac{E}{R}$$

where I represents current in amperes,

E represents emf in volts,

R represents resistance in ohms.

Note that electrons (electric charges) flow in an electric circuit. On the other hand, voltage does not flow; resistance does not flow. Electric current is defined as the *rate* of charge flow. A current of 1 ampere consists of 6.24×10^{18} electrons (6,240,000,000,000,000,000 electrons). Therefore, a current of 1 ampere denotes the passage of 1 coulomb past a point in 1 second. Although we often speak of current flow, we really refer to charge flow, because current denotes the *rate of charge flow*. If electrons flow at the rate of 3.12×10^{18} electrons per second, the current value is $1/2$ ampere.

Electric and Magnetic Circuits

To measure the current in a circuit, we connect an ammeter into the circuit as shown in Figure 1-22. It is a basic law of electricity that the current is the same at any point in the circuit. Therefore, the ammeter indicates the amount of current that flows in *each turn* of the electromagnet. Each turn produces a certain amount of magnetism in the core. As we would expect, the amount of magnetism that is produced by each turn of wire depends on the amount of current in the wire. Therefore, we describe an electromagnet in terms of *ampere-turns*. If the ammeter in Figure 1-22 indicates a current of 1 ampere, each turn on the coil represents 1 ampere-turn.

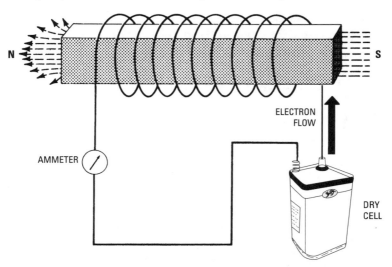

Figure 1-22 Measurement of current in the circuit.

The total number of ampere-turns on a coil is equal to the number of amperes times the number of turns. For example, with a current of 1 ampere in Figure 1-22, there will be a total of 9 ampere-turns. Next, let us suppose that we increase the current in Figure 1-22 to 2 amperes. Now 2 amperes of current flows in each turn of the coil, and we have a total of 18 ampere-turns. Since the strength of the electromagnet is proportional to the number of its ampere-turns, *we double the strength of the electromagnet by doubling the amount of current.*

Since each ampere-turn produces a certain amount of magnetism, the ampere-turn is taken as the unit of *magnetomotive force.* Magnetomotive force is measured in ampere-turns. We often compare magnetomotive force to electromotive force. In other words, electromotive force produces current in a wire, and magnetomotive force produces magnetic flux lines in a core. We will find that any core, such as air or iron, opposes the production of magnetic flux lines. This opposition is called the reluctance of the core. We often compare reluctance with resistance. In other words, production of magnetic flux lines is opposed by reluctance, and production of electric current is opposed by resistance. Therefore, we can also compare magnetic flux lines to electric current.

The foregoing comparisons lead us to the idea of a *magnetic circuit,* as shown in Figure 1-23. This diagram shows both an electric circuit and a magnetic circuit. The electric circuit consists of the 1-volt battery and the 1-ohm coil through which 1 ampere of current

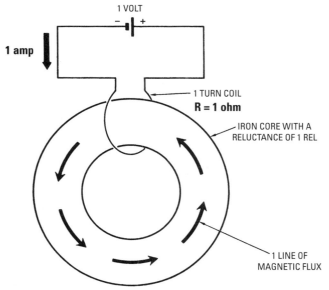

Figure 1-23 The basic magnetic circuit.

flows. The magnetic circuit in this example consists of a circular iron core. Reluctance is measured in *rels*. This core has a reluctance of 1 rel. A basic law of electromagnetism states that 1 ampere-turn produces 1 line of magnetic flux in a reluctance of 1 rel. Therefore, we write a law for magnetic circuits that is much the same as Ohm's law for electric circuits:

$$\text{magnetic flux} = \frac{\text{magnetomotive force}}{\text{reluctance}}$$

or,

$$\text{flux lines} = \frac{\text{ampere} - \text{turns}}{\text{rels}}$$

Next, suppose that we increase the current in Figure 1-23 to 2 amperes. Then, two lines of magnetic flux will be produced in the iron core. Another example is shown in Figure 1-24, where 5 amperes of current flows through 10 turns, providing a magnetomotive force of 50 ampere-turns. The number of flux lines that will be produced in Figure 1-24 depends on the reluctance of the magnetic circuit. Note that a continuous iron magnetic circuit is provided for the flux lines in Figure 1-23; the magnetic circuit in Figure 1-24 is more complicated, however, because part of the magnetic circuit is iron and the other part is air. From the previous discussion of electromagnets, we would expect air to have a much greater reluctance than iron.

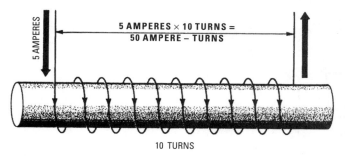

Figure 1-24 An electromagnet with 50 ampere-turns of magnetomotive force.

A volume of air that is 1 in. square and 3.19 in. long has a reluctance of 1 rel. This is 1500 times the reluctance of a typical iron core of the same size. If we place a closed iron core in a solenoid, the number of flux lines will increase 1500 times in a typical experiment.

However, the following are several facts that we must keep in mind when working with iron cores:

1. Different types of iron have different amounts of reluctance.
2. The reluctance of iron changes as the magnetomotive force is changed.
3. Since air has a much greater reluctance than iron, a magnetic circuit with a large air gap acts practically the same as an air core.

For example, cast iron has about 6 times as much reluctance as annealed sheet steel. To show the change in the reluctance of iron when the magnetomotive force is changed, we use charts such as that shown in Figure 1-25. The chart shows a magnetization curve. Note that the scales on the chart are marked off in terms of *magnetizing force* per inch and magnetic flux lines per square inch. Magnetizing force is equal to magnetomotive force per inch of core length. For example, let us consider the iron core shown in Figure 1-26. The length of this magnetic circuit is 16 in. If we apply a magnetomotive

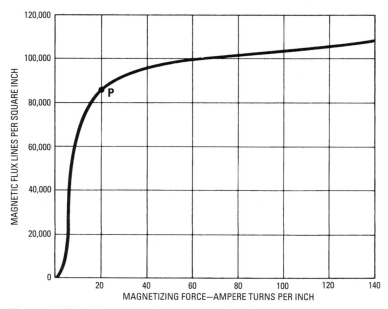

Figure 1-25 Magnetizing force versus flux density for annealed sheet steel.

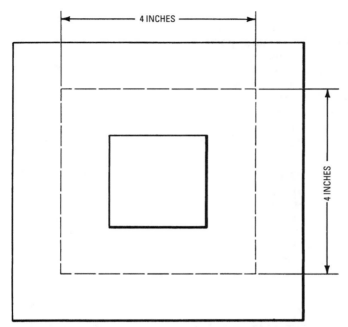

Figure 1-26 The length of this magnetic circuit is 16 in.

force of 320 ampere-turns to this core, we will have a magnetizing force of 20 ampere-turns per inch.

Let us find how many flux lines per square inch will be produced in the core (Figure 1-26) when a magnetizing force of 20 ampere-turns per inch is applied. At point P on the magnetization curve in Figure 1-25, we see that there will be about 83,000 lines per square inch produced in the core. Now, if the core of Figure 1-26 has a cross-sectional area of 1 square inch, there will be 83,000 lines of magnetic flux in the core. Or, if the core has a cross-sectional area of 2 square inches, there will be 166,000 lines of magnetic flux in the core. Again, if the core has a cross-sectional area of $1/2$ square inch, there will be 41,500 lines of magnetic flux in the core.

The number of magnetic flux lines per square inch is generally called the *flux density* in the core; flux density is represented by the letter B, and magnetizing force is represented by the letter H. Thus, a magnetization curve such as shown in Figure 1-25 is usually called a *B-H curve*. Note that the flux density in this example increases rapidly from 0 to 20 ampere-turns per inch. At higher values of magnetizing force, the B-H curve flattens off. This flattened-off portion of the curve is called the *saturation interval*. The curve will finally become horizontal, and the iron core will then have the same reluctance as air. When an iron core is completely saturated, we can remove the iron core from the electromagnet, and its magnetic field strength will remain the same.

Figure 1-27 shows some comparative B-H curves. We observe that an electromagnet with an annealed sheet-steel core will have a much stronger magnetic field than if cast iron is used for a core. Note also that if we wish to make a very strong electromagnet, it

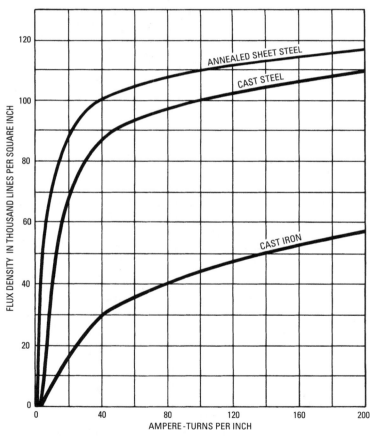

Figure 1-27 Comparative B-H curves.

is better to use a large cross-sectional area in the core instead of a large number of ampere-turns per inch. In other words, iron and steel start to saturate when more than 20 ampere-turns per inch are used, and core saturation corresponds to wasted electric current. Therefore, efficient operation requires that we use no more than 20 ampere-turns per inch and then make the cross-sectional area of the core as large as required to obtain the desired number of magnetic flux lines.

Understanding Electric Circuits

Any circuit must contain a voltage source to be of practical use. Source voltages may be very high, moderate, or very low. A dry cell is a familiar 1.5-volt source, as was shown in Figure 1-14. When a higher voltage is required, cells can be connected in *series* to form a *battery*, as shown in Figure 1-28. Note that the negative terminal of one cell is connected to the positive terminal of the next cell. This series connection causes the cell voltages to be added. Since there are four cells in the example, the battery voltage will be approximately 6 volts. New dry cells usually have an emf of slightly more than 1.5 volts. As a cell ages, its emf decreases.

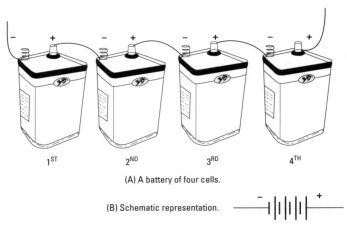

Figure 1-28 Illustrating series-connected dry cells.

Current does not flow in a circuit unless the circuit is *closed*. Figure 1-29 shows the difference between a closed circuit and an open circuit. A flashlight bulb might draw 0.25 ampere from a 1.5-volt source. Let us apply Ohm's law to find the resistance of the bulb.

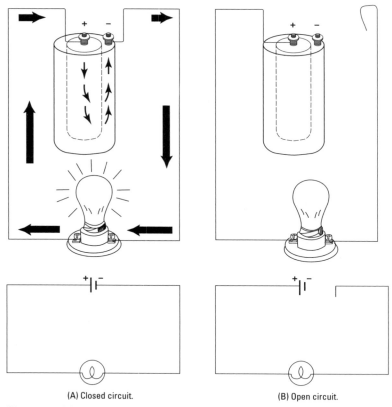

(A) Closed circuit.   (B) Open circuit.

Figure 1-29 A simple electric circuit.

It is easy to see that there are three possible arrangements of Ohm's law, as shown in Figure 1-30. Since we are interested in finding the resistance of the bulb, we will use the second arrangement of Ohm's law. Accordingly, the resistance of the bulb is 1.5/0.25, or 6 ohms. Next, let us observe how the other two arrangements of Ohm's law are used.

In case we know the applied voltage (1.5 volts) and the resistance of the bulb (6 ohms), we will use the first arrangement of Ohm's law to find the current in the circuit. Thus, the current is equal to 1.5/6, or 0.25 ampere. On the other hand, in case we know the current through the bulb (0.25 ampere) and the resistance of the bulb (6 ohms), we will use the third arrangement of Ohm's law

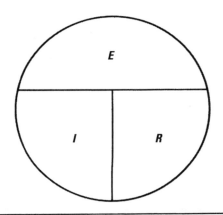

OHM'S LAW
CURRENT = ELECTROMOTIVE FORCE / RESISTANCE $I = \dfrac{E}{R}$ AMPERES = $\dfrac{\text{VOLTS}}{\text{OHMS}}$
RESISTANCE = ELECTROMOTIVE FORCE / CURRENT $R = \dfrac{E}{I}$ OHMS = $\dfrac{\text{VOLTS}}{\text{AMPERES}}$
ELECTROMOTIVE FORCE = CURRENT × RESISTANCE $E = IR$ VOLTS = AMPERES × OHMS

Figure 1-30 Ohm's law in diagram form.

to find the applied voltage. Thus, the applied voltage is equal to 0.25 × 6, or 1.5 volts. Figure 1-30 shows Ohm's law in diagram form. To find one of the quantities, first cover that quantity with a finger. The location of the other two letters in the circle will then show whether to divide or multiply. For example, to find I, we cover I and observe that E is to be divided by R. Again, to find E, we cover E and observe that I is to be multiplied by R.

A circuit can be opened by disconnecting a wire as shown in Figure 1-29B. To conveniently open and close a circuit, a switch is

28 Chapter 1

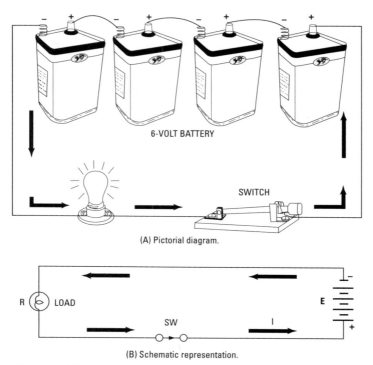

Figure 1-31 A simple electric circuit.

used as shown in Figure 1-31. This type of switch is called a knife switch. Note that the lamp is called a *load resistance,* or simply a *load.* An electrical load in a circuit changes electricity into light and heat, as in this example; in other circuits, the load may change electricity into mechanical power or some other form of power. Let us consider the action of the *fuse* shown in Figure 1-32. A fuse is a type of automatic safety switch; the fuse blows and opens the circuit in case the load (R) should become short-circuited and draw excessive current from the battery.

Fuses are made from thin strips or wire of aluminum or other metal. The resistance of a fuse is comparatively low, but because of its small cross section, the fuse heats up and melts if a certain amount of current flows through it. For example, the fuse shown in Figure 1-32A has a resistance of 1 ohm (Ω). The load R has a resistance of 29 ohms, making a total of 30 ohms of circuit resistance. Since 6 volts is applied, 0.2 ampere will flow in accordance with Ohm's

Magnetism and Electricity   29

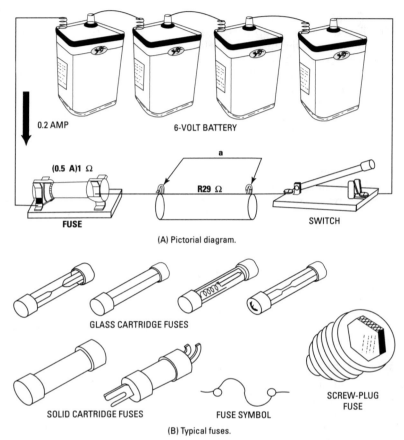

Figure 1-32   A fused circuit.

law. This fuse is made so that it will blow at a current of 0.5 ampere. Therefore, the fuse does not blow as long as R is not short-circuited.

We could short-circuit R by placing a screwdriver across the terminals. In such a case, the total circuit resistance would be reduced to 1 ohm, and 6 amperes of current would flow. Therefore, the fuse would immediately blow out and automatically open the circuit. Thereby, the battery is protected from damage due to excessive current drain. Note that after the fuse blows, the voltage between its terminals will be 6 volts. In normal circuit operation, the voltage across the load-resistor terminals (a) is 5.8 volts, in accordance with Ohm's law. The voltage across the fuse terminals is 0.2 volt, in accordance with Ohm's law. We call the voltage between the load-resistor terminals the *voltage drop* across the resistor; similarly, we call the voltage between the fuse terminals the *voltage drop* across the fuse.

When we speak of a voltage drop, we simply mean that this amount of voltage would be measured by a voltmeter connected across the terminals of resistance. It follows from the foregoing example that *Ohm's law applies to any part of a circuit, as well as to the complete circuit.* In other words, insofar as the load resistor is concerned in Figure 1-32A, 0.2 ampere is flowing through 29 ohms; therefore, the IR (current × resistance) drop across the load resistor is 5.8 volts. Insofar as the fuse is concerned, 0.2 ampere is flowing through 1 ohm; therefore, the IR drop across the fuse is 0.2 volt.

We have noted previously that both electromotive force (emf) and voltage are measured in volts. Electromotive force is often called a voltage. Nevertheless, there is a basic distinction between an emf and a voltage. For example, a dry cell is a chemical *source* of voltage, and we speak of the *emf produced by the source* of electricity. This emf is measured in volts. If a resistor is connected across a source of electricity, the current produces a voltage drop across the resistance in accordance with Ohm's law. This voltage drop is not called an emf but is called a voltage and is measured in volts. Even though the terms are normally used interchangeably, we should properly speak of the emf of a voltage source but of a voltage produced across a load resistor.

Kirchhoff's Voltage Law

The foregoing example also illustrates another basic law of electric circuits, called Kirchhoff's voltage law. This law states that *the sum of the voltage drops around a circuit is equal to the source voltage.* Note that the sum of the voltage drops across the load resistor and the fuse is equal to 6 volts (5.8 + 0.2) and that the source voltage (battery voltage) is also equal to 6 volts. We recognize that Kirchhoff's voltage law is simply a summary of Ohm's law as applied to all the resistance in a circuit. Although we do not need to use Kirchhoff's voltage law in describing the action of simple circuits, this law will be found very useful in solving complicated circuits.

Electrical Power

There are many forms of power. For example, an electric motor produces a certain amount of mechanical power, usually measured in horsepower. An electric heater produces heat (thermal) power. An electric light bulb produces both heat power and light power (usually measured in candlepower). Electrical power is measured in *watts;* electrical power is equal to volts times amperes. Thus we write

$$\text{watt} = \text{volts} \times \text{amperes}$$

or,

$$P = EI$$

With reference to Figure 1-32A, the battery supplies $6 \times 0.2 = 1.2$ watts to the circuit, the load resistor R takes $5.8 \times 0.2 = 1.16$ watts, and the fuse takes $0.2 \times 0.2 = 0.04$ watt. Note that the power taken by both resistances is equal to $1.16 + 0.04 = 1.2$ watts. In other words, the power supplied by the battery is exactly equal to the power taken by the circuit resistance. This fact leads us to another basic law called the *law of conservation of energy.* This law states that *energy cannot be created or destroyed but only changed into some other form of energy.* This is the same as saying that power can be changed only into some other form of power, because energy is equal to power multiplied by time.

In the example of Figure 1-32, electrical energy is changed into heat energy (or electrical power is changed into heat power) by the load resistor and the fuse. Since the load resistor takes 1.16 watts of electrical power, it produces 1.16 watts of heat power. With reference to Figure 1-31, light is measured in candlepower. An ordinary electric-light bulb produces approximately 1 candlepower for each watt of electrical power. For example, a 60-watt lamp normally takes 60 watts of electrical power and produces about 60 candlepower of light.

Since power is equal to IE, and $I = E/R$, we can write $P = E^2/R$. Since power is equal to IE, and $E = IR$, we can write $P = I^2R$. Thus, by substitution from Ohm's law into the basic power law and by rearranging these equations, we obtain the 12 important electrical formulas shown in Figure 1-33. In summary, these formulas state

$$I = \frac{E}{R} = \frac{P}{E} = \sqrt{\frac{P}{R}}$$

$$E = IR = \frac{P}{I} = \sqrt{PR}$$

$$R = \frac{E}{I} = \frac{P}{I} = \frac{E^2}{P}$$

$$P = IE = I^2R = \frac{E^2}{R}$$

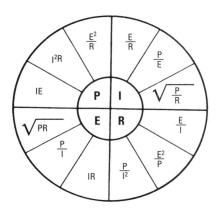

Figure 1-33 Summary of basic formulas.

These formulas are important because we may be working with a circuit in which the voltage and resistance are known, and we need to find the amount of current flow. Again, we may know the voltage and power in the circuit and need to find the amount of current flow. Or we may know the resistance and power in the circuit and need to find the amount of current flow. Similarly, we can find the voltage of a circuit if we know the current and resistance, or the current and the power, or the resistance and the power. We can find the resistance in a circuit if we know the voltage and current, or the current and power, or the voltage and power. We can find the power in a circuit if we know the voltage and current, or the current and resistance, or the voltage and resistance.

Next, let us consider the measurement of electrical energy. We measure electrical energy in watt-seconds, watt-hours, or kilowatt-hours. A watt-second is equal to the electrical energy produced by 1 watt in 1 second. In turn, a watt-hour is equal to the electrical energy produced by 1 watt during 1 hour; therefore, a watt-hour is equal to 3600 watt-seconds. A kilowatt-hour is equal to 1000 watt-hours. An instrument used to measure electrical energy is called a watt-hour meter. We are quite familiar with watt-hour meters because they are installed by public utility companies in every home, office, and shop.

Quick-Check Instruments for Troubleshooting

Electrical troubleshooting requires various tests. However, practical electricians generally make quick-checks with continuity testers and voltage indicators. Shop work often requires the use of voltmeters, ammeters, ohmmeters, wattmeters, and/or a continuity tester, which consists of a battery and some sort of signaling load (flashlight bulbs and small bells or chimes being the most common), along with a pair of test leads. Such devices are commonly used when connections are to be checked in a wiring system, or when a broken place in a wire is to be located. If the test leads are applied across a closed circuit, the bell rings. Note that if a poor connection is being checked, the bell may or may not ring, depending on the resistance of the bad connection.

A simple voltage indicator consists of a neon bulb with a built-in dropping resistor and a pair of test leads. This type of tester is usually preferred when circuits are being checked in a wiring system. If the test leads are applied across a circuit in which 70 volts or more are present, the neon bulb will glow. The bulb will glow much brighter across a 240-volt circuit than across a 120-volt circuit. However, the brightness of the glow cannot be used to estimate the voltage value accurately. A neon tester is essentially a quick-checker. However, it is a very practical tester in general troubleshooting procedures. Many of the tests formerly performed in a laborious fashion by these instruments are now superseded by multitesting instruments, such as the one shown in Figure 1-21, some of which have digital and analog displays.

Summary

A lodestone is a natural magnet consisting chiefly of a magnetic oxide of iron called magnetite. All magnets have a north and south pole. Magnetic lines of force are invisible but are always continuous, always forming a closed path. Like poles repel, unlike poles attract.

Iron and steel are the only materials that can be magnetized to any practical extent. Certain alloys, such as Alnico, can be strongly magnetized and are called permanent magnets. Permanent magnets are generally in the form of a horseshoe or bar. The space around the poles of a magnet is described as a magnetic field and is represented by magnetic lines of force. The total number of magnetic force lines surrounding a magnet is called its total magnetic flux.

Electromagnetism is the production of magnetism by an electric current. Electric current is a flow of electrons, which can be compared to the flow of water in a pipe. Electric current is electron flow; to make electrons flow, pressure must be applied to the end of the wire. This pressure applied is called the electromotive force or voltage. Electromotive force (emf) is measured in volts.

To fully understand electromagnetism or the basic laws of electricity, we must use Ohm's law. This is a law that states the relationship between voltage, current, and resistance. Voltage is the electrical pressure that causes electrons to move through a wire. Electron flow is called an electric current. Current is the movement of electrons through a conductor. A wire (or conductor) opposes the electron flow because of resistance.

Test Questions

1. What is a lodestone?
2. Do unlike magnetic poles attract or repel each other?
3. Why does a compass needle point in a north-south direction?
4. How does a permanent magnet differ from a temporary magnet?
5. Can a north pole exist without a south pole?
6. Are magnetic lines of force directed into or out of the north pole or magnet?
7. What is the definition of electromagnetism?
8. How do metals conduct electricity?
9. Is a dry cell a source of magnetism or of electricity?
10. In what way is electromotive force, or voltage, an electrical pressure?
11. Will a voltmeter measure the voltage or the current of a dry cell?
12. Why is a compass needle deflected in the vicinity of a current-carrying wire?
13. How is an electromagnet constructed?
14. Can you explain why a soft-iron core increases the strength of an electromagnet?
15. What is the name of the law that relates voltage, current, and resistance?
16. Is a magnetic circuit the same thing as an electric circuit?
17. How is an ampere-turn defined?
18. Does an ammeter measure current or voltage?
19. Can you state a law for magnetic circuits that is similar to Ohm's law?
20. Why are dry cells connected in series?
21. What is the meaning of a closed circuit? An open circuit?
22. In what way can a fuse be compared to a switch?

Chapter 2

Conductors and Insulators

A conductor is a substance that carries electric current. An insulator is a substance that does not carry electric current. Because no conductor is perfect, and because any conductor has at least a small amount of resistance, it is better to define a conductor as a substance with a very low resistance. We will also find that no insulator is perfect, and because no insulator has an infinite resistance, it is better to define an insulator as a substance with a very high resistance. Therefore, conductors, resistors, and insulators are all basically resistive substances. However, they are classified into different groups because a practical conductor has extremely low resistance, a load resistor has moderate resistance, and a good insulator has extremely high resistance.

Classes of Conductors

The substances listed in Table 2-1 have different conductivities. The best conductors are listed in the first column in the order of decreasing conductivity. For example, silver is the best conductor, lead has less conductivity, carbon has still less conductivity, moist earth is a poorer conductor than carbon, and slate has such a high resistance that it is called an insulator. Of the insulators listed, dry air is the best. A high vacuum is a better insulator than dry air; however, a vacuum can be used only in special devices such as rectifier tubes. Therefore, we will be concerned in this chapter only with

Table 2-1   Conductors and Insulators

Good Conductors	Fair Conductors	Insulators
Silver	Charcoal and coke	Slate
Copper	Carbon	Oils
Aluminum	Acid solutions	Porcelain
Zinc	Sea water	Dry paper
Brass	Saline solutions	Silk
Platinum	Metallic ores	Sealing wax
Iron	Living vegetable	Ebonite
Nickel	substances	Mica
Tin	Moist earth	Glass
Lead		Dry air

Note: *In each column, the best conductor is at the top, the best insulator at the bottom.*

35

the more common insulators used in electrical work. We will find that conductors are usually combined with insulators; for example, a conducting wire is either covered with an insulating substance or the conductor is fastened to insulating supports.

Conducting Wire

Wires used as electrical conductors are generally made of copper; however, aluminum is used to some extent. Silver is seldom used because of its high cost. Most wire is round, although square and rectangular forms are used in some applications. The basic description of a round wire is its diameter. This means the diameter of the wire itself, disregarding any insulation that might be used to cover the wire. The diameter of a wire can be measured accurately with a micrometer, such as illustrated in Figure 2-1.

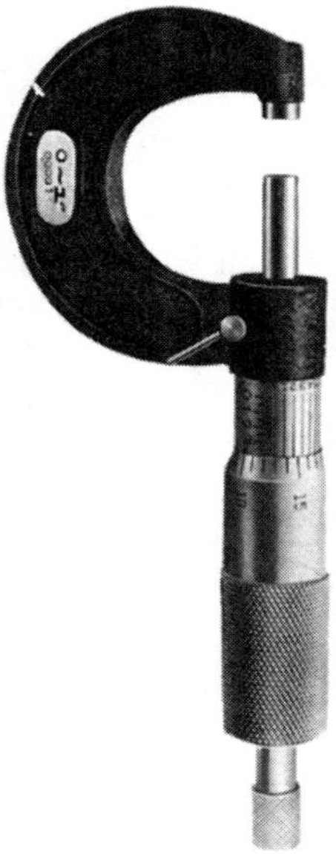

Electricians use the word *mil* to describe the diameter of a wire. A mil is 1/1000 inch. Thus, if a micrometer shows that a wire has a diameter of 0.001 inch, we say that its diameter is 1 mil. Electricians also use the term *circular mil* to describe the cross-sectional area of round wire. If a wire has a diameter of 1 mil, it is said to have a cross-sectional area of 1 circular mil. A circular mil is abbreviated CM. This is a convenient way to describe area, because circular mils are equal to mils squared. For example, if a round wire has a diameter of 2 mils, its cross-sectional area is equal to 2^2, or 4 circular mils.

Because we occasionally use square or rectangular wires, we must also have a suitable way of describing the cross-sectional area of these conductors. Electricians use the term *square mil* to describe the cross-sectional area of a square or rectangular wire. If a square wire is 1 mil on a side, it is said to have a cross-sectional area of 1 square mil. It follows that if a square wire is 2 mils on a side, it has a cross-sectional area of 4 square mils. Again, if a rectangular conductor is 3 mils wide and 2 mils thick, it has a cross-sectional area of 6 square mils. Figure 2-2 shows a comparison of the circular mil and the square mil.

Figure 2-1 A micrometer.
(Courtesy Edmund Scientific Co.).

Conductors and Insulators 37

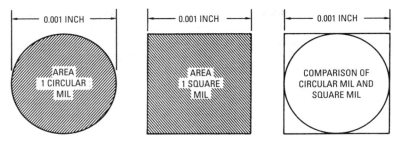

Figure 2-2 Enlarged view of circular mil and square mil.

The question may be asked why we make a comparison of the circular mil and the square mil. If a square conductor is to be replaced by a round conductor, or vice versa, we must use the same cross-sectional area to obtain the same conductivity. Therefore, electricians need to know how to change circular mils into square mils, and vice versa. Note that the number of circular mils multiplied by 0.7854 gives the number of square mils. Or, the number of square mils multiplied by 1.273 gives the number of circular mils.

Thus, we write the following formulas:

$$\text{Square mils} = \text{Circular mils} \times 0.7854$$

$$\text{Circular mils} = \text{Square mils} \times 1.273$$

Circular Mil-Foot

In practical electrical work, we describe the resistance of a wire in terms of ohms per circular mil-foot. A *mil-foot* means a circular wire 1 foot in length and 1 mil in diameter. The resistance of a mil-foot of copper wire is about 10.4 ohms. Figure 2-3 illustrates a circular mil-foot of copper wire. Note that the resistance of this wire is given for 20° centigrade. We will find that the resistance of a wire increases as its temperature increases. Therefore, the resistance of a circular mil-foot of wire is always listed in electrical handbooks at 20°C or 0°C. Of course, the wire resistance is somewhat less at 0° than at 20°.

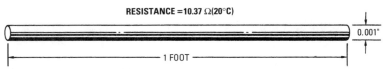

Figure 2-3 Illustrating dimensions and resistance of a circular mil-foot of copper.

38 Chapter 2

When we know the resistance of a wire per circular mil-foot, we can easily find the resistance for any cross section and length. The resistance per circular mil-foot is called rho (ρ) in electrical handbooks. (See Table 2-4.) In turn, we find the resistance of a wire of a given cross section and length by means of the formula:

$$R = \frac{\rho \times L}{CM}$$

or,

$$R = \frac{\rho \times L}{d^2}$$

where L is in feet,

　　CM is in circular mils,

　　d is in mils,

　　R is in ohms.

The foregoing formula gives the wire resistance at the temperature noted in the handbook for *rho*. We will learn how to make temperature corrections in a later part of this chapter. Figure 2-4 shows how the resistance of a wire changes with a change in its cross-sectional area.

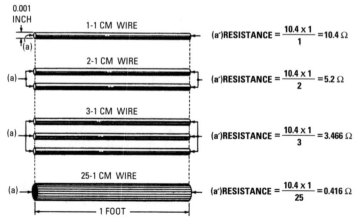

Figure 2-4 Illustrating how the resistance of a conductor decreases with an increase of the area through which the current flows.

Conductors and Insulators **39**

American Wire Gauge
Electricians usually check wire sizes with an American Standard Wire Gauge, as illustrated in Figure 2-5. The American Wire Gauge is also called a Brown & Sharpe Gauge. For brevity, the term AWG or B&S is used. Gauge numbers from 0000 to 40 are listed in Table 2-2, with corresponding diameters in mils, cross-sectional areas in circular mils and square inches, resistance in ohms per 1000 feet at 25°C and at 65°C, resistance in ohms per mile at 25°C, and pounds per 1000 feet. These wire characteristics have been determined by the National Bureau of Standards.

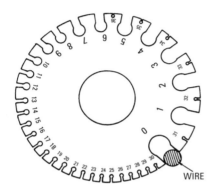

Figure 2-5 A standard wire gauge.

In selecting a wire size for a given installation, electricians are bound by the *National Electrical Code*,* which establishes the allowable current capacity for insulated wires. Table 2-3 lists the

---

*The *National Electrical Code* was originally drawn in 1897 as the result of the united efforts of various insurance, electrical, architectural, and allied interests. This original Code was prepared by the National Conference on Standard Electrical Rules, composed of delegates from various interested national associations. However, since 1911, the National Fire Protection Association has been the sponsor of the *National Electrical Code*. The *NEC* establishes the minimum standards for wiring plans and installation practices in the United States. It has been adopted by the National Board of Fire Underwriters as the NBFU Regulations and has been approved by the American Standards Association as an ASA Standard. The *NEC* is also the basis of most municipal electrical ordinances; however, local ordinances do not necessarily follow the *NEC* exactly. (Some are more stringent.) The NBFU has also established the Underwriter's Laboratories, Inc., which inspects electrical fittings, materials, and appliances to establish compliance with standard test specifications. Electricians may obtain the *National Electrical Code* and the List of Inspected Electrical Appliances from a fire underwriter's office or from the National Fire Protection Association, Inc., Batterymarch Park, Quincy, MA 02269, or from the Underwriters' Laboratories, Inc., 333 Pfingsten, Rd., Northbrook, IL 60062. Also see *Audel's Guide to the 2002 National Electrical Code.*

Table 2-2 Standard Annealed Solid Copper Wire (American Wire Gauge—B & S)

Gauge number	Diameter (mils)	Cross section		Ohms per 1000 ft.		Ohms per mile 25°C (=77°F)	Pounds per 1000 ft
		Circular mils	Square inches	25°C (=77°F)	65°C (=149°F)		
0000	460.0	212,000.0	0.166	0.0500	0.0577	0.264	641.0
000	410.0	168,000.0	0.132	0.0630	0.0727	0.333	508.0
00	365.0	133,000.0	0.105	0.0795	0.0917	0.420	403.0
0	325.0	106,000.0	0.0829	0.100	0.116	0.528	319.0
1	289.0	83,700.0	0.0657	0.126	0.146	0.665	253.0
2	258.0	66,400.0	0.0521	0.159	0.184	0.839	201.0
3	229.0	52,600.0	0.0413	0.201	0.232	1.061	159.0
4	204.0	41,700.0	0.0328	0.253	0.292	1.335	126.0
5	182.0	33,100.0	0.0260	0.319	0.369	1.685	100.0
6	162.0	26,300.0	0.0206	0.403	0.465	2.13	79.5
7	144.0	20,800.0	0.0164	0.508	0.586	2.68	63.0
8	128.0	16,500.0	0.0130	0.641	0.739	3.38	50.0
9	114.0	13,100.0	0.0103	0.808	0.932	4.27	39.6
10	102.0	10,400.0	0.00815	1.02	1.18	5.38	31.4
11	91.0	8230.0	0.00647	1.28	1.48	6.75	24.9
12	81.0	6530.0	0.00513	1.62	1.87	8.55	19.8
13	72.0	5180.0	0.00407	2.04	2.36	10.77	15.7
14	64.0	4110.0	0.00323	2.58	2.97	13.62	12.4
15	57.0	3260.0	0.00256	3.25	3.75	17.16	9.86
16	51.0	2580.0	0.00203	4.09	4.73	21.6	7.82

17	45.0	2050.0	0.00161	5.16	5.96	27.2	6.20
18	40.0	1620.0	0.00128	6.51	7.51	34.4	4.92
19	36.0	1290.0	0.00101	8.21	9.48	43.3	3.90
20	32.0	1020.0	0.000802	10.4	11.9	54.9	3.09
21	28.5	810.0	0.000636	13.1	15.1	69.1	2.45
22	25.3	642.0	0.000505	16.5	19.0	87.1	1.94
23	22.6	509.0	0.000400	20.8	24.0	109.8	1.54
24	20.1	404.0	0.000317	26.2	30.2	138.3	1.22
25	17.9	320.0	0.000252	33.0	38.1	174.1	0.970
26	15.9	254.0	0.000200	41.6	48.0	220.0	0.769
27	14.2	202.0	0.000158	52.5	60.6	277.0	0.610
28	12.6	160.0	0.000126	66.2	76.4	350.0	0.484
29	11.3	127.0	0.0000995	83.4	96.3	440.0	0.384
30	10.0	101.0	0.0000789	105.0	121.0	554.0	0.304
31	8.9	79.7	0.0000626	133.0	153.0	702.0	0.241
32	8.0	63.2	0.0000496	167.0	193.0	882.0	0.191
33	7.1	50.1	0.0000394	211.0	243.0	1114.0	0.152
34	6.3	39.8	0.0000312	266.0	307.0	1404.0	0.120
35	5.6	31.5	0.0000248	335.0	387.0	1769.0	0.0954
36	5.0	25.0	0.0000196	423.0	488.0	2230.0	0.0757
37	4.5	19.8	0.0000156	533.0	616.0	2810.0	0.0600
38	4.0	15.7	0.0000123	673.0	776.0	3550.0	0.0476
39	3.5	12.5	0.0000098	848.0	979.0	4480.0	0.0377
40	3.1	9.9	0.0000078	1070.0	1230.0	5650.0	0.0299

Table 2-3 Current-Carrying Capacities (in Amperes) of Single Copper Conductors at Ambient Temperature of Below 30°C

Size	Rubber or thermoplastic	Thermoplastic asbestos, var-cam, or asbestos var-cam	Impregnated asbestos	Asbestos	Slow-burning or weatherproof
0000	300	385	475	510	370
000	260	330	410	430	320
00	225	285	355	370	275
0	195	245	305	325	235
1	165	210	265	280	205
2	140	180	225	240	175
3	120	155	195	210	150
4	105	135	170	180	130
6	80	100	125	135	100
8	55	70	90	100	70
10	40	55	70	75	55
12	25	40	50	55	40
14	20	30	40	45	30

maximum current (in amperes) that is permitted to flow in various sizes of wire. Rubber-insulated wire tends to heat up more than varnished cambric-insulated wire, and therefore less current is allowed in rubber-insulated wire. Other insulating materials permit more rapid escape of heat and are thus allowed to carry more current.

Stranded Wires

A stranded wire consists of a group of wires, which are usually twisted to form a metallic string. Stranding improves the flexibility of a wire. Figure 2-6 shows a typical 37-strand conductor and how the total circular-mil cross section is found. We multiply the circular-mil area of each strand by the number of strands to find the total circular-mil cross section. An insulated stranded wire is called a cord. The current capacity of a cord is determined by its insulation and total circular-mil area, as for an unstranded wire.

Figure 2-6 Stranded conductor.

0.002 INCH
37-STRAND CONDUCTOR

DIAMETER OF EACH STRAND = 0.002 INCH
DIAMETER OF EACH STRAND = 2 MILS
CIRCULAR MIL AREA OF EACH STRAND = D^2 = 4 CM
TOTAL CM AREA OF CONDUCTOR = 4 × 37 = 148 CM

Aluminum Wire

Although aluminum is lighter and is easier to bend than copper, a No. 14 aluminum wire has greater resistance than an equal length of No. 14 copper wire. Therefore, larger-diameter aluminum wire must be used for a given load or current demand. As an illustration, we would usually install No. 12 aluminum wire instead of No. 14 copper wire, and we would generally install No. 4 aluminum wire instead of No. 6 copper wire. In other words, as a rough rule of thumb, we select aluminum wire two sizes larger than corresponding copper wire. However, electricians are bound by the *National Electrical Code,* to which reference should be made in particular situations. Again, local electrical codes are sometimes more stringent than the *NEC.* Aluminum wiring is forbidden in many localities, because improper installation is suspected of causing fires in some homes.

Line Drop

Since a wire has resistance, there is a voltage drop from the source end to the load end of a line. For example, if 120 V were applied to the source end, a line drop of 20 V would reduce the voltage at the load to 100 V. This might be an excessive voltage drop; an electric-light bulb that normally operates at 120 V will be dim if operated at 100 V. To increase the load voltage, a smaller load can be used to reduce the IR drop in the line. However, if a smaller load cannot be used, we must increase the wire size. In practical situations, a reasonable compromise is determined by the electrician so that the line drop is tolerable without incurring undue cost as a result of using necessarily large wire.

The electrician first selects a wire size that has permissible current capacity at maximum load (maximum current flow). Then he checks to determine whether the voltage drop will be objectionable in view of the total length of the line. If he finds that the line voltage drop will be excessive, he selects a larger size of wire to reduce the voltage drop as required. Line drop becomes the most important consideration that must be taken into account when a long line supplies a heavy load. Sometimes installations may be in locations where the ambient (surrounding) temperature is comparatively high, as in a furnace room. In such cases, proper allowance must be made for external heat on the allowable current flow, and each case has its own specific limitations. Maximum allowable operating temperatures are specified by the *National Electrical Code*.

Temperature Coefficient of Resistance

We know that the resistance of a wire increases as the temperature increases. In Table 2-4 we see the *specific resistance* or *resistivity* in ohms for a unit volume (the circular mil-foot) of various metals. These specific resistance or rho values are given at 20°C. The amount of increase in the resistance of a 1-ohm sample of a conductor per degree rise in temperature above 0°C is called its *temperature coefficient of resistance*. For copper, the temperature coefficient is approximately 0.00427. Other metals have temperature coefficients from 0.003 to 0.006.

A copper wire that has a resistance of 50 ohms at 0°C will have an increase in resistance of 50×0.00427, or 0.214 ohms for the entire length of wire for each degree of temperature rise above 0°C. At 20°C, the increase in resistance is about 20×0.214, or 4.28 ohms.

Table 2-4 Specific Resistance

	Specific resistance at 20°C	
Substance	Centimeter cube (microhms)	Circular mil-foot (ohms), or ρ
Silver	1.629	9.8
Copper (drawn)	1.724	10.37
Gold	2.44	14.7
Aluminum	2.828	17.02
Carbon (amorphous)	3.8 to 4.1	
Tungsten	5.51	33.2
Brass	7.0	42.1
Steel (soft)	15.9	95.8
Nichrome	109.0	660.0

Thus, the total resistance at 20°C is 50 + 4.28 = 54.28 ohms. Since we must often change from Fahrenheit to Celsius temperatures, Figure 2-7 shows the relations between these thermometer scales. To change a Fahrenheit reading to Celsius, we use the following formula:

$$C = \frac{5}{9}(F - 32)$$

Earth (Ground) Conduction

Moist soil is a fairly good conductor, although dry soil is a poor conductor. Therefore, moist soil can be used in case of extreme necessity for part of a circuit. For example, electric fences for cattle (explained in detail in a later chapter) must employ a ground return circuit. In general, electricians avoid using ground circuits when possible, because there is considerable voltage drop along a ground circuit as compared with a wire circuit. To understand the principle of a ground circuit, a simple arrangement consisting of a switch, dry cell, and electric bell is shown in Figure 2-8.

Conduction of Electricity by Air

Air is a very good insulator under most conditions. However, we know that electricity applied to a spark plug engine causes a spark to jump between the points of the plug. The air between the points becomes a conductor while the spark is jumping the gap. Another

46 Chapter 2

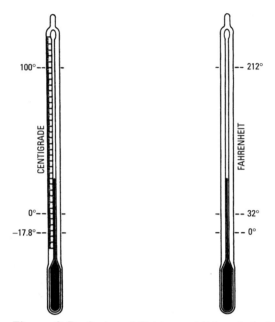

Figure 2-7 Scales of Celsius and Fahrenheit thermometers.

practical example is seen when lightning strikes from a thundercloud to the earth. The air in the path of a lightning stroke becomes a temporary conductor. We will find that air suddenly changes from an insulator to a conductor at a certain critical voltage called the *sparking voltage*. Figure 2-9 shows the voltage required to produce a spark between sharp metal points separated by various distances.

When a spark jumps an air gap, the resistance of the air in the path of the spark changes from an extremely high resistance to a very low resistance, such as 1 ohm or less. As soon as the spark has passed, the air ceases to be a conductor and again becomes a good insulator. It may surprise us to learn that electric currents in air are basically the same as electric currents in metals. The only difference is that the atoms in air are farther apart than the atoms in a metal. Therefore, we must apply more voltage across an air gap before electrons are transferred from one atom to the next.

Electricians work with both sparks and arcs. For example, automotive electricians works with sparks when they adjust and test the spark plugs in an engine. On the other hand, they work with

Conductors and Insulators 47

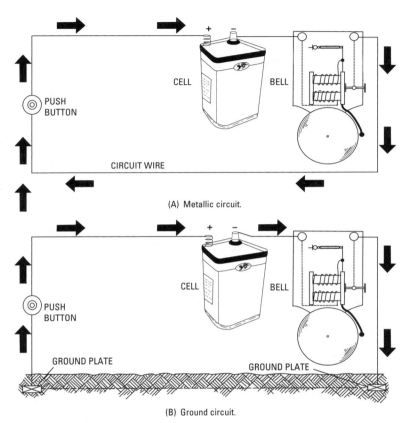

Figure 2-8 A simple bell circuit.

arcs when they weld a crack in a radiator support. The difference between a spark and an arc is that the electrodes of an arc are operated at a very high temperature, sufficient to melt iron. When air is heated to a very high temperature, its resistance becomes very low—10 ohms, for example. Therefore, an arc operates at a typical voltage of 50 V, whereas a spark operates at a typical voltage of 20,000 V.

An arc cannot start until the electrodes are heated to a very high temperature. In practice, an electrician does this by touching one electrode to the other. A heavy current flows at the point of contact, and electric power is changed into heat. The electrodes become red

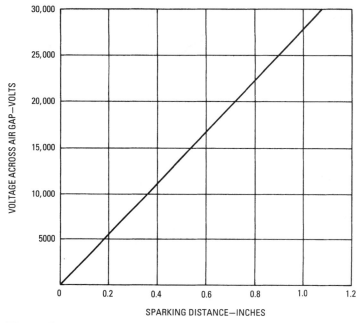

Figure 2-9 Voltage required to produce a spark between sharp metal points.

hot at the point of contact and can then be separated, and current will continue to flow in the extremely hot air between the electrodes. A very intense light is also produced by an arc, which will ruin the eyes unless dark goggles are worn. An arc welder operates with a spacing of about 1/4 in. between iron electrodes. The voltage across the arc is about 50 V and the current is about 6 A at this spacing. As the electrode spacing is changed, the voltage and current also change, as shown in Figure 2-10.

Electricians who work on large searchlights are concerned with carbon arcs, as shown in Figure 2-11. A carbon arc produces more flame than an iron arc because of vaporized carbon mixed with the extremely hot air between the electrodes. The carbon electrodes are gradually consumed by vaporization, and an automatic mechanism is provided to maintain a fixed spacing between them. When a carbon electrode is nearly consumed, it must be replaced. All arcs generate intense *ultraviolet light*. Although ultraviolet light is

Conductors and Insulators   49

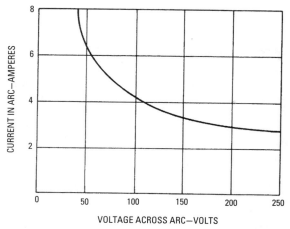

Figure 2-10   Relation of current to voltage in an arc.

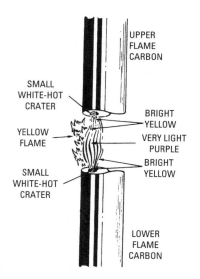

Figure 2-11   The carbon arc.

invisible, even moderate amounts will destroy the retina of the eye. Therefore, no arc is safe to look at unless dark goggles are worn.

Gases other than air are also in wide use as conductors of electric current. For example, neon gas is used at low pressure in glass tubes to provide arc conduction and generation of light for use in opposite polarity. Details of mercury-vapor arc conduction are explained in a later chapter.

Conduction of Electricity by Liquids

We know that moist soil is a fairly good conductor of electricity. It is not actually the water in the soil that provides conduction, because chemically pure water is a good insulator. Electrical conduction is provided by dissolved minerals in the water. The minerals are said to be in *water solution*. Atoms of minerals in water solution (sometimes also called an *aqueous* solution) have positive or negative electric charges, and these electrically charged atoms move toward electrodes placed in the solution. Positively charged atoms move toward the negative electrode as shown in Figure 2-8B, and negatively charged atoms move toward the positive electrode. This movement of charged atoms is an electric current.

Electricians who work with electroplating equipment are concerned with the conduction of electricity by liquids with some type of salt in solution. For example, suppose we wish to electroplate a fork with copper, as shown in Figure 2-12. A solution of blue vitriol (copper sulfate) in water is used. The fork forms the negative electrode (cathode), and a copper plate forms the positive electrode (anode). Copper sulfate dissolves to form positively charged copper atoms. Therefore, the charged copper atoms are attracted to the negative fork, and a film of metallic copper is deposited on the fork.

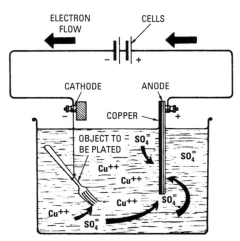

Figure 2-12 How a cell may be wired for electroplating.

Conductors and Insulators 51

(A) Porcelain tube.

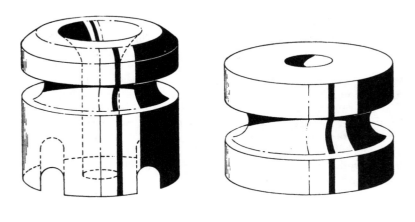

(B) Knob insulator. (C) Spool insulator.

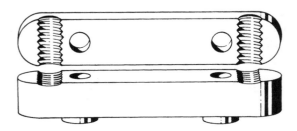

(D) Two-wire porcelain cleat.

Figure 2-13 Various types of porcelain insulators.

If we wish to silver-plate a fork, we will use a solution of silver cyanide. Silver cyanide is a poisonous silver salt that dissolves in water to form positively charged silver atoms. Thus the electroplating process is basically the same as in the case of a copper sulfate solution. If we wish to gold-plate a fork, we will use a solution of gold cyanide. Gold cyanide is also a poisonous salt. Therefore, appropriate care must be observed by electricians who work with electroplating equipment containing cyanides. Even a small amount of cyanide can be deadly if taken into the human body.

Insulators for Support of Wires

Many types of insulators are used to support electrical wires. Porcelain insulators such as those shown in Figure 2-13 are often used when the wiring is open and/or poorly insulated, such as in farm buildings. In homes, these insulators are found in old knob-and-tube wiring. Porcelain tubes are sometimes used to insulate wires that are run through holes in walls or studding (see Figure 2-14). Knob insulators are mounted on walls, ceilings, or beams by nails or screws. The groove around a knob insulator is used to bind the line wire securely in place. A spool insulator is used to insulate a wire from a wall. The insulator is typically suspended by a length of stout wire running through the center of the insulator and is secured to a screw eye in the wall. In turn, the line wire is bent around the groove

Figure 2-14 Porcelain tubes inserted in wall studs.

on the insulator and secured in place. Figure 2-15 shows how spool insulators may be used to insulate a line wire from a tree.

Two-wire porcelain cleats are secured to wood surfaces by means of screws or nails. The line wires are gripped by the grooves in the cleats and thereby prevented from sliding. Cleat-wiring installations can be made in a comparatively short time. However, the type of insulators used in various situations and locations are governed by regulations that must be observed by the electrician. Details are discussed later under the topics of house wiring and power wiring.

Classes of Insulation

Substances used as insulators in practical electrical work are classified into four groups, as follows:

>**Class A Insulation.** Class A insulation consists of (a) cotton, silk, paper, and materials similar to paper when impregnated or immersed in an insulating liquid; (b) molded or laminated materials with cellulose filler, phenolic resins, or similar resins; (c) films or sheets of cellulose acetate or similar cellulose products; and (d) varnishes or enamel applied to conductors.
>
>**Class B Insulation.** Class B insulation consists of mica, or fiberglass, all with a binder.
>
>**Class C Insulation.** Class C insulation consists entirely of mica, porcelain, glass, quartz, or similar materials.
>
>**Class O Insulation.** Class O insulation consists of cotton, silk, paper, or similar materials that are *not* impregnated or immersed in an insulating liquid.

We say that an insulation is *impregnated* when the air spaces within the insulation are filled up by an impregnating substance such as paraffin. An impregnating substance is itself a good insulator.

Insulation Resistance

Since no insulator is perfect, the insulation resistance of an insulated conductor may be measured in megohms. An instrument called a *megger* (proper name *megohm meter*) is used when it is desirable or necessary to measure insulation resistance. A megger is a type of ohmmeter that measures resistance in megohms (millions of ohms); it can measure up to several thousand megohms. Any insulating substance will break down at some value of high voltage, and the insulation of commercial insulated conductors is rated (guaranteed) to withstand a certain value of voltage. Therefore, meggers are designed to measure insulation resistance at high voltages.

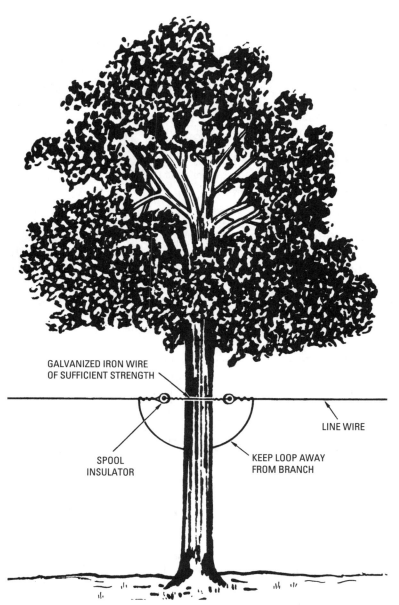

Figure 2-15 Insulation of a line wire from a tree.

Plastic-Insulated Sheathed Cables

Plastic-insulated sheathed cables such as those illustrated in Figure 2-16 are used extensively for residential and commercial wiring. The indoor-type plastic sheathed cable has a tough and flexible flat outer jacket. The wires inside are insulated with polyvinyl chloride (PVC). Two-conductor cable is used in old work, when the existing cable is of the two-conductor type. New work, consisting of wiring in new buildings, must use grounded-type receptacles; therefore, with-ground type cable, sometimes labeled W/G, must be utilized. There are three main types of nonmetallic sheathed cable. Type NM can be used only indoors in dry locations. Type NMC can be used in damp locations indoors. Only Type UF can be used outdoors under the ground. It can usually be buried without conduit (protective metallic tubing) unless there is a possibility of mechanical damage. Type UF cable resists moisture, acids, and corrosion. It can be run through masonry or between studding. If installed for outdoor circuits, a with-ground type of dual-purpose cable must be used. See the chapters on home wiring for details of various types of cable.

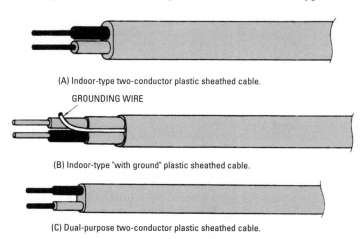

(A) Indoor-type two-conductor plastic sheathed cable.

(B) Indoor-type "with ground" plastic sheathed cable.

(C) Dual-purpose two-conductor plastic sheathed cable.

Figure 2-16 Typical plastic-insulated sheathed cable.

Summary

A conductor is a material or substance that carries electric current. No conductor is perfect, and because all conductors have some small amount of resistance, it is better to define a conductor as a substance with a very low resistance. An insulator, such as slate, paper, and glass, is a substance that does not permit electric current to flow easily.

Wires used as electrical conductors are generally made of copper or aluminum. Most wire is round, and the diameter can be measured accurately with a micrometer. Electricians use the word *mil* to describe the diameter of wire. Circular mil is used to describe the cross-sectional area of round wire. Square or rectangular wire is used in some applications, and its cross-sectional area is measured in square mils.

In practical electrical work, we describe the resistance of a wire in terms of ohms per circular wire one foot in length and one mil in diameter. When we know the resistance of a wire per circular mil-foot, we can easily find the resistance for any cross section and length. The resistance per circular mil-foot is called *rho*.

Because a wire has resistance, there is a voltage drop from the source end of the wire to the load end of the wire. Proper wire size is selected to permit current capacity at maximum load to flow. If the line voltage drop will be excessive, a larger-size wire is used to reduce the voltage drop. Line drop becomes the most important consideration that must be taken into account.

Test Questions

1. What is the definition of a conductor? Of an insulator?
2. Give an example of a good conductor, a fair conductor, a partial conductor, and an insulator.
3. What is a micrometer?
4. Define a mil.
5. How are mils related to circular mils?
6. Compare a circular mil to a square mil.
7. Why are electricians concerned with circular mils and square mils?
8. Explain the meaning of circular mil-foot of wire.
9. Does the resistance of copper wire increase or decrease as the temperature increases?
10. Is the AWG wire gauge the same as the B&S gauge?
11. How is the resistance of a wire related to its cross-sectional area? To its length?
12. Describe the use of an American Standard Wire Gauge.
13. What is meant by the current-carrying capacity of a given wire size?

Conductors and Insulators 57

14. Why are stranded conductors used in lamp cords?
15. Explain the meaning of line drop.
16. Is the resistance of a circular mil-foot (specific resistance) of copper wire greater or less than that of aluminum wire?
17. Give a general comparison of the Fahrenheit and centigrade thermometer scales.
18. How do we change a Fahrenheit temperature into a centigrade temperature?
19. Describe the meaning of a ground-return circuit.
20. About how much voltage can be withstood by needle points $1/2$ inch apart without breakdown and sparking through the air between the points?
21. In what way does an arc differ from a spark?
22. How does a salt solution conduct electricity?
23. Give several examples of insulators used to support line wires.

Chapter 3

Electric Circuits

We have learned that an electric circuit is a closed path for current flow. Electricians are concerned with many types of circuits. It has been noted in the first chapter that a *series circuit* is a circuit that supplies electricity to one or more loads; all devices in a series circuit are connected end-to-end in a closed path, and the same amount of current flows through each device. We are now ready to consider some more facts about series circuits that are of practical importance to an electrician.

Picture Diagrams and Schematic Diagrams

Figure 3-1 shows a picture diagram of two electric lamps connected in series with a battery. We occasionally use picture diagrams, but we generally work from schematic diagrams. A schematic diagram shows an electric circuit by means of *graphical symbols* instead of outline pictures. For example, Figure 3-2 shows two schematic diagrams. Standard electrical symbols are used to represent a battery, a

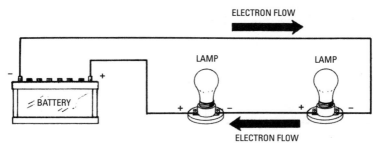

Figure 3-1 Picture diagram of two electric lamps connected in series with a battery.

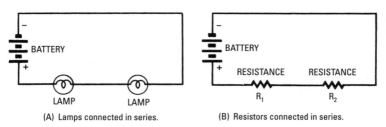

(A) Lamps connected in series.   (B) Resistors connected in series.

Figure 3-2 Schematic diagram.

59

lamp, and a resistor. Next, let us observe that the schematic diagram in Figure 3-2A can be represented by an *equivalent circuit*, as seen in Figure 3-2B. An equivalent circuit has the same electrical properties as the original circuit, but an equivalent circuit does not serve the same purpose. Let us see what this means.

If R_1 and R_2 in Figure 3-2B have the same amounts of resistance as the lamps in Figure 3-2A, and if the batteries in both circuits have the same voltage, it is clear that each circuit will draw the same amount of current. However, these circuits do not serve the same purpose because the circuit in Figure 3-2A produces light and heat, while the circuit in Figure 3-2B produces heat only. Nevertheless, the equivalent circuit is useful because it is the first step in *reducing* the original series circuit into a simplified circuit.

It is evident that the load in Figure 3-2B consists of resistance R_1 plus resistance R_2. Therefore, we can combine R_1 and R_2 into a single resistor, as seen in Figure 3-3. For example, if each lamp in Figure 3-2A has a resistance of 25 ohms, then $R_1 = 25$ ohms and $R_2 = 25$ ohms in Figure 3-2B. In turn, $R_L = 50$ ohms in Figure 3-3. The circuit in Figure 3-3 represents the final reduction, or the simplest possible equivalent circuit for the lamp circuit in Figure 3-2A. Equivalent circuits are useful because they help us to better understand the operation of complicated circuits.

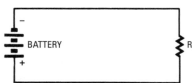

Figure 3-3 Equivalent circuit of resistor connected in series.

Voltage Polarities in Series Circuits

Electricians must understand voltage polarities in series circuits. Therefore, let us observe the polarities shown in Figure 3-1. Electrons flow from the negative terminal of the battery to the right-hand lamp in the diagram. Of course, there is a *voltage drop* across this lamp. Since the right-hand terminal of the lamp is connected to the negative terminal of the battery, and the left-hand terminal of the lamp eventually returns to the positive terminal of the battery, it is obvious that the right-hand terminal of the lamp has negative polarity with respect to its left-hand terminal.

The same amount of current flowing through the right-hand lamp also flows through the left-hand lamp in Figure 3-1. Using the same reasoning as before, we see that the right-hand terminal of this lamp must be negative with respect to its left-hand terminal. The voltage drops across the two lamps are in *series-aiding*, and these two voltage drops add up to the same amount of voltage as we find across the battery terminals. This is an example of *Kirchhoff's voltage law*, which was noted in the first chapter.

To show why we need to observe the polarity of a voltage, let us consider the connection of a voltmeter across a battery, as shown with an old-style voltmeter in Figure 3-4. To measure the battery voltage, we must connect the positive terminal of the voltmeter to the positive terminal of the battery, and we must connect the negative terminal of the voltmeter to the negative terminal of the battery. The pointer will move up-scale on the voltmeter. If we make a mistake and reverse the polarity of the meter connections, the pointer will not move up-scale on the voltmeter; instead, the pointer will move off-scale to the left. Therefore, voltmeters have their positive and negative terminals indicated.

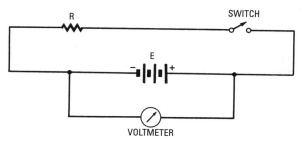

Figure 3-4 A voltmeter connected across a battery in a series circuit.

An ammeter also has its positive and negative terminals indicated; an ammeter must be connected into a circuit in proper polarity, as shown in Figure 3-5. The voltmeter is connected to read the voltage drop across R. Note carefully that the ammeter must *not* be connected across R; in effect, the battery would thereby be short-circuited through the ammeter. The ammeter would probably be damaged. Even if the ammeter did not burn out, the short circuit would soon ruin the battery. Therefore, we must keep the following rules in mind:

* An ammeter is always connected in *series* with a circuit.

* A voltmeter is always connected *across* the battery, resistor, or other device to measure a voltage drop.

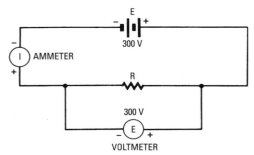

Figure 3-5 An ammeter and voltmeter connected into a circuit.

Voltage Measurements with Respect to Ground

A series circuit with a ground return is shown in Figure 3-6. The circuit consists of battery E and resistors R_1, R_2, and R_3. Voltmeter V_1 measures the voltage drop across R_1; V_2 is the voltage across the *battery and* R_1. Note carefully that to find voltage E, we *subtract* voltage V_1 from voltage V_2; we write:

$$E = V_2 - V_1$$

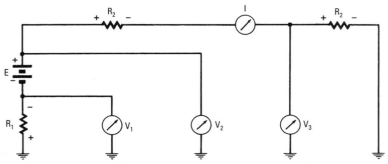

Figure 3-6 Voltage measurements with respect to ground.

Next, we observe that the voltage drop across R_2 is found by subtracting V_3 from V_2. Finally, V_3 is the voltage drop across R_3. For example, if $R_1 = 1$ ohm, $R_2 = 2$ ohms, $R_3 = 3$ ohms, and $E = 6$ volts, the current flow $I = 1$ ampere. In turn, $V_1 = 1$ volt, $V_2 = 5$ volts, and $V_3 = 3$ volts. Note that we can disregard the voltage drop across ammeter I in Figure 3-6 because the resistance of an ammeter is very small.

Note also in Figure 3-6 that the letters E and V are used to indicate voltages. We commonly use E to indicate a *source voltage* and use V to indicate a *voltage drop*. This is helpful, because it prevents confusion when we are working out a circuit problem. However, many electricians simply use E to indicate any voltage in a circuit, and other electricians use V to indicate any voltage in a circuit. It does not make any difference whether we write V or E or both, as long as we remember which voltage the letter stands for.

Resistance of a Battery

Any battery has a certain amount of resistance, called its *internal resistance*. In many circuits, we can disregard the internal resistance of a battery without getting into practical difficulties. On the other hand, we will find various circuits in which it is necessary to take the internal resistance of a battery into account. Figure 3-7A illustrates the internal construction of a dry cell. The paste contains a chemical solution called an *electrolyte*. This electrolyte has a small amount of resistance in a brand new cell, and in a nearly dead cell the resistance will be very great. Therefore, any cell can be represented by the equivalent circuit shown in Figure 3-7B. The cell operates electrically as if a perfect cell E were connected in series with an internal resistance R_{in}.

When a cell is connected to a load resistor, as shown in Figure 3-7C, the voltage across R_L will be less than E. There is a voltage drop inside the cell equal to IR_{in}. Therefore, the voltage across R_L is equal to $E - IR_{in}$. We call E the *electromotive force*, or *emf* of the cell. On the other hand, the voltage drop across R_L in Figure 3-7C is equal to the *terminal voltage* of the cell. We will find that E remains about the same (1.5 volts) even in a nearly dead cell. However, as noted previously, the resistance of R_{in} becomes greater as the cell becomes weaker. It follows that the terminal voltage of a cell is the same as its emf when no current is being drawn, even if the cell is nearly dead.

Therefore, we cannot test a cell properly with a voltmeter alone, because a voltmeter draws a very small amount of current. Instead, we must connect a load resistance across the cell as shown in Figure 3-7C and measure the voltage drop across the load resistor. A large dry cell has more electrode area than a small flashlight cell. A large cell can normally supply more current than a small cell. This means that a large cell should be tested with a load resistor that has a comparatively small value. Standard battery testers consist of a voltmeter and a dozen load resistors with different resistance values.

The voltage-selector switch on a battery tester connects a load resistor across a battery that provides a normal load or current drain. For example, an ordinary No. 6 dry cell should maintain a terminal

64 Chapter 3

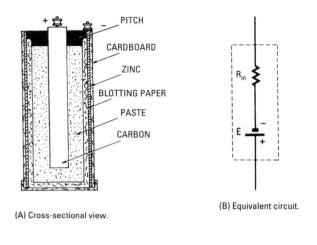

(A) Cross-sectional view.

(B) Equivalent circuit.

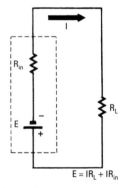

(C) Load resistor connected to equivalent circuit.

Figure 3-7 Internal resistance of a dry cell.

voltage of 75 percent of 1.5 volts, or at least $1\frac{1}{8}$ volts at a current drain of 10 amperes. This is called the *intermittent duty* rating of a dry cell. A dry cell is intended to supply a current of 10 amperes for only a short time and at widely separated intervals. On the other hand, the *constant-current* rating of a dry cell is the amount of current that the cell is intended to supply steadily over a period of several hours. In the case of a No. 6 dry cell, the constant-current rating is $\frac{1}{8}$ ampere.

Ordinary flashlights use a size-D dry cell; since it is much smaller than a No. 6 cell, a size-D cell has an intermittent duty rating of

2 amperes. Hearing aids use very small cells, which can supply much less current, either on intermittent duty or in steady service. The internal resistance of a No. 6 dry cell in good condition is about 0.1 ohm. As the cell becomes weaker, its internal resistance gradually increases. When a cell is so weak that it reads "bad" on a battery tester, its internal resistance has increased greatly. However, the emf of a bad cell will be practically the same as that of a brand new cell.

Efficiency and Load Power

Efficiency means output power divided by input power. For example, if we get half as much power out of a machine as we put into it, we say that the efficiency of the machine is 50 percent. The same principle applies to electric circuits. The *power* that we get out of the circuit shown in Figure 3-8 is the *power* in the lamp filament R_L; the power that goes into the circuit is supplied by the cell. We know that the power in the lamp is equal to VI watts, and the power supplied by the cell is equal to EI watts. Therefore, the efficiency of the circuit is equal to VI/EI, or V/E. In terms of percentage, the efficiency of the circuit is equal to $100 V/E$ percent.

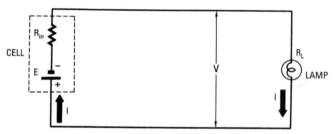

Figure 3-8 The circuit efficiency depends on the value of R_{in}.

The efficiency of the circuit in Figure 3-8 could be 100 percent only if R_{in} were zero. Since R_{in} can never be zero, the circuit efficiency must always be less than 100 percent. We will find that the efficiency of a circuit is greatest when the load is very light—that is, when a very small amount of current is drawn by the load. Let us see why this is so. If R_L has a very high value, then IR_L is much greater than IR_{in}. This is just another way of saying that V is large when R_L is large. Since E remains the same, it follows that the circuit efficiency V/E is high when R_L has a high value.

Let us consider how we can get the greatest power out of the circuit shown in Figure 3-8. We will find the value of R_L that makes VI as large as possible. It can be shown that the lamp will have the largest number of watts when $R_L = R_{in}$. This is a surprising answer at first glance. Let us note the following facts:

- If R_L were zero, maximum current would flow in the circuit. However, V would then be zero, and the power in the load would be zero.
- If R_L were infinite, maximum voltage would be dropped across the load. However, I would then be zero, and the power in the load would be zero.
- The load power has its greatest value when the load resistance is equal to the internal resistance of the cell.

A practical example is shown in Figure 3-9. The source voltage is 100 volts, and the source resistance is 5 ohms. As the resistance of the load is changed, the load voltage, circuit current, load power, and circuit efficiency change as shown in Figure 3-9B. When the load resistance is equal to the source resistance (5 ohms), the load power has its greatest value. If the load resistance is less or greater than 5 ohms, the load power decreases. We say that *maximum power transfer* occurs when the load resistance is equal to the source resistance. Note that the circuit efficiency is only 50 percent when maximum power transfer is obtained. Figure 3-9C shows how the efficiency, load voltage, circuit current, and load power change as the load resistance is changed.

Circuit Voltages in Opposition

Up to now, we have considered only cells connected in series-aiding. However, in practical electrical work, cell voltages may be connected in series-opposing. Figure 3-10 shows a 1.5-volt source connected in series-opposing with a 3-volt source. The 1.5 volts subtracts from the 3 volts, and the voltmeter reads 1.5 volts. Note the voltmeter polarity. Automobile electricians are concerned with series-opposing voltages in storage-battery charging circuits. Therefore, let us briefly consider the properties of storage batteries and battery-charging circuits.

A storage battery is also called a secondary battery, whereas a dry cell is called a primary battery. Secondary cells are different from primary cells in that a secondary cell can be recharged, whereas a primary cell cannot. The basic storage cell consists of a pair of lead plates immersed in a solution of sulfuric acid. Figure 3-11 depicts a simple storage cell and charging circuit. An electric generator is used to supply the charging voltage and current. Before the storage cell is charged, it has very little resistance, and the circuit current is practically 3 amperes.

Electric Circuits **67**

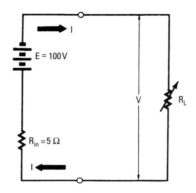

R_L	V	I	P_L	% EFF
0	0	20	0	0
1	16.6	16.6	267.6	16.6
2	28.6	14.3	409	28.6
3	37.5	12.5	468.8	37.5
4	44.4	11.1	492.8	44.4
5	50	10	500	50
6	54.5	9.1	495.4	54.5
7	58.1	8.3	482.2	58.1
8	61.6	7.7	474.3	61.6
9	63.9	7.1	453.7	63.9
10	66	6.6	435.6	66
20	80	4	320	80
30	87	2.9	252	87
40	88	2.2	193.6	88
50	91	1.82	165	91

E = OPEN-CIRCUIT VOLTAGE OF SOURCE
R_{in} = INTERNAL RESISTANCE OF SOURCE
V = TERMINAL VOLTAGE
R_L = RESISTANCE OF LOAD
P_L = POWER USED IN LOAD
I = CURRENT FROM SOURCE
% EFF = PERCENTAGE OF EFFICIENCY

(A) Circuit.   (B) Chart.

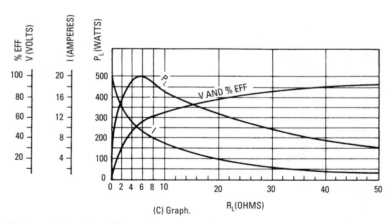

(C) Graph.

Figure 3-9 The effect of source resistance on power output.

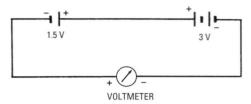

Figure 3-10 Circuit voltages in opposition.

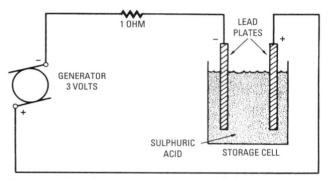

Figure 3-11 A simple storage cell and charging circuit.

As the charging process continues in Figure 3-11, chemical changes take place at the surfaces of the lead plates, and the storage cell builds up a voltage between its terminals. This voltage has the polarity shown in the diagram and opposes the generator voltage. Therefore, the charging current becomes less as the storage cell charges. Figure 3-12 shows how the terminal voltage of a storage cell increases during charge. The terminal voltage rises to 2.2 volts, then remains comparatively constant for a time, and finally rises to a maximum of 2.6 volts. At this point, the effective voltage in the circuit of Figure 3-11 is $3 - 2.6 = 0.4$ volt.

When a storage cell is removed from the charging circuit, its terminal voltage falls to approximately 2 volts; the cell can then be used as a voltage source until the chemical film on its plates has been used up. The voltage of the cell remains at practically 2 volts until it is almost discharged, and then it falls rapidly. After the cell is discharged, it can be recharged as has been explained. We measure the *capacity* of a storage cell in ampere-hours. Its ampere-hour capacity is equal to the number of amperes supplied by the cell on discharge multiplied by the number of hours that the cell

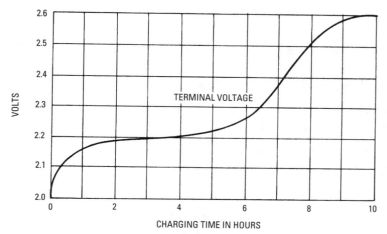

Figure 3-12 Terminal voltage versus charging time of a lead-acid storage cell.

can supply current. Commercial storage cells and batteries have the construction shown in Figure 3-13.

Another type of storage cell, called the *nickel-cadmium* cell, is in wide use. This type of cell has a terminal voltage of approximately 1.2 volts; however, it is not as heavy and has a longer life than a lead storage cell. A nickel-cadmium cell also requires less attention and care; it can be completely discharged and left uncharged for an indefinite time. This abusive treatment would ruin a lead cell. Scientists are continuously searching for new types of batteries for use in electric-powered automobiles and solar energy.

Principles of Parallel Circuits

Electricians work extensively with parallel circuits. In a parallel circuit, each load is connected in a *branch* across the voltage source, as shown in Figure 3-14. Therefore, there are as many paths for current flow as there are branches. The loads in a parallel circuit are sometimes said to be connected in *multiple*. We know that when more loads are connected into a series circuit, the total circuit resistance increases. On the other hand, when more loads are connected into a parallel circuit, the total resistance decreases.

We observe in Figure 3-14 that the voltage across all branches of a parallel circuit (from *a* to *b*) is the same because all branches are connected directly to the voltage source E. Each load resistor draws

70 Chapter 3

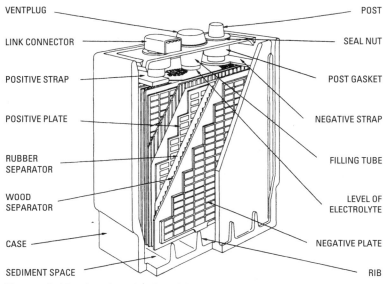

Figure 3-13 Lead-acid cell and battery.

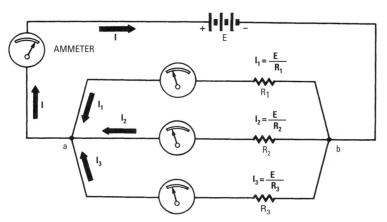

Figure 3-14 Illustrating a circuit with three resistances connected in parallel.

current independently of the other load resistors. Each branch current depends only on the load resistance in that branch. Therefore, we find the current in each branch by dividing its load resistance into the source voltage. The application of Ohm's law in each branch is shown in Figure 3-14. Since the source voltage E is applied across each branch, it follows that we may write:

$$E = I_1 R_1 = I_2 R_2 = I_3 R_3$$

Also, the three branch currents must be supplied by the total current I in Figure 3-14. Therefore, we write:

$$I = I_1 + I_2 + I_3$$

or,

$$I = \frac{E}{R_1} + \frac{E}{R_2} + \frac{E}{R_3}$$

Next, let us find an equivalent circuit for the parallel circuit in Figure 3-14. We will replace R_1, R_2, and R_3 with an equivalent resistor R_{eq}. Since $I = E/R_{eq}$, we will write:

$$\frac{E}{R_{eq}} = \frac{E}{R_1} + \frac{E}{R_2} + \frac{E}{R_3}$$

We observe that E cancels out in the foregoing formula, leaving:

$$\frac{1}{R_{eq}} = \frac{1}{R_1} + \frac{1}{R_2} + \frac{1}{R_3}$$

To show how the foregoing formula is used, let us take a practical example. Suppose that $R_1 = 5$ ohms, $R_2 = 10$ ohms, and $R_3 = 30$ ohms in Figure 3-14. Then, we find the value of R_{eq} as follows:

$$\frac{1}{R_{eq}} = \frac{1}{5} + \frac{1}{10} + \frac{1}{30} = \frac{10}{30} = \frac{1}{3}$$

or,

$$\frac{1}{R_{eq}} = 0.2 + 0.1 + 0.033 = 0.333$$

Therefore,

$R_{eq} = 3$ ohms, approximately

Our equivalent circuit is drawn as shown in Figure 3-15, and the value of R_{eq} is approximately 3 ohms. This is the method used by most electricians to find the equivalent resistance of several resistors connected in parallel. It makes no difference how many branches there may be in a parallel circuit; we simply add up the reciprocals of all the load resistors to find the reciprocal of the equivalent resistance. Then, the equivalent resistance is the denominator of this fraction.

Figure 3-15 Equivalent circuit for Figure 3-14.

Shortcuts for Parallel Circuits

Electricians use certain shortcuts in solving parallel circuits. The first shortcut applies to any number of parallel resistors, provided each resistor has the same resistance value. In this case, the equivalent resistance is found by dividing the resistance of one resistor by the number of resistors that are connected in parallel. For example, suppose that we have five 10-ohm resistors connected in parallel as shown in Figure 3-16. In turn, we write:

$$R_{eq} = \frac{10}{5} = 2 \text{ ohms}$$

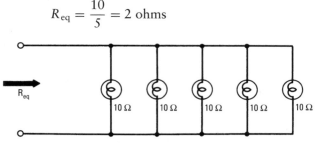

Figure 3-16 Five 10-ohm loads connected in parallel.

The second shortcut applies when we have two resistors with different values connected in parallel. In this case, the equivalent resistance is equal to the product of the two resistances divided by their sum. For example, if we have two resistors with values of 3 ohms and 6 ohms connected in parallel as shown in Figure 3-17, their equivalent resistance is found as follows:

$$R_{eq} = \frac{R_1 R_2}{R_1 + R_2} = \frac{3 \times 6}{3 + 6} = \frac{18}{9} = 2 \text{ ohms}$$

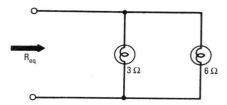

Figure 3-17 Two loads of unequal value connected in parallel.

An easy way to find the resistance of a parallel circuit is shown in Figure 3-18. This is called the leaning-ladder method; it is a graphical solution. In this example, $R_1 = 6$ ohms and $R_2 = 12$ ohms. We draw the corresponding lines as shown and note that their intersection occurs at the 4-ohm level. Therefore, the resistance of the parallel combination of R_1 and R_2 is 4 ohms. Note that the values indicated in Figure 3-18 may all be multiplied by 10, as shown in Figure 3-19. In turn, we can solve parallel circuits that have higher resistance values. Thus, if $R_1 = 100$ ohms and $R_2 = 25$ ohms, the resistance of the parallel combination is 20 ohms. Note that the values indicated in Figure 3-19 may all be multiplied by 10 to solve parallel circuits that have still higher resistance values.

It is also easy to find the resistance of three parallel-connected resistors, as shown in Figure 3-20. In this example, $R_1 = 10$ ohms, $R_2 = 8$ ohms, and $R_3 = 6$ ohms. We make the graphical solution in two steps. First, we draw the lines for R_1 and R_2, which intersect at point P. Then, we draw a line from O on OP to 6 on the left-hand vertical axis. This gives us the point of intersection P_1, showing that the parallel resistance of R_1, R_2, and R_3 is approximately 2.55 ohms. The same method may be extended to any number of resistors connected in parallel.

Conductance Values

Some electricians prefer to work with conductance values instead of resistance values when solving parallel circuits. Conductance is equal to $1/R$. We measure conductance in mhos, whereas we measure resistance in ohms. In other words, conductance is the reciprocal of resistance, and mho is the reciprocal of ohm. We let G stand for conductance, whereas we let R stand for resistance. The equivalent or total conductance of a parallel circuit is equal to the reciprocal of its equivalent or total resistance. Thus, we write:

$$G_{eq} = \frac{1}{R_{eq}}$$

74 Chapter 3

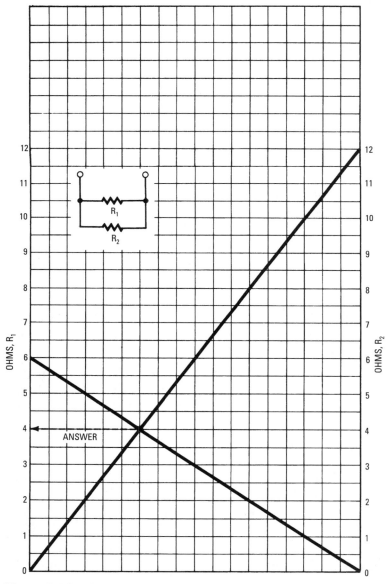

Figure 3-18 Graphic solution for resistors in parallel.

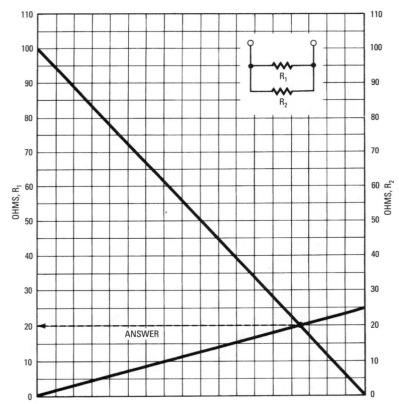

Figure 3-19 Another parallel circuit solution.

or,
$$R_{eq} = \frac{1}{G_{eq}}$$

It is easy to see that the equivalent conductance of a parallel circuit is equal to the sum of its branch conductances. For example, the equivalent conductance of the circuit shown in Figure 3-21 is written:

$$G_{eq} = G_1 + G_2 + G_3 + G_4 + G_5$$

76 Chapter 3

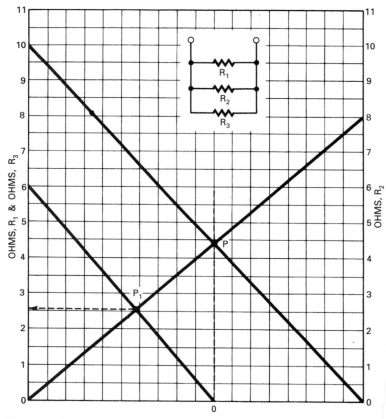

Figure 3-20 Graphic solution for three resistors connected in parallel.

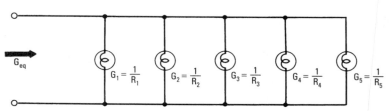

Figure 3-21 Five conductances connected in parallel.

Since G and R are reciprocals of each other, their Ohm's law equivalents are also reciprocals of each other. In turn, we write:

$$R = \frac{E}{I}$$

$$G = \frac{I}{E}$$

Therefore, Ohm's law is written with conductance values as follows:

$$I = EG$$

$$E = \frac{I}{G}$$

It seems more difficult at first to use conductance values instead of resistance values in solving parallel circuits. However, after we become familiar with the use of conductance values, we find it easier to solve complicated parallel circuits.

If an electrician prefers to use a graphical solution to convert from resistance to conductance, he may use the method shown in Figure 3-22. Note that the horizontal axis is laid off in steps from 0 to 1; the vertical axis is laid off in steps from 0 to 9. A line has been drawn at the unit level at the vertical axis for reference. To find

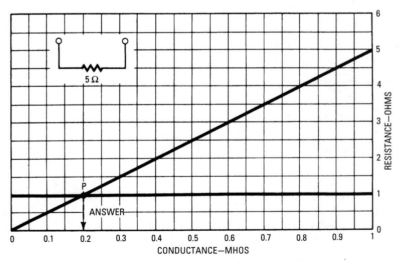

Figure 3-22 Conversion of 5 ohms to 0.2 mho.

the conductance of 5 ohms, we draw a line from 0 to 5, as shown. This line intersects the unit level at point P, which shows that the conductance value is 0.2 mho. Of course, we could also convert from conductance to resistance values by reversing the graphical procedure.

Note that the values along the horizontal axis may be divided by 10, and the values along the vertical axis multiplied by 10. Thus, a resistance of 50 ohms corresponds to a conductance of 0.02 mho. Or, the values along the horizontal axis may be divided by 100, and the values along the vertical axis multiplied by 100. For example, a resistance of 500 ohms has a conductance of 0.002 mho. This procedure is illustrated in Figure 3-23, wherein horizontal values have been divided by 10 and vertical values multiplied by 10. In Figure 3-24, horizontal values have been divided by 100 and vertical values multiplied by 100.

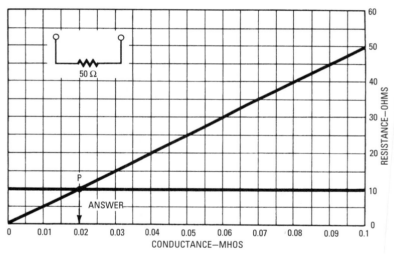

Figure 3-23 Conversion of 50 ohms to 0.02 mho.

Kirchhoff's Current Law for Parallel Circuits

Kirchhoff's current law for parallel circuits states that *at any junction of conductors, the algebraic sum of the currents is zero.* This is just another way of saying that there must be just as much electricity leaving a junction as there is electricity entering the junction. For example, let us consider junction *a* in Figure 3-25. We assume

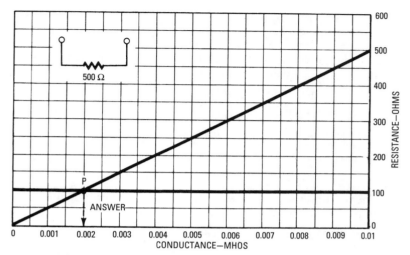

Figure 3-24 Conversion of 500 ohms to 0.002 mho.

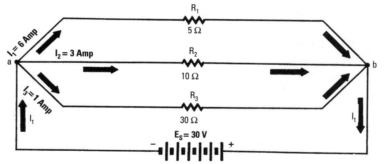

Figure 3-25 Resistors in parallel.

that the current flowing toward junction a is positive and that the currents flowing away from junction a are negative. We consider that I_t is positive and that I_1, I_2, and I_3 are negative. In turn, we write Kirchhoff's current law:

$$+I_1 - I_1 - I_2 - I_3 = 0$$

Or, for the example shown in Figure 3-25, we write:

$$+10 - 6 - 3 - 1 = 0$$

The importance of Kirchhoff's current law will be shown later by means of practical examples. We will find that, just as in a series circuit, the total power consumed in a parallel circuit is equal to the sum of the power values in each resistor. For example, the power P_1 in R_1 of the circuit in Figure 3-25 is written:

$$P_1 = EI_1 = 30 \times 6 = 180 \text{ watts}$$

Next, power P_2 in R_2 is found:

$$P_2 = EI_2 = 30 \times 3 = 90 \text{ watts}$$

Similarly, power P_3 in R_3 is found:

$$P_3 = EI_3 = 30 \times 1 = 30 \text{ watts}$$

The total power consumed by the parallel circuit in Figure 3-25 is simply the sum of the power values in each branch:

$$P_1 = P_1 + P_2 + P_3 = 180 + 90 + 30 = 300 \text{ watts}$$

We can check this answer by using the total current value in the power formula:

$$P_1 = EI_1 = 30 \times 10 = 300 \text{ watts}$$

Practical Problems in Parallel Circuits

Let us consider the parallel circuit shown in Figure 3-26. This circuit has two branches, a and b. Branch a has three lamps in parallel. L_1 takes a power of 50 watts, L_2 takes 25 watts, and L_3 takes 75 watts. Branch b also has three lamps in parallel; L_4 takes 150 watts, L_5 takes 200 watts, and L_6 takes 250 watts. The source voltage is 100 volts. Our problem is to do the following:

1. Find the current in each lamp.
2. Find the resistance of each lamp.
3. Find the current in branch a.
4. Find the current in branch b.
5. Find the total circuit current.
6. Find the total circuit resistance.
7. Find the total power in the circuit.

This problem is solved by means of Ohm's law, the power law, and Kirchhoff's current law. We proceed as follows:

1. The current in L_1 is $I_1 = P_1/E_s = 50/100 = 0.5$ ampere. Similarly, $I_2 = 0.25$ ampere, $I_3 = 0.75$ ampere, $I_4 = 1.5$ amperes, $I_5 = 2$ amperes, and $I_6 = 2.5$ amperes.

Electric Circuits 81

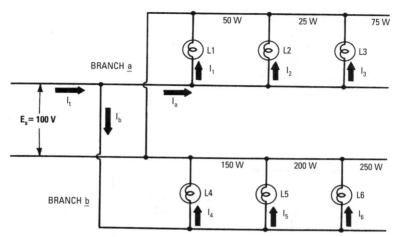

Figure 3-26 A typical parallel circuit.

2. The resistance of L_1 is $R_1 = E_s/I_1 = 100/0.5 = 200$ ohms. Similarly, $R_2 = 400$ ohms, $R_3 = 133$ ohms, $R_4 = 66.7$ ohms, $R_5 = 50$ ohms, and $R_6 = 40$ ohms.
3. The current in branch a is $I_1 + I_2 + I_3 = 0.5 + 0.25 + 0.75 = 1.5$ amperes.
4. The current in branch b is $I_4 + I_5 + I_6 = 1.5 + 2.0 + 2.5 = 6$ amperes.
5. The total circuit current is $I_a + I_b = 1.5 + 6.0 = 7.5$ amperes.
6. The total circuit resistance is $R_t = E_t/I_t = 100/7.5 = 13.3$ ohms.
7. The total power supplied to the circuit is $50 + 25 + 75 + 150 + 200 + 250 = 750$ watts. To check the total power in the circuit, note that we can write:

$$P_t = EI_t = 100 \times 7.5 = 750 \text{ watts}$$

Line Drop in Parallel Circuits

We have learned in our discussion of series circuits that line drop is sometimes excessive in a long line, even though the current-carrying capacity of the wire is not exceeded. Electricians have the same basic problem in various parallel-circuit installations. Line drop is of concern when supplying power to electric lamps, because a small reduction in voltage to a lamp results in greatly reduced light output.

For example, if a tungsten lamp is operated at 5 percent less than normal voltage, its light output is 17 percent less than normal. Therefore, unless the line drop is quite small, the lamps in Figure 3-27 will become dimmer as we proceed from the main line down to the end of the branch line. One way to avoid this difficulty is to use large-diameter wire in the branch line.

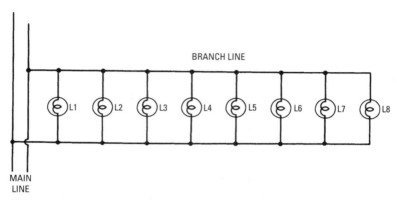

Figure 3-27 Variation in lamp brightness due to branch line drop.

However, large-diameter wire is comparatively expensive, and we prefer to use as small a diameter as possible as long as the current-carrying capacity of the wire is sufficient. The line drop that causes the lamps to become dimmer as we go down the branch line can be made much less evident by using the *return-loop* system of wiring, as shown in Figure 3-28. We can see that the distance of any lamp from the main line is the same in the return-loop system. Therefore, the lamps in this branch circuit have practically the same brightness, particularly in a case where the lamps are spaced equal distances apart. The chief disadvantage of the return-loop system is that it is a three-wire system.

Note, however, that in halls and auditoriums the lamps are often arranged around a square, rectangle, or circle. Figure 3-29 shows a circular arrangement of lamps as used in an auditorium or on some theater facades. This is called a *reentrant system*, and it requires only two wires. All of the lamps have practically the same brightness, particularly when the lamps are spaced equal distances apart. Therefore, the reentrant system is used in a branch line whenever this is possible. The length of wire used in the reentrant system is the same as in the simple arrangement of

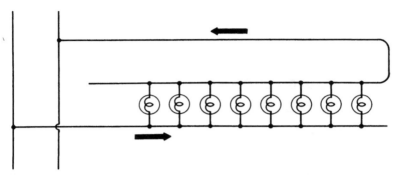

Figure 3-28 Return-loop system.

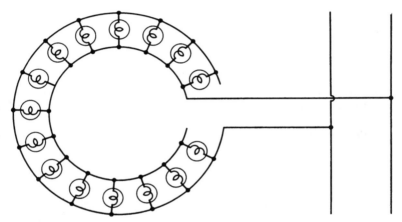

Figure 3-29 Reentrant arrangement of lamps.

Figure 3-27; however, smaller-diameter wire can be used in the reentrant system.

To get a certain chosen amount of voltage drop between the main line (service point) and the middle lamp of a reentrant system, we use a wire of such size that the total current supplied to the lamps would give the chosen voltage drop over a length $L_1 + L_2$ of wire, where L_1 is the length of wire that carries all of the current, and L_2 is $3/8$ of the length of wire in which the current is less than the total amount. Let us take a practical example of a reentrant row of two hundred 25-watt tungsten lamps. We will assume that the row is 300 feet long; one end of the row in this example is 60 feet, and the other end of the row is 40 feet from the main line, or service point. Assuming that the voltage at the service point is 115 volts, we will find the size of wire that is necessary to give 110 volts at the middle lamp of the row.

$$I = 200 \times 25/110 = 45.4 \text{ amperes}$$

$$L_1 = 60 + 40 = 100 \text{ feet}$$

$$L_2 = \frac{3}{8} \times 600 = 225 \text{ feet}$$

Now, to find the size of wire that is required, we note that the area of the wire in circular mils is given by the following formula:

$$A = \frac{10.8LI}{E_d} \text{ circular mils}$$

where A is the cross-sectional area of the wire in circular mils,
 L is the length of wire (out and back) in feet,
 I is the current in amperes,
 E_d is the permissible voltage drop.

Therefore, the size of wire required in this example will be:

$$A = \frac{10.8 \times 325 \times 45.4}{115 - 110} = 31,870 \text{ circular mils}$$

The nearest gauge is a No. 5 wire, which has a cross-sectional area of 33,100 circular mils. Since the odd-numbered sizes of wire are seldom carried in stock by local supply houses, we would probably use either No. 4 or No. 6 wire in this example.

Parallel Connection of Cells

Cells may be connected in parallel, as shown in Figure 3-30, when the load draws a greater amount of current than can be supplied by a single cell. As a practical example, let us assume that dry cells are to be used to supply 1.5 volts to a load that draws a steady current of $1/2$ ampere. A No. 6 dry cell will supply $1/8$ ampere over an extended time, such as several hours. To meet this requirement, we will connect cells in parallel, as shown in Figure 3-31A. In a parallel connection, all positive cell terminals are connected to one side of the line, and all negative cell terminals are connected to the other side of the line. The voltage across the line is the same as the voltage

Electric Circuits **85**

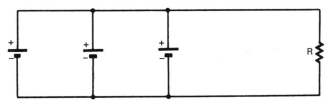

Figure 3-30 Circuit arrangement with cells connected in parallel.

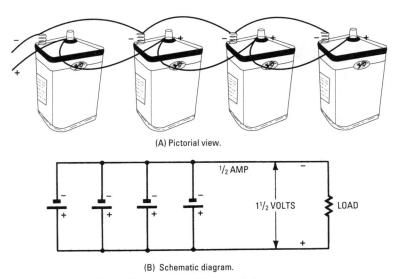

(A) Pictorial view.

(B) Schematic diagram.

Figure 3-31 Dry cells connected in parallel.

of one cell, or 1.5 volts. However, each cell can contribute its maximum allowable current of $1/8$ ampere to the line.

Since the load in Figure 3-31B draws $1/2$ ampere, we use four cells in this example. Each cell contributes $1/8$ ampere to the load. Storage cells may also be connected in parallel when a comparatively heavy current is demanded by a load. If we use lead-acid storage cells, the voltage across the load will be approximately 2 volts. It is interesting to note that a commercial lead-acid storage cell contains several negative plates and an equal number of positive plates connected in parallel, as seen in Figure 3-32. When cells are connected in parallel, the result is the same as if we used a single cell with a much larger electrode area.

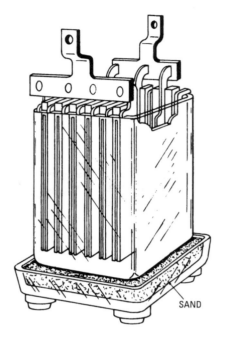

Figure 3-32 Glass jar of a storage cell supported on sand.

Summary

An electric circuit is a closed path for current flow. Current flows from negative to positive and must be observed when connecting a meter in the circuit. An ammeter is *always* connected in series with a circuit. A voltmeter is *always* connected across a battery, resistor, or any component in the circuit to be measured.

A battery is just one of many components that has a certain amount of resistance, called internal resistance. In many circuits, we can disregard the internal resistance without getting into practical difficulties. Efficiency means output power divided by the input power. For example, if a lamp had no resistance, the efficiency of the circuit would be 100 percent. We will find that the efficiency of a circuit is greatest when the load is very light—when a very small amount of current is drawn by the load.

In parallel circuits, each load is connected in a branch across the voltage source. Therefore, there are as many paths for current flow as there are branches. The loads in a parallel circuit are sometimes said to be connected in multiple. When more loads are connected into a series circuit, the total circuit resistance increases. When more loads are connected into a parallel circuit, the total resistance decreases.

Kirchhoff's law states that there must be just as much electricity leaving a junction as there is electricity entering the junction. This law is important because it helps us find the amount of branch current that flows in a circuit.

Test Questions

1. What is the definition of a series circuit?
2. Explain the meaning of an equivalent circuit.
3. Do resistances add or subtract in a series circuit?
4. What does Kirchhoff's voltage law state?
5. Discuss the meaning of the polarity of a voltmeter. Of an ammeter.
6. How are voltages measured with respect to ground?
7. Why does a dry cell have internal resistance?
8. Explain the difference between the emf and the terminal voltage of a dry cell.
9. How does a battery tester operate?
10. What is the meaning of circuit efficiency?
11. Discuss how we obtain maximum power in a load connected to a voltage source that has internal resistance.
12. If a voltage source is supplying maximum power to a load, what is the circuit efficiency?
13. Give an example of circuit voltages operating in series-opposing.
14. What is the definition of a parallel circuit?
15. How do we find an equivalent circuit that draws the same current as a parallel circuit?
16. Explain two shortcuts that can be used to solve certain parallel-circuit problems.
17. What is the meaning of conductance?
18. Do conductances add or subtract in a parallel circuit?
19. How does the light output from a lamp change with a change in voltage?
20. Explain what is meant by a return-loop system.

21. Explain what is meant by a reentrant system.
22. What is the advantage of a reentrant system?
23. Why do we connect cells in parallel?
24. How are the positive terminals and the negative terminals of cells connected in a parallel arrangement?
25. Explain why parallel-connected cells operate in the same way as a single cell with a much larger electrode area.

Chapter 4

Series-Parallel Circuits

Electricians often work with series-parallel circuits. For example, the circuit shown in Figure 4-1A looks like a parallel circuit. If we use a battery that has a very small internal resistance, such as a storage battery, we can say that Figure 4-1A is a parallel circuit for all practical purposes. On the other hand, if we use a battery consisting of small penlight cells, the internal resistance of the battery will have to be taken into account. In other words, the complete equivalent circuit is shown in Figure 4-1B. We call this a *series-parallel circuit* because some parts of the circuit are connected in series and other parts are connected in parallel.

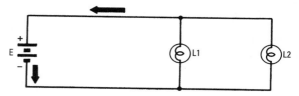

(A) Lamps in parallel connected in series with battery.

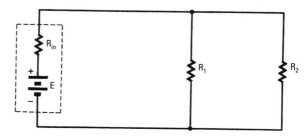

(B) Complete equivalent circuit.

Figure 4-1 Series-parallel circuits.

Current Flow in a Series-Parallel Circuit

A series-parallel circuit has branch currents and a total current in the same way that a parallel circuit does. For example, Figure 4-2 shows the total current and the branch currents in a simple series-parallel circuit. Let us see how we can find the amounts of these currents in a practical arrangement. If the battery in Figure 4-2 supplies 22 volts

89

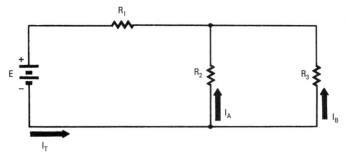

Figure 4-2 The total current and the branch currents.

and the resistances of R_1, R_2, and R_3 are 1 ohms, 2 ohms, and 3 ohms, respectively, we reason as follows:

1. R_2 and R_3 are connected in parallel; therefore, their equivalent resistance is 6/5, or 1.2 ohms.
2. R_1 is connected in series with the equivalent resistance of R_2 and R_3; therefore, the total circuit resistance is $1 + 1.2$ ohms, or 2.2 ohms.
3. Ohm's law states that $I = E/R$; therefore, I_T is equal to 22/2.2, or 10 amperes.
4. Ohm's law also states that $E = IR$; therefore, the voltage drop across R_1 is equal to 10×1, or 10 volts.
5. The voltage applied to R_2 and R_3 is equal to E minus the voltage drop across R_1; therefore, the voltage applied to R_2 and R_3 is 22/2.2, or 10 amperes.
6. Since 12 volts is applied across R_2, I_A is equal to 12/2, or 6 amperes; similarly, I_B is equal to 12/3, or 4 amperes.

It is important to observe that the total current in Figure 4-2 is equal to the sum of the branch currents I_A and I_B. That is, we write:

$$I_T = I_A + I_B$$

or,

$$10 \text{ amperes} = 6 + 4 \text{ amperes}$$

Kirchhoff's Current Law

The foregoing example illustrates an important law of electricity called Kirchhoff's current law. This law states that *the sum of the* currents leaving a junction is equal to the current entering the junction. A junction is a circuit point at which a current splits up, or branches, as shown at P in Figure 4-3. In this diagram, current I_T is entering junction P, and currents I_A and I_B are leaving junction P. We have seen that I_T is equal to the sum of I_A and I_B.

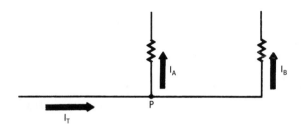

Figure 4-3 Illustrating a junction at point P.

Kirchhoff's current law is important because it helps us to find the amount of branch current that flows in a circuit such as the one shown in Figure 4-4. In this example, we are given the series resistance, the voltage drop across it, and three of the branch currents; our problem is to find the fourth branch current (I_B). We must start by finding the total current. It follows from Ohm's law that we can write

$$I_T = \frac{10}{5} = 2 \text{ amperes}$$

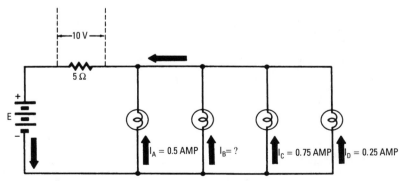

Figure 4-4 A series-parallel circuit in which I_B is unknown.

Since we know the total current, we can use Kirchhoff's current law to find I_B:

$$I_A + I_B + I_C + I_D = I_T$$

or,

$$0.5 + I_B + 0.75 + 0.25 = 2$$

Therefore,

$$I_B = 2 - 0.5 - 0.75 - 0.25 = 0.5 \text{ amperes}$$

Series-Parallel Connection of Cells

We know that cells can be connected in series to obtain a greater source voltage; we also know that cells can be connected in parallel to obtain a greater current capacity. In turn, we will find that cells can be connected in series-parallel to obtain a greater source voltage with a greater current capacity. There are two ways that this can be done, as shown in Figure 4-5. We can connect series groups of cells in parallel, or we can connect groups of cells in series. One

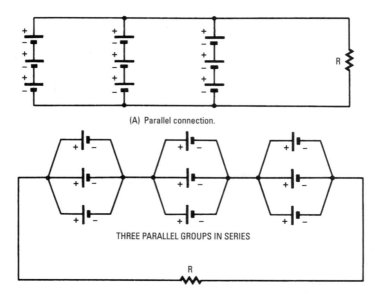

(A) Parallel connection.

THREE PARALLEL GROUPS IN SERIES

(B) Series connection.

Figure 4-5 Connecting dry cells in series-parallel.

circuit operates exactly the same as the other; therefore, it makes no difference in practice which method of connection we use.

Line Drop in Series-Parallel Circuits

The circuit shown in Figure 4-6 looks like a parallel circuit, but it is actually a series-parallel circuit. We observe that 117 V is applied at the input of the line, but there is 114.4 V at the end of the line. Therefore, the resistance of the 500-foot line is causing a line drop of 2.6 V. In this example, the wire size for the line is to be found; the wire must have a sufficiently large diameter so that the line drop will not be greater than 2.6 V. We start by finding the resistance of the line. The lamps draw a total current of 12 amperes. Accordingly, we apply Ohm's law:

$$R = \frac{E}{I} = \frac{2.6}{12} = 0.217 \text{ ohm}$$

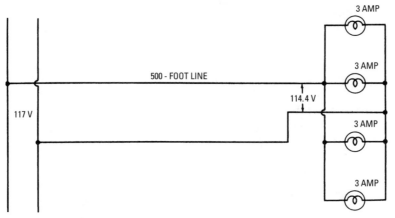

Figure 4-6 Finding wire size for 500-foot line.

This is the total resistance of the 500-foot line, which consists of 1000 feet of wire. We know that copper wire has a resistance of 10.4 ohms per circular mil-foot, that is, rho is equal to 10.4. Therefore, we write:

$$A = \frac{\rho 2L}{R} = \frac{10.4 \times 2 \times 500}{0.217} = 47,926 \text{ circular mils}$$

Since No. 4 wire has a cross-sectional area of 41,700 CM and No. 3 wire has 52,600 CM, we would select the No. 3 gauge. Finally, we must check on the current capacity of No. 3 wire; since this gauge of rubber-insulated wire is permitted to carry 80 amperes, according to the *National Electrical Code*,* our selection is suitable from the standpoints of both line drop and current-carrying capacity.

Use of a Wattmeter

We know that power is measured with a wattmeter. A wattmeter combines a voltmeter and an ammeter in one unit, and the scale of the wattmeter reads power values in watts. If we use a voltmeter and an ammeter to measure power, the instruments are connected to a line as shown in Figure 4-7A. The ammeter measures the current in the line, and the voltmeter measures the voltage across the load. In turn, we multiply volts by amperes to find the power in watts.

A voltmeter and an ammeter each have two terminals. However, since a wattmeter is a combination voltmeter and ammeter, the wattmeter has four terminals. Two of the wattmeter terminals are voltage terminals, and the other two are current terminals. As shown in Figure 4-7B, we connect the voltage terminals *across* the line, and we connect the current terminals *in series* with the line. It is very important not to make a mistake in connecting a wattmeter to a line; if the current terminals were connected across the line by mistake, the wattmeter would be damaged.

Circuit Reduction

The reduction of a series-parallel circuit to a simple equivalent circuit is based on the methods used for series circuits and for parallel circuits. For example, reduction of the circuit shown at A-1 in Figure 4-8 is very simple if we keep the following facts in mind:

- Any number of resistances connected in series can be replaced by a single resistance with a value equal to the sum of the individual resistances.

- Any number of resistances connected in parallel can be replaced by a single resistance with a value equal to the reciprocal of the sum of the reciprocals of the individual resistances.

---

*The rules laid down in the *National Electrical Code* are enforced by means of local ordinances in individual municipalities; these ordinances govern the installation of electric wiring. Note that the detailed requirements set forth in local ordinances may differ to some extent from the letter of the *NEC*. Local ordinances are the ones that will be enforced upon you; however, if there are no local ordinances, the electrician should follow the requirements laid down in the most recent issue of the *National Electrical Code*. An installation that is in accordance with the code merely ensures minimum risk from fire or accident hazards. It does not provide assurance that the installation will operate completely or efficiently.

Series-Parallel Circuits 95

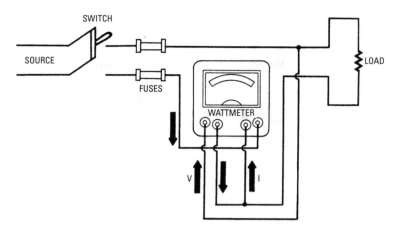

(A) With voltmeter and ammeter.

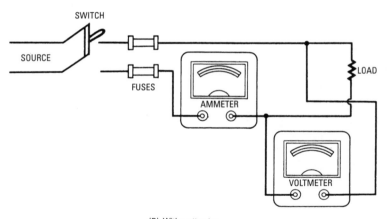

(B) With wattmeter.

Figure 4-7 Measuring power.

For example, if we have three resistances connected in series, we write:

$$R_{eq} = R_1 + R_2 + R_3$$

On the other hand, if we have three resistances connected in parallel, we write:

$$R_{eq} = \frac{1}{\frac{1}{R_1} + \frac{1}{R_2} + \frac{1}{R_3}}$$

We observe that circuit A-1 in Figure 4-8 consists of resistors R_a and R_b in series and that this series combination is in parallel with R_d. In turn, this series-parallel combination is connected in series with R_c, and the resulting combination is connected in parallel with R_f. To find the equivalent circuit, we first replace R_a and R_b with their equivalent resistance R_g, as shown in A-2. The next step is to combine R_g and R_d, replacing them with their equivalent resistance R_h, as seen in A-3. We then replace R_c and R_h with their equivalent resistance R_j, as depicted in A-4. Finally, we replace R_j and R_f with their equivalent resistance and obtain the equivalent resistance R_k, as shown in A-5.

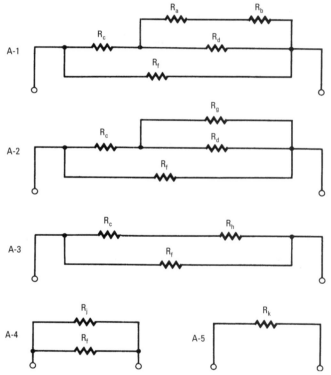

Figure 4-8 Reduction of a series-parallel circuit to an equivalent resistance.

Electricians use different ways to reduce resistances in parallel to a single equivalent resistance; the method that is used is simply

Series-Parallel Circuits 97

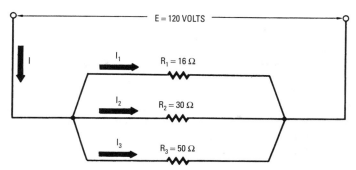

Figure 4-9 Reducing resistance in parallel circuits.

a matter of personal preference. For example, let us consider the circuit shown in Figure 4-9. We can find the equivalent resistance by any of the following methods:

Reduction by Pairs. In reduction by pairs, we take the resistors two at a time. Thus, we may select R_1 and R_2 and find their equivalent resistance:

$$R_{eq} = \frac{R_1 R_2}{R_1 + R_2} = \frac{16 \times 30}{16 + 30} = \frac{480}{46} = 10.434 \text{ ohms}$$

Next, we will take R_{eq} and R_3 and find their equivalent resistance:

$$R_{eq} = \frac{R_{eq} R_3}{R_{eq} + R_3} = \frac{1200}{139} = \frac{521.7}{60.434} = 8.63 \text{ ohms, approx.}$$

Reduction by Product-and-Sum Formula. If we prefer, we can take all three resistors at the same time in the product-and-sum formula. Accordingly, we write:

$$R_{eq} = \frac{R_1 R_2 R_3}{R_1 R_2 + R_1 R_3 + R_2 R_3} = \frac{16 \times 30 \times 50}{480 + 800 + 1500}$$

$$= \frac{24,000}{2780} = \frac{1200}{139} = 8.63 \text{ ohms, approx.}$$

Reduction by Reciprocals. To use the method of reduction by reciprocals, we write the reciprocal of the equivalent resistance as the sum of the reciprocals of the individual resistances:

$$\frac{1}{R_{eq}} = \frac{1}{R_1} + \frac{1}{R_2} + \frac{1}{R_3} = \frac{1}{16} + \frac{1}{30} + \frac{1}{50}$$

or,

$$\frac{1}{R_{eq}} = \frac{75}{1200} + \frac{40}{1200} + \frac{24}{1200} = \frac{139}{1200}$$

$$\frac{1}{R_{eq}} = \frac{1200}{139} = 8.63 \text{ ohms, approx.}$$

Reduction by Conductances. In reduction by conductances, we write the equivalent conductance as the sum of the individual conductances:

$$G_{eq} = G_1 + G_2 + G_3 = 0.0625 + 0.0333 + 0.02$$
$$= 0.1158 \text{ mhos, approx.}$$

or,

$$R_{eq} = \frac{1}{G_{eq}} = \frac{1}{0.1158} = 8.63 \text{ ohms, approx.}$$

Power in a Series-Parallel Circuit

The power consumed by the loads in a series-parallel circuit is equal to the sum of the power values consumed by each load. For example, Figure 4-10 shows a 3000-watt (3 kW) load consisting of a heater, hot plate, and flatiron, each of which consumes 1000 watts.

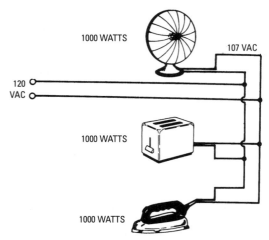

Figure 4-10 Power consumed by appliances and line resistance.

However, the total power in this circuit also includes the power loss to the line due to line drop. The line drop in this example is 117 − 107, or 10 volts. To find the power loss in the line, we must first find the current drawn by the load. Accordingly, we write

$$107 I = 3000 \text{ watts}$$

or,

$$I = \frac{3000}{107} = 28 \text{ amperes, approx.}$$

The power loss in the line is equal to the line drop multiplied by the line current:

$$P_{\text{line}} = 10 \times 28 = 280 \text{ watts, approx.}$$

Therefore, the power consumption of the circuit shown in Figure 4-10 is equal to 3280 watts. The power input to the line is the product of the input voltage and the line current:

$$P_{\text{in}} = 117 \times 28 = 3276 \text{ watts, approx.}$$

Note that we find a value of 3276 watts for the power input but that we found a value of 3280 watts for the circuit power. The reason is that both of these answers are approximate because we rounded off the decimals in our arithmetic.

Next, the *efficiency* of the circuit shown in Figure 4-10 is equal to the load power divided by the input power:

$$\text{Efficiency} = \frac{3000}{3276} = 0.91 = 91\%, \text{ approx.}$$

Horsepower

Figure 4-11 shows a series-parallel circuit in which the power in the loads is given in horsepower. *One horsepower is equal to 746 watts.* Therefore, the total load power in this example is 746 watts. In turn, the line current will be approximately 7 amperes, and the power input to the line will be approximately 819 watts. The power lost due to line drop is approximately 70 watts. This is a very simple example in which we have assumed that the motors in the washer and freezer are 100 percent efficient. However, no motor is 100 percent efficient; therefore, a 1-horsepower motor load will actually consume somewhat more that 746 watts from the line. (For details of motor power consumption, reference should be made to a specialized motor book.) Note that the illustration in Figure 4-11 is used to show horsepower only and does not intend to give misleading wiring advice.

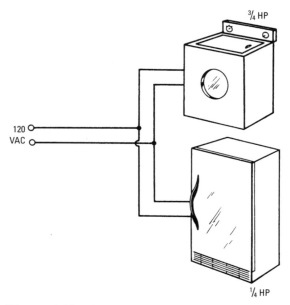

Figure 4-11 A series-parallel circuit with loads measured in horsepower. This figure is for illustration only. In the home, a clothes washer and a freezer should each have its own circuit.

Three-Wire Distribution Circuit

A series-parallel circuit of particular importance to electricians is called a three-wire distribution circuit. This type of circuit often operates at a total voltage of 240 volts, while the loads operate at 120 volts. As shown in Figure 4-12, a *positive feeder*, a *negative feeder*, and a neutral wire are used in the basic arrangement. The loads are connected between the negative feeder and the neutral, and between the positive feeder and the neutral. When the loads are *unbalanced* (unequal), the neutral wire carries a current equal to the difference of the currents in the positive and negative feeders.

In the example of Figure 4-12, load L_1 draws 10 amperes, load L_2 draws 4 amperes, and the neutral wire carries a current of $10 - 4 = 6$ amperes. The direction of current flow in the neutral wire is always the same as that of the smaller of the currents in the positive and negative feeders. Note that the current flow is to the left in the positive feeder and that this current is smaller than the current in the negative feeder. In turn, the current in the neutral wire is in the same direction as in the positive feeder.

Series-Parallel Circuits 101

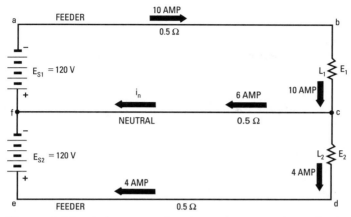

Figure 4-12 A three-wire distribution circuit with two loads.

With respect to junction c in Figure 4-12, we observe that the algebraic sum of the currents entering and leaving the junction is zero:

$$+10 - 4 - 6 = 0$$

Next, let us see how to find the load voltage E_1 in Figure 4-12. Since the algebraic sum of the voltages around the circuit $fabcf$ must be zero, we write:

$$+120 - 10 \times 0.5 - E_1 - 6 \times 0.5 = 0$$

or,

$$E_1 = 120 - 5 - 3 = 112 \text{ volts}$$

Thus, the voltage across load L_1 is 112 volts. The source voltage is 120 volts, and the line drop is 8 volts. In other words, the line drop in the negative feeder is equal to the line resistance (0.5 ohm) times the line current (10 amperes), or 5 volts. The line drop in the neutral wire is equal to the line resistance (0.5 ohm) times the line current (6 amperes), or 3 volts. Thus, the total line drop between E_{S1} and L_1 is 8 volts.

Load voltage E_2 in Figure 4-12 is found in a similar manner. We write:

$$+120 + 6 \times 0.5 = E_2 - 4 \times 0.5 = 0$$

or,

$$E_2 = 120 + 3 - 2 = 121 \text{ volts}$$

102 Chapter 4

In tracing the circuit from f to c, note that we proceed against the direction of the current arrow, and therefore the IR drop of 6×0.5 volts has a plus sign. The load voltage (121 volts) is 1 volt greater than the source voltage (120 volts). The total source voltage is 240 volts, and the total load voltage is $112 + 121 = 233$ volts. Observe that this total load voltage is also equal to the difference between the total source voltage and the sum of the voltage drops in the positive and negative feeders, or $240 - (2 + 5) = 233$ volts.

When we have a balanced load on the positive and negative sides of a three-wire system, the current in the neutral is zero, and the currents in the feeders are equal. However, when the loads are unbalanced, the unbalance current flows in the neutral. Therefore, the voltage decreases on the heavily loaded side, and the voltage increases on the lightly loaded side. We see that if the neutral wire had zero resistance, there would be no unbalance in the load voltages. For this reason, it is desirable to use a low-resistance neutral wire when unbalanced loads are present in a system.

A more complicated three-wire circuit is shown in Figure 4-13. The source voltage is 120 volts between each outside wire and the neutral wire. Load currents in the upper side of the system are 10, 4, and 8 amperes, respectively, for L_1, L_2, and L_3 in this example. In the lower side of the system, the load currents for loads L_1 and L_5 are 12 and 6 amperes, respectively. To find the load voltages, we must find the currents in each outside wire and in the neutral wire.

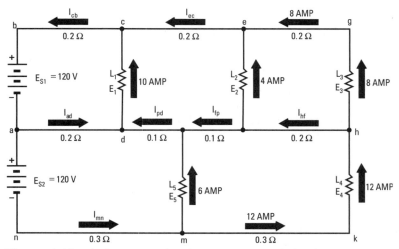

Figure 4-13 A three-wire distribution circuit with five loads.

Series-Parallel Circuits

Since the resistances of these wires are given, the voltage drops and the load voltages can be found after the currents are found.

To find the currents, it is best to start at the load farthest from the source in Figure 4-13. Electrons flow out of the negative terminal at n and return to the positive terminal at b. It is standard practice to assume that currents flowing toward a junction are positive and that currents flowing away from a junction are negative. Now, let us apply Kirchhoff's current law at junction h; the neutral current I_n (flowing from h to f) is evidently:

$$12 - 8 - I_{hf} = 0$$

or,

$$I_{hf} = 12 - 8 = 4 \text{ amperes}$$

In the same way, at junctions f, e, p, m, d, and c, we write at f,

$$4 - 4 - I_{fp} = 0$$

or,

$$I_{fp} = 4 - 4 = 0 \text{ amperes}$$

at e,

$$4 + 8 - I_{ec} = 0$$

or,

$$I_{ec} = 4 + 8 = 12 \text{ amperes}$$

at p,

$$6 + 0 - I_{pd} = 0$$

or,

$$I_{pd} = 6 + 0 = 6 \text{ amperes}$$

at m,

$$+I_{mn} - 6 - 12 = 0$$

or,

$$I_{mn} = 6 + 12 = 18 \text{ amperes}$$

at d,

$$I_{ad} + 6 - 10 = 0$$

or,

$$I_{ad} = -6 + 10 = 4 \text{ amperes}$$

at c,
$$I_{cb} + 10 + 12 = 0$$
or,
$$I_c = -10 - 12 = 22 \text{ amperes}$$

Therefore E_{S1} in Figure 4-13 supplies 22 amperes and E_{S2} supplies 18 amperes. The electron flow in all parts of the lower wire is outward from the source, and the electron flow in all parts of the upper wire is back toward the source. The current in the neutral wire is always equal to the difference in the currents in the two outside wires, and the electron flow is in the direction of the smaller of these two currents.

In Figure 4-13, the neutral current in section ad is 3 amperes, which is the difference between 19 amperes and 22 amperes; also, it is in the direction of the smaller current in section mn. The neutral current in section pd is 6 amperes, which is the difference between 18 amperes and 12 amperes; it is in the same direction as the 12 amperes in section ec. The neutral current in section fp is zero because the current in each outside wire in that section is 12 amperes. The neutral current in section hf is 4 amperes, which is the difference between 12 amperes and 8 amperes; it is in the direction of the smaller outside current in section ge.

To find the load voltages in Figure 4-13, we apply Kirchhoff's voltage law to the various individual circuits. Thus, to find the voltage E_1 across L_1, the algebraic sum of the voltages around the circuit $abcda$ is equated to zero. Starting at a,
$$-120 + 22 \times 0.2 + E_1 \times 0.2 = 0$$
or,
$$E_1 = 120 - 4.4 - 0.8 = 114.8 \text{ volts}$$

To find load voltage E_2, we trace circuit $dcefpd$. Starting at d,
$$-114.8 + 12 \times 0.2 + E_2 + 0 \times 0.1 - 6 \times 0.1 = 0$$
or,
$$E_2 = 114.8 - 2.4 - 0 + 0.6 = 113 \text{ volts}$$

To find voltage E_3, we trace circuit $feghf$. Starting at f,
$$-113 + 8 \times 0.2 + E_3 - 4 \times 0.2 = 0$$
or,
$$E_3 = 113 - 1.6 + 0.8 = 112.2 \text{ volts}$$

To find load voltage E_4, we trace circuit *nadpfhkmn*. Starting at *n*,

$$-120 - 4 \times 0.2 + 6 \times 0.1 + 0 \times 0.1 + 4 \times 0.2 + E_4$$
$$+12 \times 0.3 + 18 \times 0.3 = 0$$

or,

$$E_4 = 120 + 0.8 - 0.6 - 0 - 0.8 - 3.6 - 5.4 = 110.4 \text{ volts}$$

To find load voltage E_5, we trace circuit *nadpmn*. Starting at *n*,

$$-120 - 4 \times 0.2 + 6 \times 0.1 + E_5 + 18 \times 0.3 = 0$$

or,

$$E_5 = 120 + 0.8 - 0.6 - 5.4 = 114.8 \text{ volts}$$

The foregoing is an example of a comparatively involved series-parallel circuit problem. However, it is a very practical type of problem for the electrician. We recognize that this problem would have been very difficult to solve if we did not understand Kirchhoff's current law and Kirchhoff's voltage law.

Summary

A series-parallel circuit has branch currents and a total current in the same way that a parallel circuit does. In other words, the total current is equal to the sum of the total branch currents.

A wattmeter combines a voltmeter and an ammeter in one unit and reads power values in watts. If the voltmeter reading and the ammeter reading were multiplied, we would have the power, which would be read in watts.

Series-connected cells will produce a greater source voltage, and parallel-connected cells will produce a greater current capacity. In turn, if cells are connected in series-parallel, we can obtain a greater source voltage with a greater current capacity.

Horsepower is a unit of power, that is, the capacity of a mechanism to do work. It is the equivalent of raising 33,000 pounds one foot in one minute, or 550 pounds one foot in one second. One horsepower is equal to 746 watts.

When we have a balanced load on the positive and negative sides of a three-wire system, the current in the neutral wire is zero, and the currents in the feeders are equal. When the loads are unbalanced, the unbalanced current flows in the neutral wire. The direction of current flow in the neutral wire is always the same as that of the smaller of the currents in the positive and negative feeders. A three-wire system often operates at a total voltage of 240 volts, while the load can operate at 120 volts.

Test Questions

1. What is the definition of a series-parallel circuit?
2. Explain the meaning of a branch current.
3. State Kirchhoff's current law.
4. What is a junction?
5. Name two ways in which cells can be connected in series-parallel.
6. Why must the branch voltages in a series-parallel arrangement of cells be equal?
7. Discuss how a wattmeter is connected into a circuit.
8. What is the meaning of reduction by pairs? By the product-and-sum formula?
9. What is the meaning of reduction by reciprocals? By conductances?
10. How do we calculate the load power in a series-parallel circuit? The total power?
11. Explain how we find the efficiency of a series-parallel circuit.
12. How many watts are there in one horsepower?
13. What is the basic difference between a three-wire distribution system and a two-wire system?
14. State the arrangement of a neutral wire in a three-wire distribution system.
15. What is the meaning of unbalanced loads?
16. Why do unbalanced loads cause current flow in a neutral wire?
17. Would unbalanced loads have unbalanced voltages if the neutral wire has zero resistance? Why?
18. What is the direction of current in the neutral wire when the loads are unbalanced?
19. Can a load voltage be greater than the source voltage when the loads are unbalanced?
20. Name the laws of electricity that we use in solving a three-wire distribution system.

Chapter 5

Electromagnetic Induction

Electricians work with electromagnetic induction in many practical situations. Various electrical devices operate on the principle of electromagnetic induction, and circuits often have a response (circuit action) based on this principle. For example, we may note the following typical devices:

Transformers   Generators   Induction motors
Watt-hour meters   Ignition coils   Magnetos
Compensators   Induction regulators Choke reactors

Principle of Electromagnetic Induction

When a permanent magnet is inserted into a coil, as shown in Figure 5-1, the voltmeter deflects. In other words, the changing magnetic field in the coil *induces* a voltage in the coil winding. When the permanent magnet is withdrawn from the coil, a voltage of opposite polarity is induced in the coiled winding. Note that if the permanent magnet is motionless in the coil, there is no induced voltage. Note also that when the magnet is moved faster, a greater voltage is induced.

It is the *relative* motion of a magnetic field and a wire (or coil) that induces a voltage in the wire. For example, if a coil is moved through

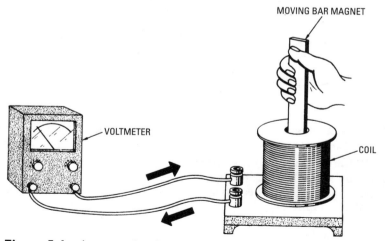

Figure 5-1   An example of electromagnetic induction.

a magnetic field, a voltage is induced in the coil. This principle is used in an electrical instrument called a *fluxmeter*, or *gaussmeter*, to measure the strength of a magnetic field. The instrument consists of a *flip coil* connected to a *galvanometer*, as shown in Figure 5-2. A flip coil consists of a few turns of wire, usually less than an inch in diameter. A galvanometer is simply a sensitive current meter. To use a fluxmeter, the electrician places the flip coil in the magnetic field to be measured and then flips the coil out of the field. In turn, the galvanometer reads the strength of the magnetic field in lines of force per square inch.

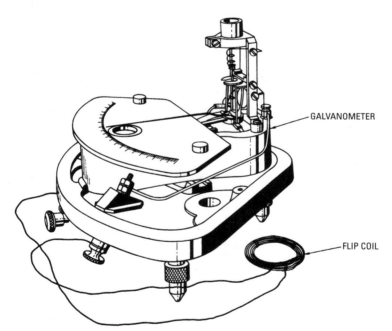

Figure 5-2 Construction of a flip coil and galvanometer.

The meter in Figure 5-2 is constructed so that the pointer remains at the point on the scale to which it is deflected when the flip coil is moved out of the magnetic field that is to be measured. Therefore, it makes no difference how fast or how slow the electrician moves the flip coil out of the magnetic field. Before another measurement of magnetic field strength is made, the pointer is adjusted back to zero on the scale.

Laws of Electromagnetic Induction

When a conductor moves with respect to a magnetic field, a voltage is induced in the conductor only if the conductor *cuts* the lines of magnetic force. To cut lines of force means to move at right angles to the lines, as shown in Figure 5-3. In this example, conductor *AB* is moving downward through the magnetic field, and this motion is at right angles to the lines of force. The lines of force are directed from the north pole to the south pole of the magnet. A voltage is induced in the conductor. A negative potential appears at *A*, and a positive potential appears at *B*. If we connect a wire from *A* to *B*, electrons flow in the wire as shown.

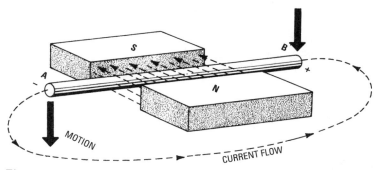

Figure 5-3 Induction of voltage in a conductor moving through a magnetic field.

Electricians need to know the polarity of the voltage induced in a conductor or the direction of induced current in the conductor (which amounts to the same thing). This is found by means of *Lenz's law*. Figure 5-4 illustrates Lenz's law. If you point in the direction of the magnetic flux with the index finger of your left hand and point your thumb in the direction that the conductor moves, your middle finger then points in the direction of electron flow in the conductor. Note that we hold the index finger, thumb, and middle finger at right angles to one another when we apply Lenz's law.

The amount of voltage that is induced in a moving conductor depends on how fast the conductor cuts lines of magnetic force. If a conductor moves at a speed such that it cuts 10^8 lines of force per second, there will be 1 volt induced in the conductor. Of course, if we use a number of conductors, the induced voltage will be multiplied by the number of conductors. For example, if the coil in Figure 5-1

110   Chapter 5

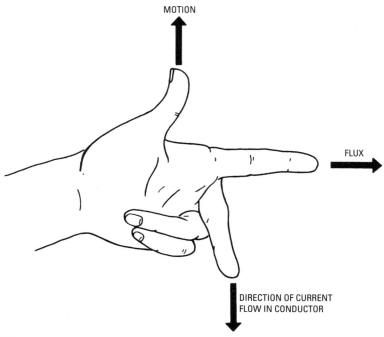

Figure 5-4   Illustrating Lenz's law.

has 100 turns, the induced voltage will be 100 times greater than if only one turn were used.

Since the amount of voltage induced in a conductor depends on how fast the conductor cuts lines of force, it might seem that the reading obtained on a fluxmeter such as the one shown in Figure 5-2 would depend on how fast the flip coil is removed from a magnetic field. However, we will find that the speed with which the flip coil is moved has no effect on the meter reading. The reason that speed makes no difference is that the galvanometer indicates the *total quantity of electricity* induced in the flip coil. This total quantity of electricity is the same whether a small current is induced over a long period of time or a large current is induced over a short period of time. A small current is produced by a small induced voltage, and a large current is produced by a large induced voltage.

Self-Induction of a Coil

Electricians often observe that a large flaming spark, or arc, appears across the switch contacts when the circuit to an electromagnet is

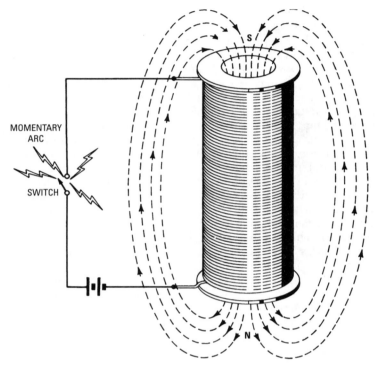

Figure 5-5 Self-induction causes arcing at the switch contacts.

opened. This circuit action is illustrated in Figure 5-5. When a larger electromagnet is used, the arc is hotter and larger. This circuit action is explained as follows:

1. When the switch is closed (Figure 5-5), current flows through the coil turns and builds up a magnetic field in the space surrounding the coil.
2. A magnetic field contains *magnetic energy*; the larger the coil, the greater is the amount of energy in its magnetic field.
3. At the instant the switch is opened, current flow stops in the coil, and the magnetic force lines quickly collapse into the coil. This is a moving magnetic field that cuts the coil turns and induces a voltage in the turns—the *voltage of self-induction*.
4. While the magnetic flux lines are collapsing, the energy in the magnetic field becomes changed into heat and light energy in the form of a momentary arc between the switch contacts.

There are some important facts concerning self-induction that we should keep in mind. The voltage of self-induction depends on the resistance of the arc path; if the switch is opened quickly, the arc path is longer and the self-induced voltage rises to a higher potential than if the switch is opened slowly. If the switch is opened with extreme rapidity, it may be easier for sparks to jump between turns of the coil than between the switch contacts. The high voltage of self-induction can break down the insulation on the coil wire and damage the magnet. Therefore, electrical equipment that uses large electromagnets carrying heavy current is often provided with a protective device to prevent the voltage of self-induction from rising to dangerous potentials.

We will find that the polarity of self-induced voltage is such that it tends to maintain current flow in the same direction as before the switch was opened. The direction of current in the arc between the switch contacts is the same as it was in the circuit before the switch was opened. This is often called the *flywheel effect* of an electromagnet. Just as a flywheel continues to turn in the same direction after an engine is turned off, an electromagnet continues to supply current in the same direction to a circuit after the switch is opened.

With reference to Figure 5-6, a lamp is connected across an electromagnet. When the switch is closed, the lamp glows dimly because the coils have comparatively low resistance and are connected in parallel with the lamp. However, when the switch is opened, the energy in the collapsing magnetic field causes the coil to apply a large

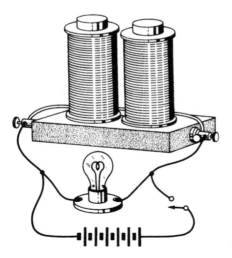

Figure 5-6 Self-induction of the electromagnet.

Electromagnetic Induction   113

momentary voltage across the lamp. Therefore, the lamp flashes brightly. If the coil is sufficiently large, the bulb will burn out when the switch is opened.

An electric bell is illustrated in Figure 5-7. The contact breaker automatically opens the circuit to the electromagnet when the armature is attracted; then, the armature springs back, and the circuit is again closed through the contacts. This opening and closing action repeats rapidly. Each time the contacts are opened, a spark jumps between the contacts due to the voltage of self-induction. To prevent the contacts from being burned away in a short time, special metals such as platinum are used. If a soft metal such as copper were used for the contacts, the life of the bell would be comparatively short.

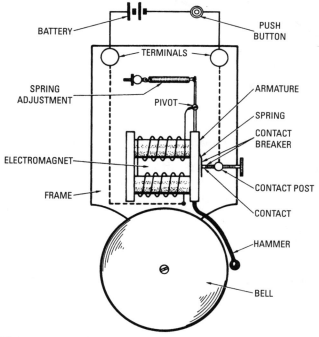

Figure 5-7   An electric bell.

Transformers

Transformers are widely used in electrical equipment. In this chapter we will consider certain types of transformers called *ignition coils*, or *spark coils*. Since a bar magnet produces an induced voltage when it is moved into or out of a coil, as shown in Figure 5-1, it follows that an electromagnet can be substituted for the bar magnet, as depicted in Figure 5-8. It also follows that we can place the electromagnet into the coil and induce a current in the coil by opening and closing the circuit of the electromagnet. *This is an example of transformer action.* A transformer has two windings, called the *primary* (P) and the *secondary* (S).

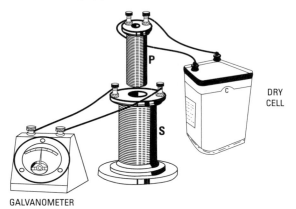

GALVANOMETER

Figure 5-8 A moving electromagnet generates a current.

When the primary circuit of a transformer is closed, as shown in Figure 5-9A, magnetic flux lines build up in the space surrounding each primary wire B. In turn, these expanding flux lines cut each secondary wire A and induce a voltage in the secondary. When the primary circuit is opened, the flux lines collapse and again cut the secondary wire; therefore, a voltage is again induced in the secondary. Note that when the switch is closed, the galvanometer deflects in one direction, but when the switch is opened, the galvanometer deflects in the other direction. The flux lines cut the secondary wire in opposite directions when the switch is opened and closed.

If the switch is held closed in Figure 5-9A, the galvanometer does not deflect. In other words, the pointer jumps on the galvanometer scale only during the instant that the magnetic field is building up or collapsing. It is easy to see that if the flux lines from a primary wire cut more than one secondary wire, more voltage will be induced in the secondary (see Figure 5-9C). Since an ignition coil in an automobile must produce a very high voltage in order to jump the gap of the spark plug, the secondary is wound with a large number

Electromagnetic Induction **115**

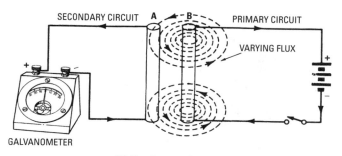

(A) Transformer action.

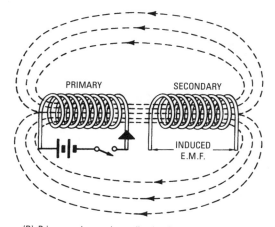

(B) Primary and secondary coils, showing magnetic flux lines.

Figure 5-9 Basic power transformer action.

of turns as compared with the primary. Figure 5-10 shows the arrangement of an ignition coil and its circuit. The primary is often called the low-tension winding, and the secondary the high-tension winding.

The ground circuit, shown by dotted lines in Figure 5-10, is familiar to every automotive electrician. The engine block and car frame serve as part of the ignition circuit. We observe that both the primary and secondary currents travel through ground circuits. If the metalwork of the car were not used as a ground path for the ignition system, the wiring would be more complicated. Note that a pair of contacts in the distributor are opened and closed by a

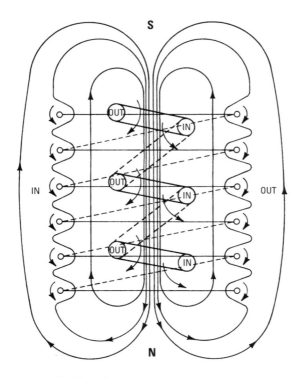

(C) Primary flux lines cutting secondary turns.

Figure 5-9 Continued.

revolving cam; these contacts operate as a switch to open and close the primary circuit.

A spark coil, or vibrator coil, is similar to an ignition coil except that the contacts are opened and closed by electromagnetic action, as seen in Figure 5-11. Note the *capacitor*, which is connected across the vibrator contacts. In actual practice, we will also find a capacitor connected across the contacts in Figure 5-10. Let us carefully observe the action of a capacitor, because an electrician is often concerned with this device. A capacitor basically consists of a pair of metal plates spaced near each other, with a sheet of insulating material, such as waxed paper, mica, or plastic, between the plates. With reference to Figure 5-12, a momentary current flows when the switch is closed. The battery forces electrons into the negative plate and takes electrons away from the positive plate. As soon as the capacitor *charges* to the same voltage as the battery, current flow stops.

Electromagnetic Induction 117

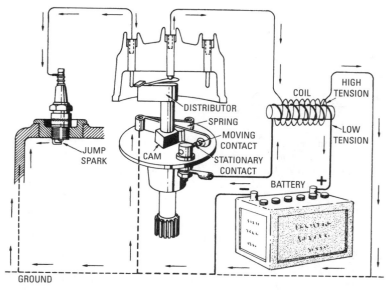

Figure 5-10 Diagram of automobile ignition system using battery, ignition coil, and high-tension distributor.

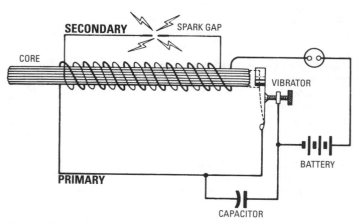

Figure 5-11 Diagram of a vibrator spark coil.

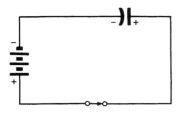

Figure 5-12 Capacitor action in a circuit.

In its charged condition, a capacitor contains electrical energy that can be returned to the circuit. For example, suppose that we open the switch in Figure 5-12. The capacitor remains charged because there is no path for escape of electrons from the negative plate to the positive plate. However, if we short-circuit the capacitor terminals (as with a screwdriver), we will observe that a momentary current flows through the short circuit and produces a snapping spark at the point of contact. The larger the area of the plates in the capacitor, and the closer together the plates are mounted, the greater is the short circuit on *discharge*.

Now, let us return briefly to Figure 5-11. It is highly desirable to prevent the vibrator contacts from sparking or arcing when the contacts open. The reason for this is that a spark or arc wastes electrical energy that would otherwise appear as induced voltage in the secondary. The capacitor prevents sparking or arcing when the contacts open because the self-induced voltage from the primary is then used to charge the capacitor instead of jumping the air gap between contacts. The self-induced voltage is stored in the capacitor. A short time after the capacitor is charged, it proceeds to discharge back through the primary winding. During the charging process, a little time is provided for the vibrator contacts to separate enough that a spark does not jump between them. Therefore, sparking is practically eliminated at the contacts, and the electricity stored in the capacitor next produces a heavy momentary current flow through the primary. In turn, the secondary voltage is much greater than if a capacitor were not used.

Previous mention has been made of electric fences, such as those installed by electricians in rural areas. An electric fence operates at sufficiently high voltage that an unpleasant shock is provided upon contact. It usually consists of a single wire supported by insulators on small posts that are spaced considerably farther apart than conventional fence posts. An important feature of the electrical system is the inclusion of sufficient series resistance that current flow is limited to only a few milliamperes. This current limitation guards against the possibility of fatal shock.

An electric fence operates generally at 7500 volts and 2 milliamperes short-circuit current. This is usually a DC voltage that is switched on and off rapidly by a mechanical or equivalent electric device. Interruption of the high voltage ensures that an animal or person contacting the electric fence will not freeze to the wire and be seriously burned or otherwise injured. The return circuit for an electric fence is provided by a ground connection. Electrical equipment for fence operation is often specified by codes, which may also require that the products have official approval.

Most of these devices are energized by a 120-volt line through an interrupter to the primary of a transformer that resembles a spark coil or ignition coil. The secondary may be wound with sufficiently fine wire that the short-circuit current is automatically limited to a small value. Otherwise, a capacitor or resistor is connected in series with the secondary circuit to limit the short-circuit current as required. In practically all installations, official inspection is required before the system can legally be placed in operation.

Advantage of an Iron Core

In Figures 5-10 and 5-11, the primary and secondary coils are wound on a single iron core. This core makes the ignition coil or spark coil more efficient than if the primary and secondary were wound on separate iron cores with an air space between them. A single iron core also makes for much greater efficiency than an air core. The reason for this improvement in efficiency is seen in Figure 5-13. If an air core or if separate iron cores are used as shown in Figure 5-13A and B, most of the magnetic force lines from the primary fail to cut the secondary coil. On the other hand, if a single iron core is used for both coils, as shown in Figure 5-13C, most of the primary magnetic force lines then cut the secondary coil.

Choke Coils

Choke coils are used in various types of electrical equipment. For example, in Figure 5-14A, we see how a choke coil is connected in a line conductor with a lightning arrester. If lightning strikes the line, a very large surge of current travels along the line. This heavy current surge would damage equipment at the end of the line (such as the generator shown in Figure 5-14B) unless a protective device called a lightning arrester is used. A lightning arrester is basically a spark gap connected between the line and ground.

Since the high voltage produced by a lightning stroke will take the easiest path to ground, the strong electrical surge tends to jump the spark gap to ground instead of flowing through the generator in Figure 5-14B. We know that it takes quite a high voltage to break

120 Chapter 5

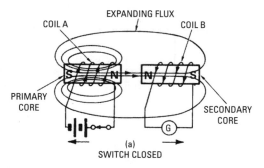

(A) The expanding flux fails to cut the secondary coil.

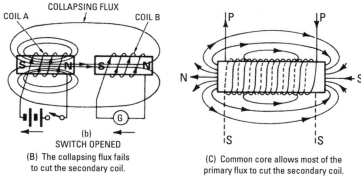

(B) The collapsing flux fails to cut the secondary coil.

(C) Common core allows most of the primary flux to cut the secondary coil.

Figure 5-13 Illustrating single and separate iron cores.

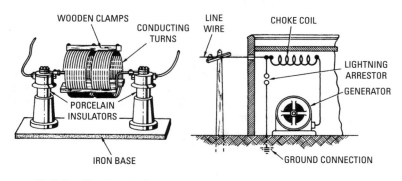

(A) Choke coil used in a circuit.

(B) Using a lightning arrester.

Figure 5-14 Illustrating the choke coil and its application.

down a spark gap, even when the gap has a comparatively small spacing. Therefore, to make sure that the air gap breaks down before the high-voltage surge reaches the generator, a choke coil is connected between the gap and the generator.

A choke coil consists of a number of turns of large-diameter wire, as shown in Figure 5-14A. Since the resistance of the choke coil is very small, the steady current from the generator flows easily through the choke coil, and there is very little voltage drop across the coil. However, when a sudden electrical surge travels down the line, the choke coil blocks the surge and makes it jump the spark gap to ground. Let us see why a choke coil stops a sudden electrical surge.

When a voltage is suddenly applied to a coil, as shown in Figure 5-15, the magnetic flux lines build up, or expand. As the flux lines build up, they cut the turns of the coil and induce an emf. Note carefully that *the induced voltage opposes the applied voltage*. This is called inductive opposition to a sudden current change; electricians usually call this opposition *inductive reactance*. Because the sudden current surge is opposed by the induced emf, it takes a certain amount of time for the surge to overcome this opposition and get through the coil.

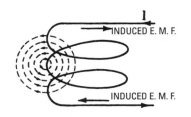

Figure 5-15 Illustrating counter electromotive force.

During the time that the current surge tries to build up a magnetic field in the choke coil, a very high voltage drop is produced across the spark gap, and the surge follows the easiest path to ground by jumping the gap. In other words, before a magnetic field can build up to any great extent in the choke coil, the lightning surge has been harmlessly passed to ground through the spark gap.

Electricians often call the induced emf in Figure 5-15 a *counter emf* (cemf), or *back emf*. Counter emf is found in any coil arrangement when voltage is suddenly applied to the coil. Therefore, it takes a certain amount of time for a surge voltage to produce current flow through a coil. On the other hand, after the switch has been closed for a short time in a circuit, as shown in Figure 5-16, the current

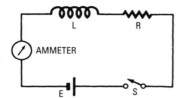

Figure 5-16 Illustrating a time delay before the ammeter indicates a current flow.

will have risen to its full value given by Ohm's law, and the current thereafter flows steadily as if the coil were not in the circuit.

Reversal of Induced Secondary Voltage

We will find that the secondary voltage reverses in polarity when the magnetic flux in a spark coil stops expanding and starts to collapse. For example, we see in Figure 5-13A that the expanding lines of force cut the secondary in a direction such that the right end of the secondary core has a south polarity. (This is shown by our left-hand rule.) Next, we see in Figure 5-13B that the collapsing lines of force cut the secondary in the opposite direction, and the right-hand end of the secondary core now has a north polarity. Therefore, the induced secondary voltage changes its polarity as the expanding magnetic field changes into a collapsing field.

Switching Surges

When a switch is closed in a circuit containing a coil, the current rises to a steady value after a short length of time. A magnetic field has then been built up in the space around the coil. Next, when the switch is opened, we know that a momentary arc is produced between the switch contacts, as shown in Figure 5-5. Electricians call this self-induced current a *switching surge*. We will now observe how much time it takes the current to build up in a coil circuit such as the one shown in Figure 5-17A when the switch is closed. If a small coil is used, the current builds up fast. On the other hand, if a large coil is used, the current builds up slowly.

The opposition that a coil has to current buildup is measured in *henrys*. Electricians call the *henry* the unit of inductance, just as they call the ohm the unit of resistance. If a coil has an inductance of 1 henry, the current will build up at the rate of 1 ampere each second, as shown in Figure 5-17B. In other words, one second after the switch is closed, the current will have built up from zero to 1 ampere. Two seconds after the switch is closed, the current will have built up to 2 amperes, and so on. If there were no resistance in the circuit, and if the battery could supply any amount of current, the

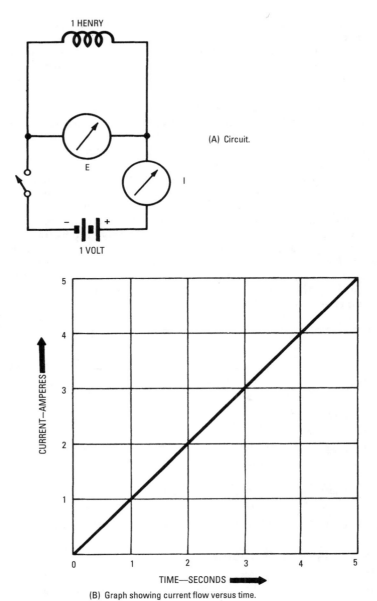

(A) Circuit.

(B) Graph showing current flow versus time.

Figure 5-17 A voltage of 1 volt applied to 1 henry of pure inductance.

circuit current would continue to build up as long as the switch remained closed.

However, we know that even large wire has a small amount of resistance and that a battery has some internal resistance. Therefore, we cannot have a circuit with zero resistance. This is just another way of saying that any coil connected to a battery is represented by a circuit containing both inductance and resistance, as shown in Figure 5-16. Therefore, the current buildup shown in Figure 5-17B will level off after enough time has passed, even if the circuit resistance is very small. This leveled-off value of current is given by Ohm's law: $I = E/R$.

Figure 5-18 shows how current builds up in large and small ignition coils. A large coil gives a hotter spark, but the contacts must be allowed to remain closed longer. A small coil gives a weaker spark, but the current buildup levels off more quickly. Therefore, to operate a large ignition coil at high speed, we must apply more voltage to the primary so that the current builds up to a sufficiently high value even though enough time is not available for the current to level off. If an ignition coil has a small primary inductance, it is called a fast coil. On the other hand, an ignition coil that has a large primary inductance is called a slow coil.

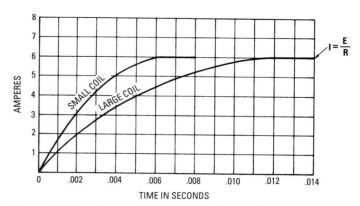

Figure 5-18 Current buildup in small and large ignition coils.

The Generator Principle

Various devices that operate on the principle of electromagnetic induction were noted at the beginning of this chapter. Although we cannot discuss all of these devices in one chapter, we will conclude

Electromagnetic Induction 125

with a brief explanation of the generator principle. Generators are familiar to automotive electricians, power-station electricians, and workers in many other branches of electricity. The basic generator principle was shown in Figure 5-3. A voltage is *generated* by induction in a wire that cuts magnetic flux lines.

A generator consists basically of conductors *rotating* in a magnetic field, as shown in Figure 5-19. For ease of explanation, a conducting loop is depicted that is divided into sections with light and dark lines. In (a), no lines of force are being cut, and no voltage is induced. However, in (b), the loop is cutting the lines of force at right angles, and the induced voltage is greatest. At (c), the loop cuts fewer lines of force, and the induced voltage decreases. At (d), the induced voltage again has its greatest value, but the polarity of the induced voltage is reversed. At (e), we have returned to the starting point, and no voltage is induced in the loop at this point in its rotation.

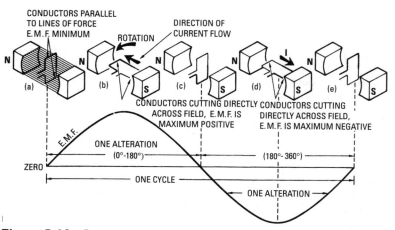

Figure 5-19 Basic electric generator action.

The induced voltage from the basic generator is supplied to an external circuit by means of sliding contacts called *brushes* and *slip rings*. Electricians call the rotating loop assembly an *armature*. The voltage supplied by the generator varies in value periodically and changes its polarity periodically as shown by the curve in Figure 5-19. This curve is called a *sine-wave alternating voltage*. If an alternating voltage is applied across a load resistor, a sine-wave alternating current (AC) flows through the resistor. The basic principles of AC current and AC circuit action are explained in the next chapter.

Summary

When a magnet is inserted into a coil that is connected to a voltmeter, the meter will deflect. The changing magnetic field in the coil induces a voltage in the coil winding. The relative motion of a magnetic field and a wire induces a voltage in the wire.

The amount of voltage induced in a moving conductor depends on how fast the conductor cuts magnetic lines of force. Since the amount of voltage induced in a conductor depends on how fast the conductor cuts lines of force, it might seem that the reading obtained on a fluxmeter would depend on how fast the flip coil is removed from a magnetic field, but this is not true. The reason that speed makes no difference is that the galvanometer indicates the total quantity of electricity that is induced in the flip coil. A flip coil is a small coil used for measuring a magnetic field. When connected to a galvanometer or other instrument, a flip coil gives an indication whenever the magnetic field of the coil or its position in the field is suddenly reversed.

Transformers play a very important part in the field of electricity. A transformer is used to convert a current or voltage from one magnitude to another, or from one type to another, by electromagnetic induction. When the primary circuit of a transformer is closed, magnetic flux lines build up in the space surrounding each primary wire. In turn, these expanding flux lines cut each secondary wire and induce a voltage in the secondary. When the primary circuit is opened, the flux lines collapse.

Primary and secondary coils are wound on a single iron core. The system is used in ignition coils and spark coils. A single iron core is more efficient than if the primary and secondary were wound on separate iron cores, because of the air space between the cores.

Test Questions

1. What is the principle of electromagnetic induction?
2. Explain Lenz's law.
3. If 1 volt is induced in a moving conductor, how many magnetic lines of force does the conductor cut each second?
4. What is meant by the self-induction of a coil?

Electromagnetic Induction 127

5. How can the voltage of self-induction break down the insulation of coil wires when the current to a large coil is suddenly stopped by opening a switch?
6. Discuss transformer action.
7. Why is the secondary of an ignition coil wound with a large number of turns?
8. How can sparking be suppressed between the vibrator contacts of a spark coil?
9. Explain how a capacitor stores electric charge when it is connected to a battery.
10. Why are the primary and secondary of an ignition coil or a spark coil wound on a single iron core?
11. Discuss the polarity of voltage induced in the secondary of an ignition coil when the magnetic flux of the primary is expanding. What happens while the flux is collapsing?
12. What is a choke coil? Give an example of its use.
13. Define counter electromotive force.
14. Why does it require a certain amount of time for a suddenly applied voltage to build up current flow through a coil?
15. If 1 volt is switched across a 1-henry coil, how fast does the current flow build up?
16. What do electricians mean by switching surge?
17. Why does the buildup of current flow through a coil level off to the value given by Ohm's law?
18. Explain the difference between a fast and a slow ignition coil.
19. State what electricians mean by the generator principle.
20. When a conducting loop rotates in a magnetic field, why does the induced voltage rise and fall during a complete revolution?
21. Why does the polarity of induced voltage in an armature loop reverse its polarity each time it completes one-half of a complete revolution?
22. Explain what is meant by a sine-wave alternating voltage.
23. How many cycles are generated by a complete revolution of an armature loop?
24. How many alterations are there in one cycle?

Chapter 6

Principles of Alternating Currents

An alternating voltage or current is usually defined as a voltage or current that changes in strength according to a sine curve. An alternating voltage reverses its polarity on each alternation, and an alternating current reverses its direction of flow on each alternation. Electricians usually speak of AC voltage and AC current. An AC generator is commonly called an *alternator*. Automotive electricians are concerned with the repair and maintenance of alternators. Figure 6-1 shows the development of an AC voltage. The point of maximum voltage occurs at 90° and is also called the *crest* voltage or *peak* voltage of the sine curve.

Frequency

The frequency of an AC voltage of current is its number of cycles per second. For example, electricity supplied by public utility companies in the United States has a frequency of 60 cycles per second (60 cps). Electricians sometimes call cycles per second by the newer name, *hertz*. Thus, 60 cycles per second is the same as 60 hertz (60 Hz). Figure 6-2 shows the relation between time and voltage at a frequency of 1 cps. Each alternation is completed in $1/2$ second. In this example, the time for one complete cycle, called the *period* of the AC voltage, is 1 second. If an AC voltage has a frequency of 60 cps, its period is $1/60$ second.

Instantaneous and Effective Voltages

An *instantaneous* voltage is the value of an AC voltage at a particular instant. For example, the maximum voltage shown in Figure 6-1 is an instantaneous voltage; the positive maximum voltage occurs at 90°, and the negative maximum voltage occurs at 270°. Electricians generally consider that the most important instantaneous AC voltages are the *maximum* (crest or peak) voltage and the *effective* voltage. We will first explain what is meant by the effective voltage and then observe why it is so important. The effective voltage is 70.7 percent of the peak voltage and occurs at 45°, as shown in Figure 6-3.

Practically all AC voltmeters read the effective value of an AC voltage. For example, the AC voltmeter illustrated in Figure 6-4 has its scale marked to indicate effective voltages. An effective AC voltage may also be called a root-mean-square (rms) voltage. Since

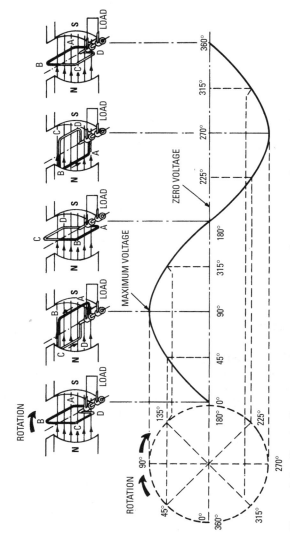

Figure 6-1 Simple illustration showing generation of a sine curve during an alternating current cycle.

Principles of Alternating Currents 131

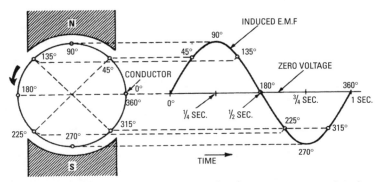

Figure 6-2 Curve showing relationship between time and induced voltage in an alternating current circuit.

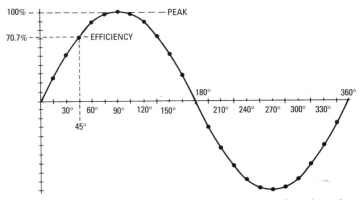

Figure 6-3 The effective value is 70.7 percent of peak and occurs at 45°.

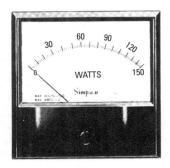

Figure 6-4 An AC voltmeter that indicates effective voltage. *Courtesy Simpson Electric Co.*

nearly all AC voltmeters indicate effective values, this particular value is always understood when an electrician speaks of 120 volts, 240 volts, etc. The word effective is understood, although it is not commonly stated in electrical diagrams and instruction sheets. If an electrician wishes to speak of a crest or peak voltage, he will state "169 crest volts" or "169 peak volts," for example. An effective voltage of 120 volts has a crest voltage of approximately 169 volts.

Let us see why effective AC volts are so important in practical work. For example, Figure 6-5 shows the arrangement of a resistance element in an electric range. A three-wire distribution system is used, with 230 volts between the feeders and 115 volts between each feeder and the neutral wire. We will find that the resistance element will have the same temperature whether we use 230 DC volts between the feeders or use 230 effective AC volts between the feeders. In other words, *a given effective AC voltage provides the same amount of power as the same value of DC voltage.*

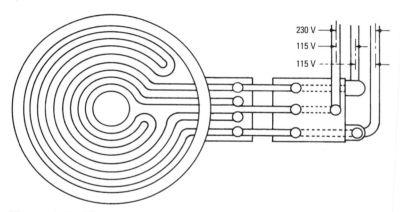

Figure 6-5 The resistance element in an electric range.

As another practical example, let us consider an electric light bulb that is rated for operation on 120 volts AC. This means that the bulb is rated for operation on 120 effective volts of AC. We will find that the bulb provides the same amount of light if it is operated on 120 volts DC. Similarly, a soldering iron, toaster, percolator, or space heater draws the same amount of power from either a 120-volt DC line or a 120-effective-volt AC line. The power in watts consumed by a resistance is equal to E^2/R, whether E is taken in DC volts or in effective AC volts.

Ohm's Law in AC Circuits

In most practical electrical work with AC circuits, we use effective values of voltage and current in Ohm's law. Just as an effective voltage is 70.7 percent of the peak voltage, an effective current is 70.7 percent of the peak current. In terms of effective (rms) values of current and voltage, we write Ohm's law:

$$I_{eff} = \frac{E_{eff}}{R}$$

or,

$$I_{rms} = \frac{E_{rms}}{R}$$

As we proceed with our discussion of AC circuits, we will simply write I and E with the understanding that the letters stand for I_{eff} and E_{eff} or for I_{rms} and E_{rms}, which are the same thing.

Electricians sometimes work with peak voltages (crest voltages). We know that an effective voltage is 0.707 of the peak voltage. Therefore, a peak voltage is 1.414 times the effective voltage. Similarly, a peak current is 1.414 times the effective current. Therefore, we may write Ohm's law for peak values as follows:

$$1.414 I_{eff} = \frac{1.414 E_{eff}}{R}$$

or,

$$I_{peak} = \frac{E_{peak}}{R}$$

Once in a while, we work with instantaneous voltages and currents. We write Ohm's law for instantaneous voltages and currents as follows:

$$i = \frac{e}{R}$$

where small letters i and e represent instantaneous values.

In other words, Ohm's law is true for effective, peak, or instantaneous values of voltage and current. The only thing that we must watch for is to use the *same kind* of values in Ohm's law. We would get incorrect answers if we used a peak voltage value and an effective current value in Ohm's law. With reference to Figure 6-3, various instantaneous values are indicated by dots along the sine curve. We see that since the instantaneous value at 45° is the same as the effective value, Ohm's law for effective values is merely a special case of Ohm's law for instantaneous values. Similarly, since the instantaneous value at 90° is the same as the peak value, Ohm's law for peak values is merely a special case of Ohm's law for instantaneous values.

It would be very tedious to draw sine waves to represent AC voltages. Therefore, electricians represent the armature of a basic AC generator (see Figure 6-1) by means of an arrow, as shown in Figure 6-6. This arrow is called a *vector*. As the arrow rotates, it generates a sine curve of voltage, just as a loop of wire rotating in a magnetic field generates a sine curve of voltage. A sine curve is usually called a sine wave because its shape suggests a water wave. Note that the length of the vector E in Figure 6-6 represents the peak voltage of the sine wave, and the dotted line, which has a length e, represents the instantaneous voltage of the sine wave.

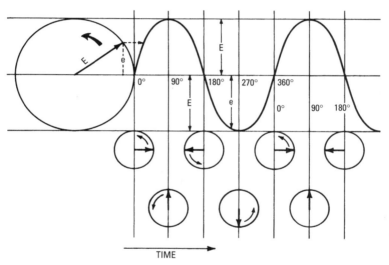

Figure 6-6 Vector representation of an AC voltage.

When an AC voltage is applied across a resistance, a sine-wave current flows through the resistance, as shown in Figure 6-7B. Note that the voltage and current vectors rotate together and that the current rises and falls in step with the voltage. For example, at time t_1, the voltage in Figure 6-7A has an instantaneous value indicated by point P_e; at this instant, the current in Figure 6-7B has an instantaneous value indicated by point P_i. Figure 6-7C shows how the voltage and current vectors rotate together.

Principles of Alternating Currents 135

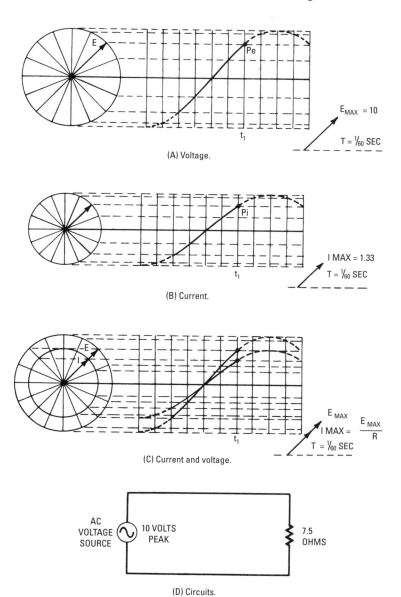

Figure 6-7 Vector representation of AC voltage and current.

In the foregoing example, the peak voltage is 10 volts, which is applied across a resistance of 7.5 ohms, as shown in Figure 6-7D. Since $I = E/R$, the peak current is 1.33 amperes. Since the frequency is assumed to be 60 cps (60 Hz), the current and voltage go through their peak positive values each 1/60 second and go through their peak negative values each 1/60 second. Thus far, we see that Ohm's law is applied to resistive AC circuits in the same way that Ohm's law is applied to resistive DC circuits.

Power Laws in Resistive AC Circuits

Let us consider the power that is consumed in the 7.5-ohm resistor in Figure 6-7. Because AC voltage and current have instantaneous, effective, and peak values, it follows that we can find three power values in the 7.5-ohm resistor. The basic power laws are written:

$$\text{Power} = EI = I^2 R = \frac{E^2}{R}$$

Now, consider the use of effective voltage and current values in the power laws. This is the most important case, because effective power corresponds to DC power, as we have learned. In other words, if we find the power consumed by a resistor when an effective voltage is applied, the same power will be consumed if we apply a DC voltage equal to the effective voltage. In terms of effective voltages and currents, the power laws are written:

$$\text{Effective power} = E_{\text{eff}} I_{\text{eff}} = I_{\text{eff}}^2 R = \frac{E_{\text{eff}}^2}{R}$$

Therefore, the effective power in the 7.5-ohm resistor (see Figure 6-7) can be found by multiplying the effective voltage by the effective current. The effective (rms) voltage is equal to 10 × 0.7071, or 7.071 volts. The effective current is equal to 1.33 × 0.7071, or 0.94 ampere. This is written:

$$\text{Watts}_{\text{eff}} = 7.071 \times 0.94 = 6.65, \text{ approximately.}$$

Of course, the power in the resistor rises and falls. The *peak power* in the resistor is given by:

$$\text{Peak power} = E_{\text{peak}} I_{\text{peak}}$$

Therefore, the peak power in the example of Figure 6-7 is found as follows:

$$\text{Watts}_{\text{peak}} = 10 \times 1.33 = 13.3, \text{ approximately.}$$

Principles of Alternating Currents   137

Note that *effective power is equal to* $1/2$ *of peak power*. Since the power rises and falls in the resistor, we have at any instant an *instantaneous power* value in the resistor:

Instantaneous power $= ei$

or,

Watts$_{\text{instantaneous}} = ei$

or,

$w = ei$

Electricians represent instantaneous power by a small letter w, just as e represents instantaneous voltage and i represents instantaneous current. It is obvious that the instantaneous power value at 45° in Figure 6-7 is the same as the effective power value and that the instantaneous power value at 90° is the same as the peak power value. Note that power does not depend on frequency; the power value is the same, regardless of frequency.

Combining AC Voltages

We know how to combine DC voltages by connecting cells in series to form a battery. We also know how to combine voltage drops when we trace around a DC circuit according to Kirchhoff's voltage law. There are certain types of electrical jobs in which an electrician needs to know how to combine AC voltages. To understand how this is done, let us consider the simple example shown in Figure 6-8. Two lamps are connected in series with an AC source. The source supplies 240 volts; the filament resistances of the lamps are the same, and there is a voltage drop of 120 volts across each lamp.

We see that Kirchhoff's voltage law is applied to the circuit in Figure 6-8 just as if it were a DC circuit. It will be helpful for us to observe how the vector voltages combine. The source voltage E_S is represented by a vector as shown in Figure 6-8B. We know that the *polarity of the load* voltage opposes the *polarity of the source* voltage at any given instant. Therefore, the load-voltage vectors are drawn as shown in Figure 6-8C. Note that each vector is half the length of the source vector, and each load-voltage vector points in the *opposite* direction to the source vector. We add the load-voltage vectors by placing them end-to-end. Observe that since $E_{L1} + E_{L2}$ has the same length as E_S but points in the opposite direction, the vectors cancel; this shows that the algebraic sum of the voltage drops around the circuit is equal to zero.

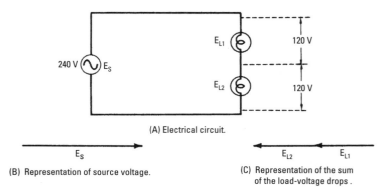

Figure 6-8 Two lamps connected in series with an AC source.

Note that E_S, E_{L1}, and E_{L2} have the same frequency in Figure 6-8. Vectors can be used to represent AC voltages, provided that the voltages are sine waves and provided that they have the same frequency. Vectors can also be used to combine sine wave voltages that have the same frequency but which are more or less out of step. For example, let us consider the two AC voltage sources shown in Figure 6-9. A pair of sine wave voltage sources E_1 and E_2 are shown in Figure 6-9A. We know that cells can be connected in series-aiding or in series-opposing. If the two AC voltage sources are connected so that the sine waves are in step, as shown in Figure 6-9B, the voltages are in series-aiding and the voltages add. On the other hand, if the sine waves are completely out of step, as shown in Figure 6-9C, the voltages are in series-opposing and the voltages subtract.

The foregoing examples are very simple. However, electricians may have to work with equipment in which a pair of sine wave voltages that have the same frequency are halfway out of step, as shown in Figure 6-10. The peaks of the two sine wave voltages are 90° apart. In this example, the AC voltage sources are represented as a pair of generators; however, we will find that other devices such as capacitors or coils produce sine waves that have peaks 90° apart. It is easier to understand such situations if we start with generators such as those shown in Figure 6-10.

The sine wave voltages generated in loops a and b (Figure 6-10A) are 90° apart because the loops are located 90° apart on the two-pole armatures. These armatures are assumed to be mounted on a common shaft in this example. Note that when loop a is cutting squarely across the magnetic field, loop b is moving parallel to the

Principles of Alternating Currents 139

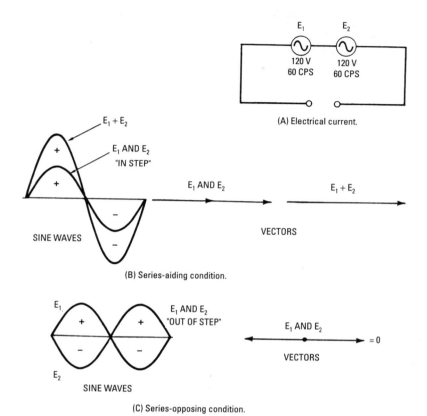

Figure 6-9 Two AC voltage sources with the same frequency, connected in series.

field and is not cutting magnetic lines of force. Therefore, the voltage in loop *a* is maximum when the voltage in loop *b* is zero. When two voltages are 90° apart, electricians say that the two voltages are 90° *out of phase*. Or, they will say that the *phase angle* between E_a and E_b is 90°, as shown in Figure 6-10B.

If loops *a* and *b* in Figure 6-10A are connected in series, and the maximum voltage generated in each loop is 10 volts, these voltages do *not* combine to give a maximum voltage of 20 volts, simply because the maximum voltages of the loops do not occur at the same time. The peak voltages are separated by 90°, or by ¼ cycle. Because

140 Chapter 6

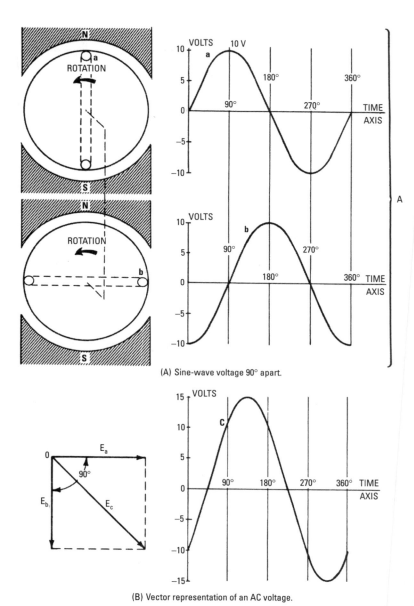

(A) Sine-wave voltage 90° apart.

(B) Vector representation of an AC voltage.

Figure 6-10 Two AC voltages of the same frequency that are halfway out of step.

Principles of Alternating Currents 141

(C) The sum of a and b.

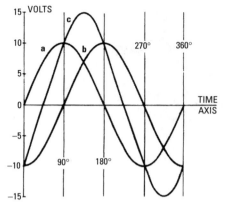

Figure 6-10 Continued.

they are out of phase, we cannot simply add the two voltages. However, we can add E_a and E_b vectorially, as shown in Figure 10B, and their vectorial sum is given by E_c.

In Figure 6-10B, E_c is the vector sum of E_a and E_b; in other words, E_c is the diagonal of a parallelogram that has sides of length E_a and E_b. In this example, the parallelogram is a square, because $E_a = E_b$. Therefore, E_c is equal to $\sqrt{2}$ times 10 volts, or approximately 14.14 volts. Let us observe Figure 6-10C. This diagram shows how the sine waves E_a and E_b add to give E_c. In other words, if we go along point-by-point and add the instantaneous values of E_a and E_b, we will get the sine wave E_c. We observe that it is much easier and quicker to find the vector sum of E_a and E_b.

Figure 6-11 shows how the square on the hypotenuse of a right-angled triangle is equal to the sum of the squares on the other two sides. This is a very important fact for us to remember, because it shows us how two AC voltages are combined when the voltages are not the same. For example, let us suppose that two AC voltages of the same frequency have peak values of 8 volts and 6 volts and are 90° out of phase. As shown in Figure 6-11, the vector sum of E_a and E_b is 10 volts. If we prefer, we can find E_c by writing:

$$E^2{}_c = E^2{}_a + E^2{}_b$$

or,

$$E_c = \sqrt{E^2{}_a + E^2{}_b}$$

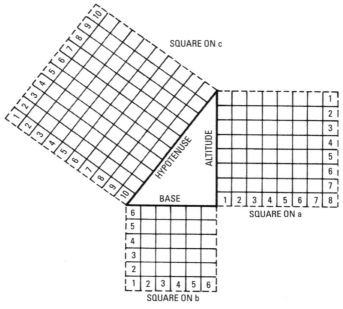

Figure 6-11 The square on the hypotenuse is equal to the sum of the squares on the sides of the right-angled triangle.

In the foregoing example, $E_a = 8$ volts and $E_b = 6$ volts. Therefore:

$$E_c = \sqrt{8^2 + 6^2} = \sqrt{64 + 36} = \sqrt{100} = 10 \text{ volts}$$

Whether we use vectors or arithmetic, we find that E_c has a peak voltage of 10 volts. To find the effective or rms value of E_c, we multiply 10 by 0.707 to obtain 7.07 volt rms. The principle of vector sums is the most important principle of AC for us to keep in mind, because we will work with vector sums again and again as we proceed with our study of practical electricity.

Transformer Action in AC Circuits

Transformers are among the most basic and useful devices used in AC circuits. We have previously considered transformer action in ignition coils and spark coils. Now, we must see how transformers operate in AC circuits. The arrangement of an elementary transformer

Principles of Alternating Currents 143

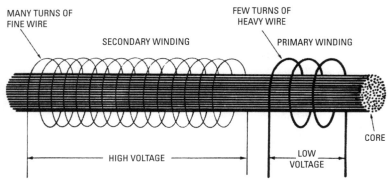

Figure 6-12 Transformer consisting of one core and primary and secondary windings.

is shown in Figure 6-12. In this example, the primary has only a few turns, while the secondary has many turns. An iron core is used to obtain good efficiency by making most of the primary magnetic flux cut the secondary. If we apply a low AC voltage to the primary of the transformer, we will get a high AC voltage from the secondary, as shown in Figure 6-13.

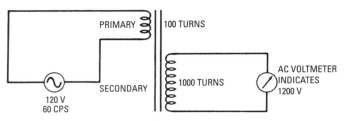

Figure 6-13 Step-up transformer action.

When the secondary has more turns than the primary, a transformer is called a *step-up* transformer. The voltage step-up ratio is equal to the turns ratio:

$$\frac{E_{\text{pri}}}{E_{\text{sec}}} = \frac{N_{\text{pri}}}{N_{\text{sec}}}$$

where E represents AC voltage,

N represents number of turns.

144 Chapter 6

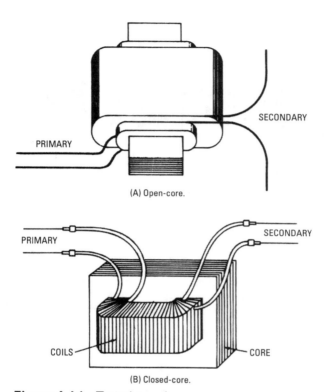

Figure 6-14 Typical transformers.

Open-core transformers, as shown in Figure 6-14A, are used in some branches of electrical work. However, closed-core transformers, as depicted in Figure 6-14B, are much more widely used, particularly when the secondary must supply considerable current. Closed-core transformers are more efficient because practically all of the primary magnetic flux cuts the secondary turns. Electricians call the construction in Figure 6-14B a *shell-type* core.

If a transformer has more turns on the primary than on the secondary, it is called a *step-down* transformer. The transformer shown in Figure 6-15 steps down the applied voltage by a 4-to-1 ratio. In this example, a 25-ohm load resistor is connected to the secondary terminals. Note carefully that a step-down transformer steps up the secondary current. On the other hand, a step-up transformer steps

Principles of Alternating Currents 145

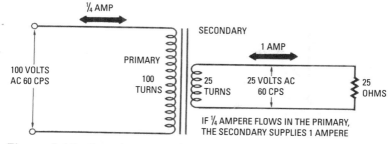

Figure 6-15 Step-down transformer.

down the secondary current. In the example shown in Figure 6-15, the primary voltage is stepped down from 100 volts to 25 volts; the primary current is stepped up from ¼ ampere to 1 ampere.

The primary-to-secondary current ratio is equal to the secondary-to-primary turns ratio:

$$\frac{I_{pri}}{I_{sec}} = \frac{N_{sec}}{N_{pri}}$$

This is just another way of saying that the primary power is equal to the secondary power. For example, in Figure 6-15, the primary power is 100 × 0.25, or 25 watts; the secondary power is 25 × 1, or 25 watts. Of course, if a transformer has poor efficiency, the secondary power will be less than the primary power. For example, let us suppose that an open-core transformer is used which is only 80 percent efficient. If the primary draws 25 watts, the secondary will supply only 20 watts. Note that shell-type transformers have a high efficiency, which may be as great as 99 percent.

Another type of closed-core transformer is shown in Figure 6-16 and is called a *core-type* transformer. The efficiency of a core-type transformer is between that of an open-core transformer and a shell-type transformer. Note in Figure 6-16 that some of the magnetic force lines stray from the iron path into the surrounding air. These stray flux lines are called *leakage flux* and cause reduced efficiency.

Let us consider the AC resistance of the primary and secondary for the loaded transformer shown in Figure 6-15. The primary draws ¼ ampere with 100 volts applied; it follows from Ohm's law that the AC resistance of the primary is equal to E/I, or 400 ohms. The secondary supplies 1 ampere at 25 volts; thus, the AC resistance is 25 ohms. This is an AC resistance ratio of 400/25, or 16-to-1. In

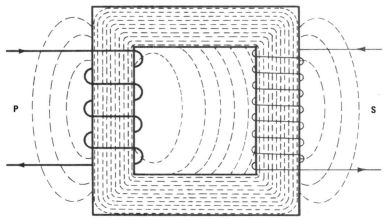

Figure 6-16 The core-type transformer.

other words, the AC resistance ratio of a transformer is equal to the *square* of the turns ratio:

$$\frac{N^2_{pri}}{N^2_{sec}} = \frac{R_{AC\ pri}}{R_{AC\ sec}}$$

We have 100 primary turns and 25 secondary turns in Figure 6-15. We write:

$$\frac{100^2}{25^2} = \frac{10,000}{625} = \frac{16}{1}$$

Electricians usually call the AC resistance ratio of a transformer its *impedance ratio*. Impedance simply means AC resistance, as will be explained in greater detail later. At this time we are chiefly concerned with the importance of the transformer impedance ratio in practical electrical work. For example, we have an AC source in Figure 6-17 with an internal resistance of 400 ohms. We will ask how maximum power will be transferred from this source to a 25-ohm load. Maximum power will be transferred if R_L is made to look like it has the same resistance as R_{in}. This is accomplished by the transformer.

Just as we found in Figure 6-15, the primary impedance is 400 ohms and the secondary impedance is 25 ohms in this example. Therefore, the source in Figure 6-17 works into a 400-ohm primary and maximum power is transferred into the primary. The load

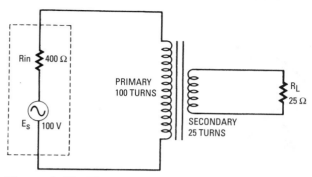

Figure 6-17 The transformer turns ratio provides maximum power from the source to the load.

works out of a 25-ohm secondary and maximum power is transferred into R_L. Therefore, a 4-to-1 turns ratio on the transformer provides maximum power transfer from the source to the load in this example. For an ideal transformer, the system efficiency would be 50 percent, because half of the available power in the primary circuit is lost in the internal resistance. In turn, if the transformer is 99 percent efficient, the system efficiency will be 49.5 percent.

It is important to remember that voltage, current, and power relations in a transformer do not depend on frequency. We could use a 60-cps source or a 1000-cps source in Figure 6-17, and the relations of voltage, current, and power would not be changed. However, a core that is very efficient at 60 cps might be less efficient at 1000 cps. We will merely observe in this chapter that the electrician always selects a transformer that is rated properly for the particular application. For example, the transformer in Figure 6-15 would be rated for a primary voltage of 100 volts, for 25 watts of power, and for an operating frequency of 60 cps. A 4-to-1 turns ratio would be specified in this application.

In various types of electrical equipment, a transformer may be wound with several secondaries to supply several loads at different voltages. Figure 6-18 illustrates small transformers that have several secondary windings. The total power supplied by such a transformer is equal to the sum of the power values supplied by the secondaries. A transformer should never be operated above its power rating. However, a transformer operates satisfactorily at any power value below its power rating. If a mistake is made and a transformer is operated considerably above its power rating, the efficiency will be

Figure 6-18 Typical transformers with more than one secondary winding. *(Courtesy United Transformer Corp)*

poor and the transformer will run hot. Insulation is likely to become burned with resulting short circuits between turns in the coils.

Core Loss and Core Lamination

Because it has a low reluctance, iron is the only suitable substance for use in transformer cores. However, iron is also a conductor of electricity. Therefore, AC current in a transformer winding induces voltage in the core and causes current to flow in the core as shown in Figure 6-19. These core currents are called *eddy currents*. Eddy currents are undesirable because they reduce the efficiency of the transformer. In other words, iron has resistance, and eddy currents produce an I^2R core loss. Power that is lost in the core cannot be transferred to the secondary winding.

If iron were an insulator, it would have no eddy current loss. Therefore, we make the iron core look like it has more resistance to eddy currents. This is done by *laminating* the core, as shown in Figures 6.19B and C. Note that the path for eddy current flow is longer in B than in A. The path for eddy current flow is even longer in C. When the total path for eddy current flow is very long, its resistance is very high; in turn, the amount of eddy current that flows is small.

We can make the eddy current value as low as may be desired by simply using more and thinner laminations. Note that laminations must be insulated from one another; otherwise, a laminated core would be the same as a solid core insofar as eddy current flow is concerned. Laminations are usually lacquered to provide insulation. A transformer core is made by stacking laminations as shown in Figure 6-20. Ignition coils and spark coils generally use cores made

Principles of Alternating Currents 149

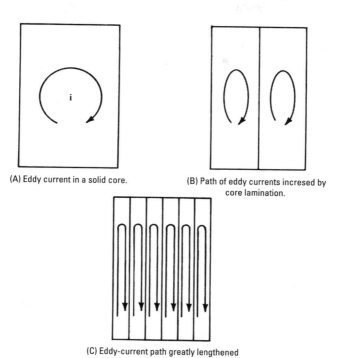

(A) Eddy current in a solid core.

(B) Path of eddy currents incresed by core lamination.

(C) Eddy-current path greatly lengthened by numerous laminations.

Figure 6-19 Transformer core action.

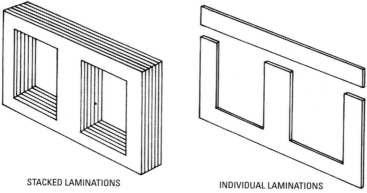

STACKED LAMINATIONS

INDIVIDUAL LAMINATIONS

Figure 6-20 A core formed by stacked laminations.

from bundles of iron wires, as seen in Figure 6-12. The iron wires are lacquered to provide insulation, and eddy currents are greatly reduced, just as in a laminated core used in transformers.

DC versus AC Resistance

The primary and secondary windings of an ideal transformer would have zero resistance. However, copper wire has a certain amount of resistance. Therefore, the primary and secondary windings have certain DC resistance values. The *winding resistance* of a coil can be measured with an ohmmeter. This DC resistance is undesirable because it produces an I^2R loss in transformer operation and reduces the efficiency of the transformer. The I^2R power loss due to the DC winding resistance of the primary and secondary coils is called the *copper loss* of the transformer. This copper loss can be reduced by winding the primary and secondary coils with larger wire. Of course, the DC resistance of the windings cannot be reduced to zero, although it might be so small that it can be neglected in practice.

Let us consider the AC resistance of the primary and secondary. When the secondary of an ideal transformer is open-circuited, the primary would have infinite reactance and would draw no current from an AC source. However, a practical transformer does not have infinite primary reactance, although this reactance is very high in the case of an efficient transformer. Therefore, the primary of an efficient transformer draws a slight amount of AC current from the source when the secondary is open-circuited. This is called the no-load current. The no-load current is not quite 90° out of phase with the primary voltage because a practical transformer has copper loss and core loss. These losses consume a small amount of power from the source.

It follows from previous discussion that if a resistance load is connected across the secondary terminals of a transformer, a substantial AC current is drawn by the primary. We say that the connection of the secondary load has caused the primary to have a lower value of AC resistance. Note that this AC resistance cannot be measured with an ohmmeter; the value of the AC resistance is simply a voltage/current ratio. It is the ratio of primary AC voltage to in-phase primary AC current. Even an ideal transformer would have a value of primary AC resistance that depends on the value of the load connected across the secondary terminals.

Summary

Voltage is continually varying in value and reversing its direction at regular intervals in an alternating-current circuit. The frequency of an AC voltage or current is its number of cycles per second, sometimes called *hertz,* which is the newer name. An instantaneous voltage is the value of an AC voltage at a particular instant.

The most important instantaneous AC voltages are the maximum (crest or peak) and the effective voltages. The effective voltage is 70.7 percent of the peak voltage. The effective voltage is also called rms (root-mean-square) voltage. Peak or crest voltage is the maximum value present in a varying or alternating voltage and can be either positive or negative.

Transformers are useful devices in AC circuits. If a secondary has more turns than the primary, it is called a step-up transformer. Open-core transformers are used in some branches of electrical work, but closed-core transformers are more efficient because the primary magnetic flux lines cut the secondary, supplying considerably more current.

The efficiency of a core-type transformer is between that of an open-core and a shell-type transformer. Some of the magnetic lines of force stray from the core into the surrounding air. These stray flux lines are called *leakage flux* and cause reduced efficiency in the transformer output.

Eddy current is current developed in the core, and it reduces the transformer voltage. Since iron is used in transformer cores, and iron has resistance, the cores are laminated to block the paths through which eddy currents can flow.

Test Questions

1. What is an alternator?
2. Explain what is meant by the frequency of an AC voltage.
3. How is the frequency of an AC voltage related to its period?
4. What do we mean by an instantaneous voltage? An effective voltage?
5. Discuss the difference between an effective voltage and a peak voltage.
6. Does Ohm's law hold true for instantaneous, effective, and peak values of AC voltages and currents?
7. What is a vector? Why is it helpful to represent AC voltages and currents by vectors?
8. Explain the difference between peak power, effective power, and instantaneous power.
9. Draw voltage vectors to illustrate Kirchhoff's voltage law for an AC series circuit.

Chapter 6

10. Show how AC voltage sources can be connected in series-aiding and in series-opposing.
11. What is meant when we say that two AC voltages are 90° out of phase?
12. If two AC voltages are equal but are 90° out of phase, what is their total voltage?
13. If two AC voltages have values of 8 volts and 6 volts and are 90° out of phase, what is their total voltage?
14. Define a step-up transformer. A step-down transformer.
15. Why does a step-up transformer step the AC voltage up and step the AC current down? (*Hint:* The primary power is equal to the secondary power.)
16. Define an open-core transformer. A closed-core transformer.
17. What is the difference between a core-type and a shell-type transformer?
18. How is a transformer used to obtain maximum power transfer between an AC source and a load?
19. State the percentage efficiency of a high-quality transformer.
20. Why does stray flux (leakage flux) reduce the efficiency of a transformer?
21. Discuss eddy currents in transformer cores.
22. What is meant by a laminated core?
23. Why must core laminations be insulated from one another?
24. Explain why laminations reduce core loss.
25. Why should a transformer be operated within its power rating?

Chapter 7

Inductive and Capacitive AC Circuits

Inductive AC circuits are used in many types of electrical equipment. For example, electricians who work in theaters and motion-picture studios are concerned with lighting-control equipment. The basic principle of lighting control is shown in Figure 7-1. To control the brightness of the lamp, it is connected in series with a coil that has adjustable inductance. The coil has its least inductance when the iron core is withdrawn; the coil has its greatest inductance when the core is fully inserted. In turn, the circuit current is at its greatest when the inductance value is at its least, and the current is at its least when the inductance is at its greatest.

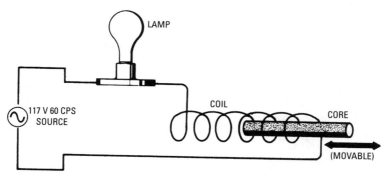

Figure 7-1 Basic lighting-control circuit.

A coil with an adjustable core is called a *variable reactor* or a *variable inductor*. Inductance opposes the flow of AC current because self-induction causes the inductance to have *reactance*. We know that inductance is measured in henrys. We will find that although reactance is different from resistance, reactance is nevertheless measured in ohms. Obviously, electricians must know the difference between reactance and resistance and must know how to find the number of ohms of reactance in a circuit. Therefore, let us consider the relation of henrys to ohms of reactance.

Inductive Circuit Action

We know that inductance opposes any change in current flow. If we increase the voltage across an inductor, an increase in current flow builds up gradually due to the self-induced voltage that opposes current change. On the other hand, if we decrease the voltage across an inductor, the current flow falls gradually due to the self-induced voltage. In other words, the most noticeable difference between a resistive circuit and an inductive circuit is in their response speeds. The current flow in a resistive circuit changes immediately if the applied voltage is changed. On the other hand, the current flow in an inductive circuit is delayed with respect to a change in applied voltage.

We know that inductance is the result of magnetic force lines cutting the coil turns when the magnetic field strength changes. Therefore, the inductance of an air core is greatly increased if we insert an iron core (see Figure 7-1). Note that if the coil wire is doubled back on itself, as shown in Figure 7-2, the magnetic fields will cancel, and the inductance will be zero. This is called a noninductive coil and is used in electrical equipment in which we wish to employ wire-wound resistors that have no inductance.

In Chapter 5 we learned that the unit of inductance is the henry and that an inductor has an inductance of 1 henry if the current flow increases at the rate of 1 ampere per second when 1 volt is applied across the inductor. We will find that the inductance of a coil increases as the square of the number of turns. For example, with reference to Figure 7-3, a coil with two layers of wire has four times as much inductance as a coil with one layer; a coil with three layers of wire has nine times as much inductance as a coil with one layer. Let us see why this is so.

If we wind a single-layer coil, we have a certain number of turns that are cut by a magnetic field of a certain strength. If we wind two layers on the coil, we have twice as many turns that are cut by a magnetic field that is twice as great. Therefore, the self-induced voltage is four times as great as in a single-layer coil. This is just another way of saying that the inductance of a coil increases very rapidly as we add more layers to its winding. We will now assume that we have constructed a 1-henry coil, and we will ask how much opposition it has to AC current flow. To simplify our question, we will assume that the coil has been wound with large wire so that its DC resistance can be neglected.

Electrical measurements with an AC voltmeter and ammeter will show that Ohm's law for an inductor is slightly more complicated than Ohm's law for DC with a resistor. We will find that Ohm's law for an inductor is written:

$$I = \frac{E}{2\pi f L} \text{ amperes}$$

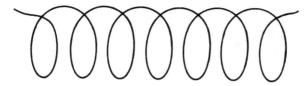

(A) Inductive coil with air core.

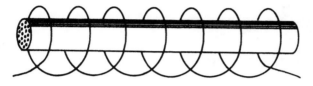

(B) Inductive coil with iron core.

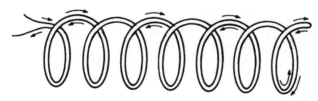

(C) Noninductive coil.

Figure 7-2 The inductance of an air-core coil is increased by an iron core; if the wire is doubled back on itself, the inductance is zero.

where I represents AC amperes,

E represents AC volts,

π is 3.1416,

f is the frequency in cps,

L is henrys of inductance.

Therefore, if we apply an AC voltage of 117 volts across an inductance of 1 henry, and the frequency is 60 cps, as shown in Figure 7-4, the AC current flow is approximately 0.31 ampere. It is helpful to write Ohm's law for an inductive circuit as follows:

$$\text{AC Current} = \frac{\text{AC Voltage}}{\text{Reactance}}$$

We observe that reactance can be compared to resistance in Ohm's law. In other words, $2\pi fL$ is measured in AC ohms. Electricians use

156 Chapter 7

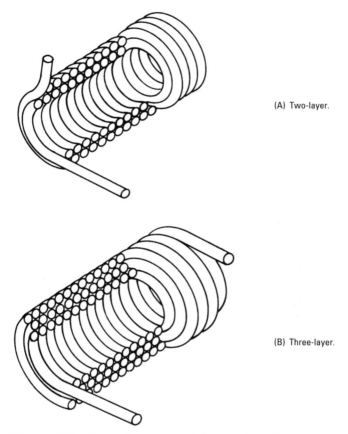

(A) Two-layer.

(B) Three-layer.

Figure 7-3 Illustrating two- and three-layer coils.

X_L to represent AC ohms. Thus, $X_L = 2\pi fL$. Therefore, Ohm's law for inductive circuits is written

$$I = \frac{E}{X_L}$$

It is obvious that if we used a 117-volt DC source in Figure 7-4 instead of a 117-volt AC source, an extremely large current would flow because the wire resistance of the 1-henry coil was assumed to be very small.

Next, let us note how the current varies in the circuit of Figure 7-4 when the frequency is changed. Ohm's law for alternating current

Inductive and Capacitive AC Circuits   157

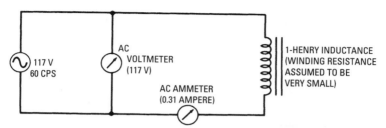

Figure 7-4 Illustration of Ohm's law for a purely inductive circuit.

in an inductive circuit tells us that if we increase the frequency from 60 cps to 120 cps, the current will be reduced by one-half. On the other hand, if we decrease the frequency from 60 cps to 30 cps, the current will be doubled.

If we keep the frequency at 60 cps but double the inductance to 2 henrys, the current will be reduced by one-half. On the other hand, if we reduce the inductance to $\frac{1}{2}$ henry, the current will be doubled. Of course, if we keep the frequency and the inductance the same, the current will double if we double the applied AC voltage.

Power in an Inductive Circuit

We know that the power in a DC resistive circuit is equal to volts multiplied by amperes. This is also true in an AC resistive circuit, as shown in Figure 7-5. On the other hand, in a circuit that has inductance only, as shown in Figure 7-6, the *true power* is zero. This fact might seem puzzling until we observe the circuit action. We recognize that the current flow through an inductor is delayed when voltage is applied due to the self-induced voltage of the inductor, which opposes the applied voltage. It can be shown that the AC current in an inductor is delayed 90° with respect to the applied voltage, as seen in Figure 7-6.

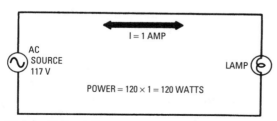

Figure 7-5 AC power in a resistive circuit.

158 Chapter 7

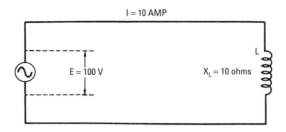

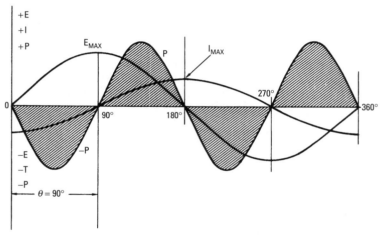

Figure 7-6 Power in an inductive circuit.

Now, if we multiply instantaneous voltages by instantaneous currents, we will find the instantaneous power. Over a complete cycle we see that half of the power is below the axis (negative power) and the other half of the power is above the axis (positive power). Therefore, the average power in an inductor is half positive and half negative; no power is consumed by the inductor. Thus, we can simply say that an inductor takes power from the source during the positive alternation and returns this power to the source during the negative alternation.

However, although no power is taken on the average by the inductor from the source in Figure 7-6, we see that power is surging back and forth in the circuit; the inductor first stores power in its magnetic field and then returns this stored power to the circuit. Thus, the circuit draws current but *consumes no power*. This is just another way of saying that the source does no work in this circuit, or that the inductor does not actually load the source. The inductor is an apparent load. Current that flows into the inductor on the positive alternation is stored, and this current is then returned to the circuit on the negative alternation.

The product of voltage and current in a purely inductive circuit is called *apparent power* because it is floating power that does no work. Apparent power is also called *reactive power* and is measured in *vars*. We write vars to represent volt-amperes reactive. In other words, we write, for Figure 7-6:

$$\text{Vars} = E_{\text{rms}} I_{\text{rms}}$$

Electricians measure vars with a varmeter, as shown in Figure 7-7. A varmeter looks like a wattmeter, but it is constructed so that it indicates apparent power instead of true power. Note that if the load were a resistance, such as a lamp, the varmeter would read zero. On the other hand, if the load were a pure inductance, the varmeter would read the product of the rms voltage and the rms current.

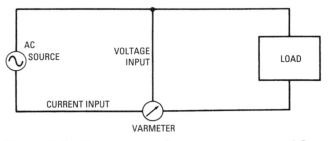

Figure 7-7 Measurement of apparent power in an AC circuit.

Resistance in an AC Circuit

We know that in a circuit containing only resistance (Figure 7-8A), the current and voltage are in phase. Therefore, there is true power in the resistor, which appears as heat. True power is also called real power. We also know that this true power in watts is equal to $E_{\text{rms}} I_{\text{rms}}$. The shaded area in Figure 7-8B represents the product of instantaneous voltages and instantaneous currents. The entire shaded area is positive and accordingly represents true power. Note that when the voltage is negative, the current is also negative, and therefore their product must be positive.

Next, we will ask what the AC circuit action will be when inductance and resistance are connected in series, as shown in Figure 7-9.

160 Chapter 7

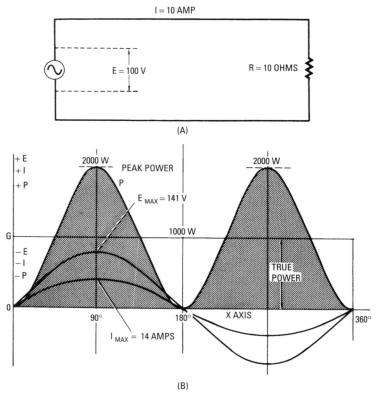

Figure 7-8 Relation between E, I, and P in a resistive circuit.

The AC current I_O flows through both the inductor and the resistor. Accordingly, a voltage drop is produced across the inductor and across the resistor. We observe that the resistive voltage drop E_R is in phase with the current but that the inductive voltage drop E_L is 90° out of phase with the current. E_R is 50 volts and E_L is 86.6 volts in this example.

Note that the applied voltage E_O in Figure 7-9 is the hypotenuse of a right triangle, the sides of which are 50 and 86.6; in turn, we write:

$$E_v = \sqrt{50^2 + 86.6^2} = 100 \text{ volts}$$

If we use a protractor in Figure 7-9B, we will find that E_O makes an angle of 60° with I_O. Electricians call this angle the *power-factor*

Inductive and Capacitive AC Circuits 161

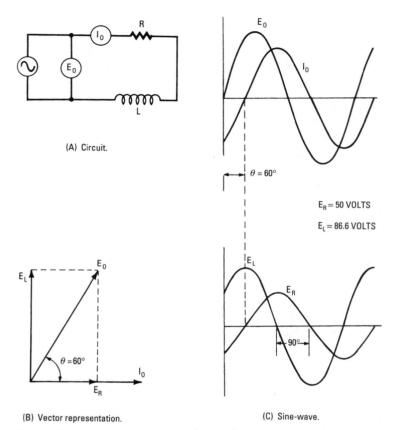

Figure 7-9 Resistance and inductive reactance in series.

angle of the circuit. It is evident that if we keep L constant and increase R, the power-factor angle will become smaller. In large factories and power plants, electricians are concerned with power-factor angles. However, details of this kind of electrical work will be explained later.

Impedance in AC Circuits

The circuit in Figure 7-9 has both resistance and inductive reactance. The total opposition to AC current flow is called *impedance*. Impedance is measured in ohms, just as resistance and reactance are measured in ohms. If an AC circuit contains reactance only, the circuit impedance is the same as the circuit reactance. Likewise, if an AC circuit contains resistance only, the circuit impedance is the same as the circuit resistance. When a circuit has both resistance and reactance, we find the circuit impedance as follows:

$$Z = \sqrt{R^2 + X^2}$$

where Z represents the ohms of impedance,

R represents the ohms of resistance,

X represents the ohms of reactance.

We can draw an impedance diagram as a right triangle, such as shown in Figure 7-10. The power-factor angle is the angle θ between R and Z in the diagram.

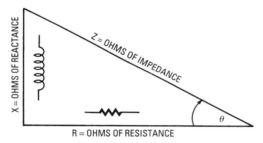

Figure 7-10 An impedance diagram.

Power in an Impedance

When inductance and resistance are connected in series with an AC source, as shown in Figure 7-11, there are three power values to be considered. We will find a true power in *watts*, a reactive power in *vars*, and an apparent power in *volt-amperes*. These power values are found as follows:

1. The impedance of the circuit is 14.14 ohms.
2. The current is 100/14.14, or 7.07 amperes.
3. The true power is I^2R, or $50 \times 10 = 500$ watts.
4. The reactive power is I^2X, or $50 \times 10 = 500$ vars.
5. The apparent power is EI, or $100 \times 7.07 = 707$ volt-amperes.

Only the true power does useful work in the circuit. For example, if the resistor represents a lamp filament, as in Figure 7-1, the true power produces light. The reactive power merely surges back and forth in the circuit. The apparent power provides reactive power and true power, as shown in the power diagram of Figure 7-12.

Inductive and Capacitive AC Circuits   163

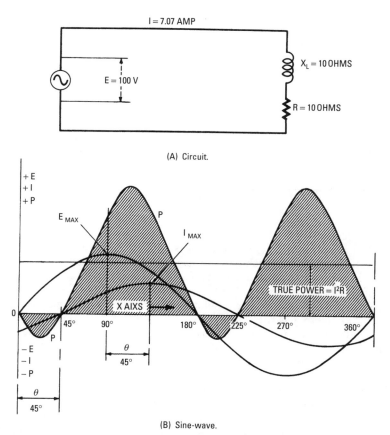

Figure 7-11   Power in a circuit containing L and R in series.

Capacitive Reactance

Electricians often work with capacitive reactance, as well as with inductive reactance. Capacitance is measured in *farads* with a capacitor tester. A 1-farad capacitor will store 1 coulomb of charge when 1 volt is applied across the capacitor. A *coulomb* is the amount of charge produced by 1 ampere flowing for 1 second. Because the farad is a very large unit of capacitance, we generally work with *microfarads;* a microfarad is one millionth of a farad (10^{-6} farad). The symbol μ or the letter m is used to represent *micro;* for example, 50 microfarads is commonly written as $50 \mu f$, $50 \mu fd$, 50 mf, or 50 mfd.

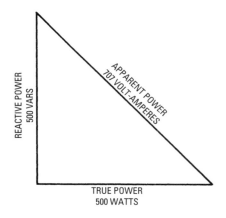

Figure 7-12 A power diagram.

We know that if an ideal inductor is connected across a DC voltage source, the inductor short-circuits the source. On the other hand, if we connect a capacitor across a DC voltage source, the capacitor is an open circuit. Although there is a momentary rush of DC current into the capacitor to charge its plates, the current flow thereafter is zero. Capacitance is the opposite of inductance in a DC circuit. We will find that they are opposites in other respects also. We often hear automotive electricians speak of *condensers* and *capacity*. However, it is now standard practice to use the terms *capacitors* and *capacitance*. It makes no difference which terms are used, except that the new terms are preferred in most offices and shops.

Figure 7-13 shows a simple experiment of basic importance. The lamp connected in series with a capacitor and a DC voltage source does not glow. On the other hand, when an AC source is used, the lamp glows. If we increase the capacitance, the lamp will glow brighter. By comparing Figure 7-13 with Figure 7-1, we will recognize that the circuit actions are opposite.

It is evident that AC current does not really flow through a capacitor. What actually happens is that the current flows into the capacitor on the first alternation and charges the capacitor plates. Electricity is then stored in the capacitor, and this electricity returns to the circuit on the second alternation as the capacitor discharges. We can compare a capacitor to an inductor in the sense that a capacitor does not consume power. In other words, the power in a capacitor is reactive power and is measured in vars. Therefore, the AC circuit in Figure 7-13 has a reactive power value in the capacitor that is measured in *vars*, a real (true) power value in the lamp that

Inductive and Capacitive AC Circuits 165

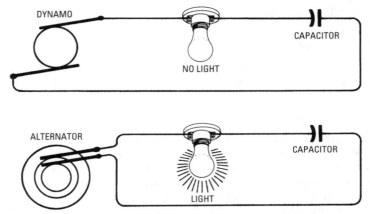

Figure 7-13   Lamp connected in series with a capacitor.

is measured in *watts*, and an apparent power value supplied by the alternator that is measured in *volt-amperes*.

Because a large capacitance stores more electricity than a small capacitance, it also has less opposition to AC current flow. The amount of AC current flow will depend, moreover, on how many times a second the capacitor is charged and discharged.

Therefore, the AC current flow increases when the frequency is increased. In its simplest form, Ohm's law for a capacitor is written:

$$I = \frac{E}{X_C}$$

where I represents the number of AC amperes,

E represents the number of AC volts,

X_C represents the number of capacitive ohms.

We know that the number of capacitive ohms (reactive ohms) will depend on the number of farads of the capacitor and on the AC frequency. In turn, we write:

$$X_C = \frac{1}{2\pi f C}$$

where X_C represents the number of capacitive ohms,

f is the frequency in cps,

C is the capacitance in farads.

Thus, Ohm's law for a capacitor may be written in the following form:

$$I = 2\pi fCE$$

For example, if a capacitor has a capacitance of 133 microfarads (133μf, or 1.33×10^{-4} farad), its reactance at 60 cps will be:

$$X_C = \frac{E}{2\pi fC} = \frac{1}{6.28 \times 60 \times 1.33 \times 10^{-4}} = 20 \text{ ohms}$$

Let us observe Figure 7-14. When a sine wave AC voltage is applied to a capacitor, the current is greatest when the voltage starts to rise from zero. This is so because there is no charge in the capacitor at this instant, and therefore there is no back voltage to oppose the flow of current. On the other hand, the current is zero when the voltage reaches its peak value. This is because the capacitor is then fully charged and its back voltage is equal and opposite to the applied peak voltage. This is just another way of saying that the current in a capacitive circuit goes through its peak value before the applied voltage goes through its peak value.

Electricians say that the current in a capacitive circuit *leads* the applied voltage in time. This is just the opposite of an inductive circuit, in which the current is delayed with respect to the applied voltage (i.e., the current *lags* the voltage in an inductive circuit). There is a 90° phase difference between voltage and current in a capacitor. However, we have a *leading phase* in a capacitor, whereas we have a *lagging phase* in an inductor.

Figure 7-15 shows the power in a capacitor, which is the product of instantaneous voltages and currents. We observe that the shaded areas represent positive power and negative power on successive alternations. In other words, the average power in the capacitor is zero. All of the capacitor power is reactive power and is measured in vars:

$$\text{Vars} = R_{\text{rms}} I_{\text{rms}}$$

Thus, a capacitor is like an inductor in that it is a *reactor* and consumes no real power. However, capacitance is the opposite of inductance in that capacitive current leads the applied voltage by 90°, whereas inductive current lags the applied voltage by 90°.

Capacitive Reactance and Resistance in Series

Figure 7-16 shows capacitance and resistance connected in series with an AC source. To find the current in this circuit, we proceed as

Inductive and Capacitive AC Circuits

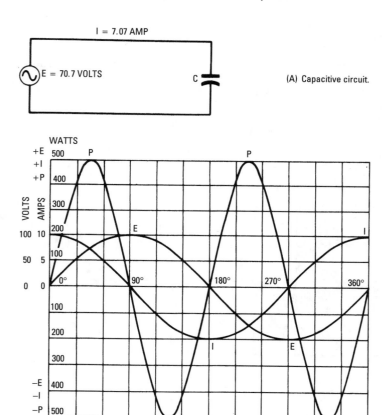

(A) Capacitive circuit.

(B) Sine curves of current, voltage, and power.

Figure 7-14 Voltage, current, and power in a capacitor.

follows:

1. The impedance of the circuit is $Z = \sqrt{R^2 + X_C^2} = 44.66$ ohms.
2. The current is $I = E/Z = 3$ amperes.
3. The resistive voltage drop is $E_R = IR = 60$ volts.
4. The capacitive voltage drop is $E_C = IX_C = 120$ volts.

168 Chapter 7

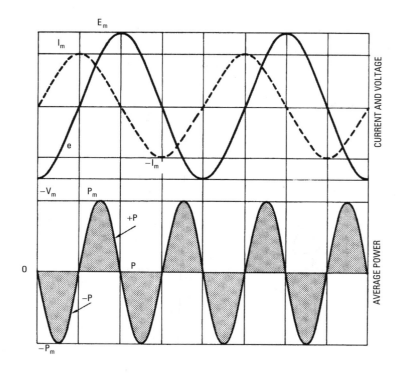

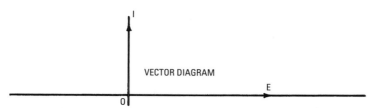

Figure 7-15 The power in a capacitor is half positive and half negative.

Next, let us find the three power values in the circuit of Figure 7-16:

1. The true power is $P = I^2 R = 180$ watts.
2. The reactive power is $P = I^2 X_C = 360$ vars.
3. The apparent power is $P = EI = 402$ volt-amperes.

Inductive and Capacitive AC Circuits   169

Electricians use many different kinds of capacitors. For example, refrigerator electricians use capacitors such as those illustrated in Figure 7-17. Very large capacitors are used in power plants, as illustrated in Figure 7-18. Capacitors built up from metal plates separated by sheets of insulation are called static capacitors. These capacitors operate at high voltage; for example, the capacitor in Figure 7-18 is operated at voltages that may be stated in *kilovolts* (a kilovolt is 1000 volts).

Figure 7-11 showed the development of power in a circuit with inductance and resistance. The development of power in a circuit with capacitance and resistance is much the same, as seen in Figure 7-19. Basically, the only difference between an RC circuit and an RL circuit is in the lead or lag of the circuit current with respect to the applied voltage in the circuit.

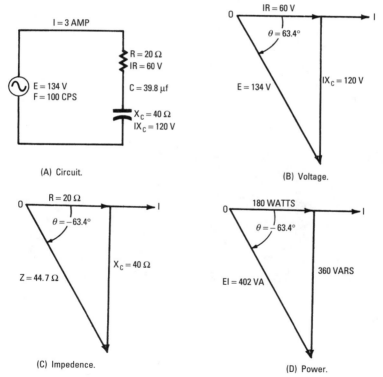

Figure 7-16 Capacitive reactance and resistance in series.

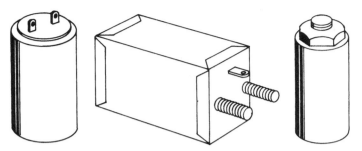

Figure 7-17 Various motor-starting capacitors.

Figure 7-18 Large static capacitors used in factories. *Courtesy Westinghouse Electric Corp.*

Inductive and Capacitive AC Circuits 171

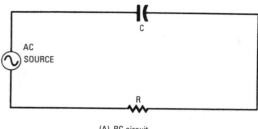

(A) RC circuit.

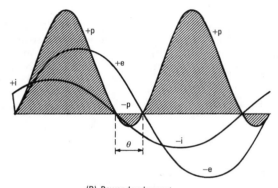

(B) Power development.

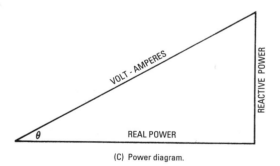

(C) Power diagram.

Figure 7-19 Power circuit with capacitance and resistance.

Inductance, Capacitance, and Resistance in Series

Electricians who work with refrigerator motors and other electrical equipment are often concerned with circuits that have inductance, capacitance, and resistance connected in series with an AC voltage source, as shown in Figure 7-20A. As we would expect, the voltage drop across the resistor is in phase with the circuit current, while the voltage drops across the inductor and capacitor are 90° out of phase with the circuit current.

Since the current is the same at any point in the circuit of Figure 7-20A, we take the current as a common reference in following the circuit action. In turn, the source voltage E is the vector sum of IR, IX_L, and IX_C. In Figure 7-20B, we see the relations of IR, IX_L, and IX_C. Note that IX_L and IX_C are each 90° away from I and point in opposite directions (IX_L and IX_C have a phase angle of 180°). Therefore, the total reactive voltage E_X is the difference between IX_L and IX_C. We write:

$$E_x = IX_L - IX_C = 45 - 15 = 30 \text{ volts}$$

Since X_L is greater than X_C in this example, the inductance dominates the circuit action, and the circuit current lags the applied voltage E. In other words, the phase angle in Figure 7-20B is lagging. The impedance diagram for the circuit shows that the impedance is 50 ohms. To find the circuit impedance by arithmetic, we write:

$$Z = \sqrt{R^2 + (X_L - X_C)^2}$$

or,

$$Z = \sqrt{40^2 + (45 - 15)^2} = \sqrt{2500} = 50 \text{ ohms}$$

We know that the circuit current in Figure 7-20A is given by $I = E/Z$:

$$I = \frac{E}{Z} = \frac{50}{50} = 1 \text{ ampere}$$

In turn, the four power values in the circuit of Figure 7-20A are as follows:

$$P_R = I^2 R = 40 \text{ watts}$$
$$P_L = I^2 X_L = 45 \text{ vars}$$
$$P_C = I^2 X_C = 15 \text{ vars}$$
$$P_{\text{apparent}} = EI = 50 \text{ volt} - \text{amperes}$$

Inductive and Capacitive AC Circuits **173**

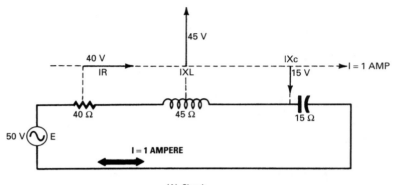

(A) Circuit.

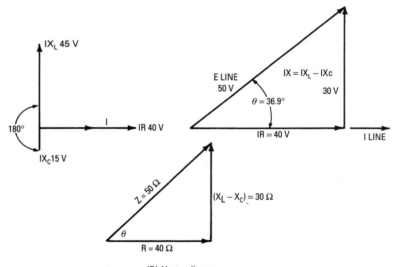

(B) Vector diagram.

Figure 7-20 Resistance, inductance, and capacitance connected in series.

Note that the difference between P_L and P_C is 30 vars. We know that apparent power is equal to:

$$P_{app} = \sqrt{\text{Real power}^2 + \text{Reactive power}^2}$$

or,

$$P_{app} = \sqrt{40^2 + 30^3} = 50 \text{ volts} - \text{amperes}$$

In large factories and power plants, electricians are particularly concerned with adjusting the capacitance in circuits such as the one shown in Figure 7-20A so that the inductive and capacitive reactances are equal. Figure 7-21 shows a circuit in which $X_L = X_C$. We recognize that X_L exactly cancels X_C. Therefore, *the circuit action is the same as it would be if there were only resistance present in the circuit.* Since there is effectively no inductance or capacitance in this circuit, the phase angle between the source voltage and the circuit current is zero. Electricians say that this kind of a circuit has been *corrected for zero power factor.*

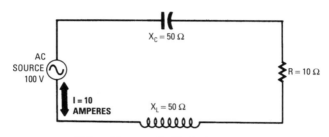

(A) X_C and X_L are equal in this circuit.

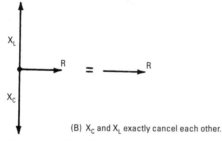

(B) X_C and X_L exactly cancel each other.

Figure 7-21 Conductive and reactance circuit.

It follows that the current in the circuit of Figure 7-21 is given by:

$$I = \frac{E}{R} = \frac{100}{10} = 10 \text{ amperes}$$

When the inductive reactance is equal to the capacitive reactance in a circuit, we see that the number of volt-amperes is the same as the resistive load power. This is a desirable condition in a power distribution system because there is no reactive current surging in the

Inductive and Capacitive AC Circuits

line to increase the line losses. Therefore, the efficiency of a power distribution system is greatest when the inductive reactance that may be present is exactly cancelled by inserting an equal amount of capacitive reactance. This is the purpose of the large capacitor illustrated in Figure 7-18.

Some electricians call the circuit in Figure 7-21 a *series-resonant circuit*. This simply means that the phase angle between the source voltage and circuit current is zero. It is important to note that there is a *voltage rise* across the capacitor and across the inductor in a series-resonant circuit. For example, in Figure 7-21, the voltage drop across the capacitor is equal to IX_C, or 500 volts. Similarly, the voltage drop across the inductor is equal to IX_L, or 500 volts. Although the source voltage is 100 volts in Figure 7-21, the voltage across L rises to 500 volts and the voltage across C rises to 500 volts. In turn, we must use inductors and capacitors that are rated for 500-volt operation. This voltage rise across an inductor and across a capacitor in a resonant LCR circuit is sometimes called the *voltage magnification* of the circuit. The voltage magnification can be very great when the load resistance is small. For example, if we used a 1-ohm load resistor in Figure 7-21, the voltage across L would be 5000 volts, and across C it would be 5000 volts. Figure 7-22 shows the instantaneous voltage, current, and power relations in a series-resonant circuit.

Inductance and Resistance in Parallel

Electricians often work with circuits that consist of inductance and resistance connected in parallel. For example, an AC line may be connected to a group of lamps and to a motor such as an induction motor. The lamps provide a resistive load, and the motor provides a load that has an inductive component. Let us consider the impedance of an RL parallel circuit, such as that shown in Figure 7-23A. The inductive reactance has a value of $2\pi fL$ ohms. To find the circuit impedance, we may write:

$$Z = \frac{R X_L}{\sqrt{R^2 + X^2_L}}$$

However, to simplify the calculation of Z, it is advisable to make use of the triangle relations shown in Figure 7-23B. We lay off the line R with a length proportional to the resistance, and we lay off the line X_L with a length proportional to $2\pi fL$. Then, we draw the line Z perpendicular to the hypotenuse of the triangle.

176 Chapter 7

The length of Z then gives the number of ohms of impedance. Note also that the angle θ is the power-factor angle of the RL parallel circuit.

When working with power-factor correction, as will be explained, electricians often find it useful to change a parallel RL circuit into an equivalent series RL circuit, as shown in Figure 7-24. In other words, when an inductance L_P is connected in parallel with a resistance R_P (see Figure 7-24A), we can make up an equivalent circuit by connecting an inductance L_S in series with a resistance R_S (see Figure 7-24B). The relations of inductance and resistance values in the series and parallel circuits are easily found by drawing the triangle relations shown in Figure 7-24C. Note that smaller values of series inductance and resistance have the same circuit action as larger values of parallel inductance and resistance.

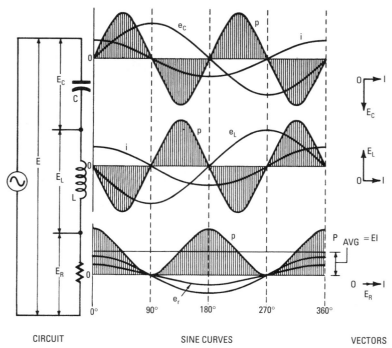

Figure 7-22 Instantaneous voltage, current, and power relations in a series-resonant circuit.

Inductive and Capacitive AC Circuits **177**

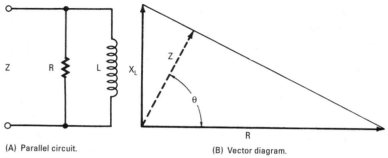

(A) Parallel circuit.   (B) Vector diagram.

Figure 7-23 Impedance parallel circuit.

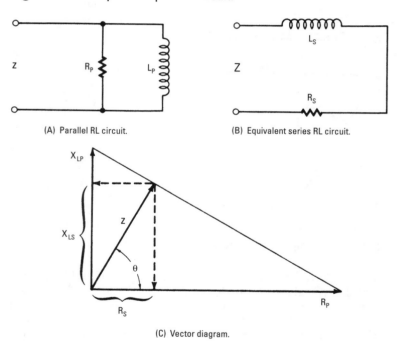

(A) Parallel RL circuit.   (B) Equivalent series RL circuit.

(C) Vector diagram.

Figure 7-24 Impedance circuit.

Capacitance and Resistance in Parallel

Electricians may also work in circuits that consist of capacitance and resistance connected in parallel. For example, an AC line may be connected to a group of lamps and to a motor such as a synchronous motor. The lamps provide a resistive load, and the motor provides a load that has a capacitive component. Let us consider the impedance of an RC parallel circuit, such as the one shown in Figure 7-25A. The capacitive reactance has a value of $1/(2\pi fC)$ ohms. To find the circuit impedance, we may write:

$$Z = \frac{R X_C}{\sqrt{R^2 + X^2_C}}$$

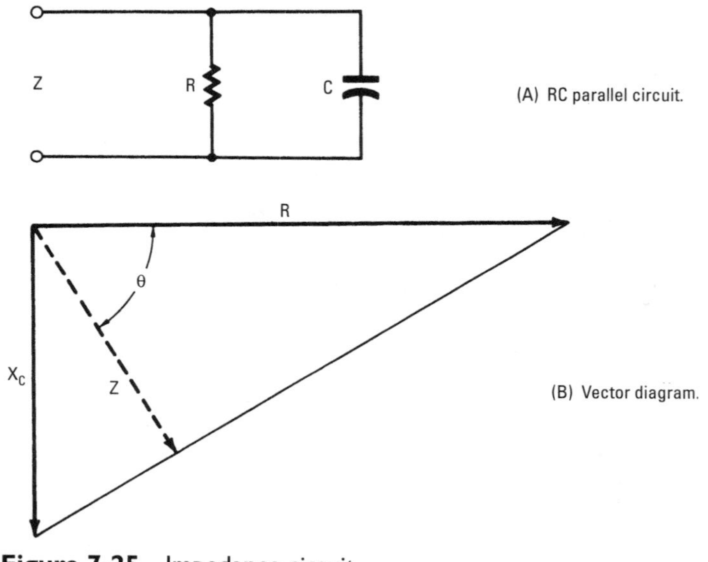

(A) RC parallel circuit.

(B) Vector diagram.

Figure 7-25 Impedance circuit.

However, to simplify the calculation of Z, it is advisable to make use of the triangle relations shown in Figure 7-25B. We lay off the line R with a length proportional to the resistance, and we lay off the line X_C with a length proportional to $1/(2\pi fC)$. Then, we draw the line Z perpendicular to the hypotenuse of the triangle. The length of Z then gives the number of ohms of impedance. Note also that the angle θ is the power-factor angle of the RC parallel circuit.

When working with power-factor correction, electricians often find it useful to change a parallel RC circuit into an equivalent series RC circuit, as shown in Figure 7-26. In other words, when a capacitance C_P is connected in parallel with a resistance R_P, we can make up an equivalent circuit by connecting a capacitance C_S

in series with a resistance R_S. The relations of capacitance and resistance values in the series and parallel circuits are easily found by drawing the triangle relations shown in Figure 7-26C.

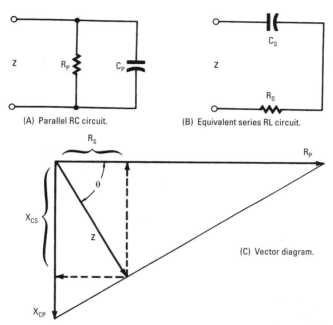

Figure 7-26 Impedance circuit.

Inductance, Capacitance, and Resistance in Parallel

Electricians in factories or shops are sometimes concerned with circuits that contain resistance, capacitance, and inductance in parallel. For example, an AC line may be connected to a group of lamps, to a synchronous motor, and to an induction motor. In this case, we are concerned with a basic RLC parallel circuit as shown in Figure 7-27. The R, X_L, and X_C vectors are drawn as shown in Figure 7-27B. Note that X_L and X_C extend in opposite directions. Therefore, X_L and X_C will tend to cancel each other, and if $X_L = X_C$, the reactances cancel completely, leaving resistance only.

When the values of X_L and X_C are equal, electricians state that the circuit in Figure 7-27A is a *parallel-resonant circuit*. This means that L and C are effectively not present, inasmuch as they cancel

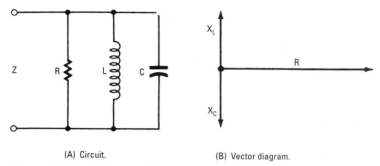

(A) Circuit.   (B) Vector diagram.

Figure 7-27 RLC parallel circuit.

each other. Therefore, current is drawn by R only. When a circuit is parallel resonant, it is also said to be *power-factor corrected,* since the power-factor angle is obviously zero in the power-factor corrected circuit, particularly when the power demand is heavy. This is because a power-factor corrected circuit has no surging line currents and therefore operates at maximum efficiency. (The I^2R line loss due to surging line currents has been eliminated.)

From a practical point of view, induction motors are in wider use than synchronous motors. If we add a suitable number of synchronous motors to a group of induction motors, we can operate the power circuit at unity power factor. This is the most common method of obtaining power-factor correction. However, we can also connect a capacitor of suitable value across an induction motor, and the motor will operate at unity power factor. Such a capacitor is often called a static capacitor, as noted previously.

When power-factor correction is partial (incomplete), there is a reactance left over in the circuit. For example, let us consider the LC parallel circuit shown in Figure 7-28A. In this example, X_L is greater than X_C, and there is a capacitive reactance left over in the circuit. The input reactance X is a capacitive reactance. To find out how much capacitive reactance is left over, we use the triangle relations shown in Figure 7-28B. Point P on the X_L axis represents the amount of inductive reactance; point Q on the X_C axis represents the amount of capacitive reactance. The lines PO and OQ intersect at point R. We draw a line RS perpendicular to the X_C axis, thereby determining point S. The length of the line segment OS denotes the values of the input reactance X in Figure 7-28A.

Note that the length of the line OO in Figure 7-28B must be 10 units. That is, we lay off this horizontal line equal to 10 reactance units. It is necessary only that we make OP proportional to the

Inductive and Capacitive AC Circuits   181

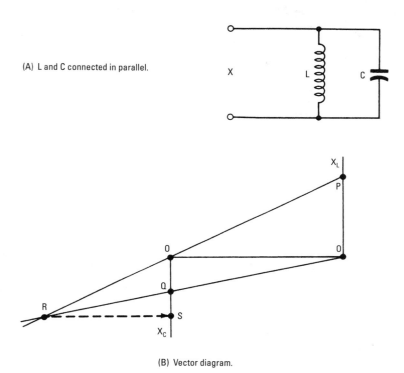

(A) L and C connected in parallel.

(B) Vector diagram.

Figure 7-28   Total reactance.

inductive reactance of L and make OQ proportional to the capacitive reactance of C. Note that if X_C should happen to be greater than X_L, the diagram would be drawn as shown in Figure 7-29. In this example, there is inductive reactance left over, and the amount of inductive reactance that is left over is given by the length of the line segment OS.

With these principles in mind, it is easy to find the amount of capacitance (or inductance) that will be required to correct the power factor in any practical circuit. For example, suppose that an induction motor consists of an effective inductance in series with an effective resistance, as depicted in Figure 7-24B. In turn, the triangle diagram in Figure 7-24C shows the amount of effective parallel inductance. To correct the power factor, we must connect a suitable amount of capacitance in parallel with the inductance. In other words, the reactance of this capacitance must be equal to the reactance of the effective parallel inductance that we have determined.

182 Chapter 7

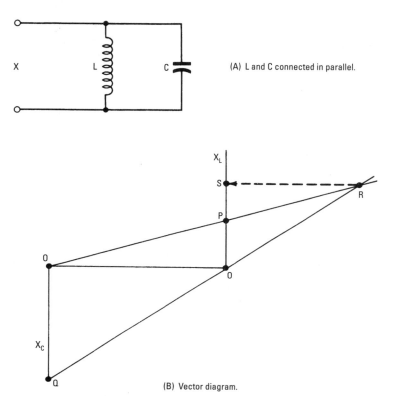

Figure 7-29 Total reactance.

Summary

Inductance, which is measured in henrys, is the property that opposes any change in the existing current. Inductance is the result of magnetic lines of force cutting the coil turns when magnetic field strength changes. Therefore, the inductance of an air-core coil is greatly increased if we insert an iron core.

The total opposition to resistance and inductive reactance is called impedance. Impedance is measured in ohms. Another type of reactance is called capacitive reactance, which is also measured in ohms. A capacitor is like an inductor in that it is a reactor and consumes no real power. Capacitance is the opposite of inductance in that capacitive current leads the applied voltage by 90°, whereas inductive current lags the applied voltage by 90°.

Inductive and Capacitive AC Circuits **183**

A series-resonant circuit is one in which an inductor and capacitor are connected in series and have values such that the inductive reactance of the inductor will be equal to the capacitive reactance of the capacitor at a particular frequency.

Test Questions

1. How can the brightness of a lamp be controlled with a variable inductance?
2. Explain the different symbols that may be used for coils and resistors by electricians.
3. How is a noninductive coil wound? Where would it be used?
4. State Ohm's law for a purely inductive circuit.
5. In what way does true power differ from reactive power?
6. Discuss how to find the reactance of a coil.
7. How do we define impedance?
8. Explain how to find the impedance of a circuit that has resistance and inductance.
9. What is meant by a power diagram?
10. Discuss how to find the reactance of a capacitor.
11. Define a henry. A farad.
12. What is the definition of phase angle?
13. Explain the difference between lagging and leading phase angles.
14. How do we draw an impedance diagram?
15. What do we call an instrument that measures real power? Reactive power?
16. How would we measure the apparent power in a reactive circuit?
17. Why does capacitive reactance tend to cancel inductive reactance in a series circuit?
18. If the inductive reactance is exactly equal to the capacitive reactance in an LCR series circuit, how do we describe the circuit?
19. What is the value of the power-factor angle in a resonant circuit?
20. Explain what electricians mean by power-factor correction.

21. Why is power-factor correction important in power distribution systems?

22. Discuss an example of voltage rise across an inductor and across a capacitor in a series-resonant circuit.

23. Why is the voltage magnification greater in a series-resonant circuit when the load resistance is reduced?

24. Why is voltage magnification of concern to the electrician?

25. What do we call an instrument that measures capacitance values?

Chapter 8
Electric Lighting

Light is a form of energy. It has wavelike properties and is called a visible form of radiant energy. Light and heat are basically very similar; however, heat has a longer wavelength than light and is invisible. Only those radiant wavelengths between the approximate limits of 0.00038 to 0.00076 millimeters are visible to the human eye. Note that a meter contains 39.37 inches and that a millimeter is 0.001 of a meter. For convenience, the wavelengths of light are usually expressed in *angstroms*. One angstrom is equal to 1/10,000,000 of a millimeter. In other words, the range of visible wavelengths is from about 3800 to 7600 angstroms. Table 8-1 shows the visible wavelengths, which are a very small segment of the entire range of radiant energy.

Table 8-1   Classifications of Radiant Energy

Frequency, cycles/second										
10^2	10^4	10^6	10^8	10^{10}	10^{12}	10^{14}	10^{16}	10^{18}	10^{20}	10^{22}
10^6	10^4	10^2	1	10^{-2}	10^{-4}	10^{-6}	10^{-8}	10^{-10}	10^{-12}	10^{-14}
Wavelength, meters										

Within the abovementioned visible range, the colors corresponding to various wavelengths are as shown in Figure 8-1. The eye has greatest sensitivity for wavelengths of about 5500 angstroms; that is, yellow can be seen under poor conditions of illumination when blue or red cannot be seen. Under dim illumination, the sensitivity curve tends to shift as shown by the shaded region in Figure 8-1. Therefore, violet disappears first and red remains visible. Yellow disappears last as the illumination becomes very dim. As each color disappears, it becomes a gray shade, and finally black.

Sources of Light

Light sources are broadly divided into natural sources such as the sun, moon, stars, aurora borealis, phosphorescent insects and fishes, and open fires. The other broad division is into sources of artificial light, such as oil and gas lamps, incandescent lamps, arc lamps, various types of gaseous glow lamps, and fluorescent lamps.

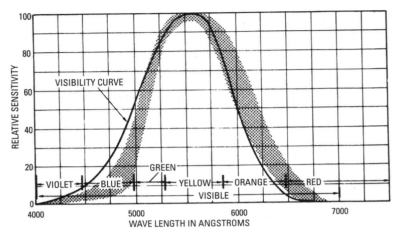

Figure 8-1 Eye sensitivity curve.

Measurement of Amount of Illumination

The original method for comparing amounts of illumination was to use the light from a standard tallow candle held 12 inches from the surface to be illuminated. This unit of illumination is called the *foot-candle*. Electricians often use a light meter to measure foot-candles, as illustrated in Figure 8-2. We are also concerned in many practical situations with the total light output that a lamp radiates in all directions. Let us consider a point source of light placed inside a sphere that has a radius of 12 inches. If the entire inner surface of the sphere is illuminated by an amount of 1 foot-candle, the total light output from the source is 1 *lumen*.

Figure 8-2 Light meter.

Electric Lighting 187

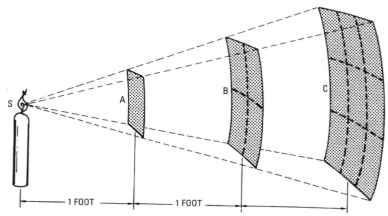

Figure 8-3 Illustrating the inverse square law.

We will find that the intensity of illumination falls off as the square of the distance from the source. With reference to Figure 8-3, note that the illumination at A is 1 foot-candle. Next, if we remove surface A and illuminate surface B, the same amount of light now illuminates 4 times as much area. Therefore, the illumination at B is ¼ foot-candle. Then, if we remove surface B and illuminate surface C, the same amount of light now illuminates 9 times as much area as at A. Therefore, the illumination at C is ⅑ foot-candle.

To get a basic idea of the amount of illumination provided by an incandescent lamp, it is helpful to note that a 60-watt lamp with a tungsten filament and an inside-frosted bulb has a brightness of about 60 foot-candles. By way of comparison, the noonday sun has a brightness of about 1,000,000 foot-candles. The moon has a brightness of approximately 5 foot-candles. In other words, these are comparative brightnesses of the light sources. Very bright light sources such as the sun or an arc lamp will damage the retina of the eye if viewed directly.

Tungsten Filaments

Since the light output from a filament depends on its temperature, only metals with high melting points are suitable for use as lamp filaments. Tungsten melts at approximately 6120°F. A typical tungsten lamp is operated at a filament temperature of 4800°F. Since tungsten gradually evaporates at this temperature, the average life of the lamp is about 1000 hours. Note that if a tungsten filament is operated

5 percent below rated voltage, the lamp life will be doubled, but the light output will be 17 percent below normal. The efficiency of the lamp (light output divided by power input) will be reduced 10 percent.

Tungsten has the best efficiency of any filament material in general use, and 20 lumens per watt is typical for large lamps. A 100-watt tungsten lamp will have an efficiency of about 15 lumens per watt. A 250-watt photoflood lamp has an efficiency of approximately 35 lumens per watt, but its average life is only 3 hours. A 6-watt lamp, such as used in electric signs, has an average life of 1500 hours, but its efficiency is less than 7 lumens per watt. Filaments in high-vacuum lamps are operated at about 4000°F; gas-filled lamp filaments operate 800° higher, and photoflood lamp filaments operate a little above 5600°F.

A gas-filled lamp tends to reduce evaporation of the tungsten filament and can therefore be operated at a higher temperature than a vacuum lamp. Most gas-filled lamps employ a mixture of nitrogen and argon. Krypton is the best gas for this purpose but is so expensive that it is used only in special-purpose lamps, such as miners' lamps. The gas pressure in a gas-filled lamp is about 20 percent below atmospheric pressure. Most filaments are *coiled* in the form of a long helical spring, which reduces the tendency of the gas to cool the filament. A helical filament is sometimes coiled into another helix of greater diameter to further reduce the cooling action of the gas.

Helical filaments also have a slower rate of tungsten evaporation. This evaporation eventually causes *bulb blackening* because the tungsten vapor condenses as a black film on the inner surface of the bulb. In a gas-filled lamp, the hot gas carries the tungsten vapor upward. Therefore, a black spot forms at the top of the bulb instead of spreading over the entire inner surface, as in a high-vacuum bulb. Chemicals called getters are often placed inside the bulb to capture tungsten vapor and thereby reduce the rate of blackening. A piece of wire mesh called a *collector grid* may also be attached to each lead-in wire to attract the particles of tungsten vapor.

Starting Surge of Current

A metal increases in resistance as its temperature increases. Figure 8-4 shows a current versus voltage curve for a tungsten filament. The current starts to rise rapidly at first and then rises more slowly as the filament becomes hotter. Figure 8-5 shows how the resistance of a tungsten filament increases as it heats up. The cold resistance of an ordinary lamp is about $1/15$ of its hot resistance. The

Electric Lighting 189

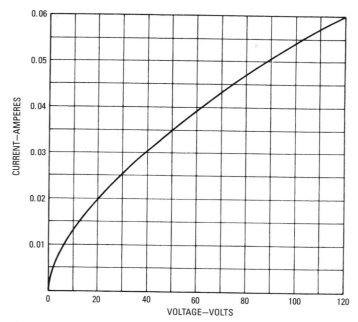

Figure 8-4 Current versus voltage curve for a tungsten filament rated for 120-volt operation.

cold resistance of a photoflood lamp is approximately $1/20$ of its hot resistance. Since it takes a certain amount of time for a filament to come up to operating temperature, a surge of current flows when the switch is turned on. This starting surge of current is 15 times as great as the normal operating current for an ordinary lamp and is 20 times as great for a photoflood lamp.

The practical electrician must use a switch that will withstand starting surges without damage. For example, in a shop or factory, a line of lamps might draw 50 amperes in normal operation. However, the starting-surge current will be 50×15, or 750 amperes. Unless the switch contacts can withstand 750 amperes without damage, the life of the switch will be comparatively short. As another example, if a line of photoflood lamps normally operates at 50 amperes, the switch contacts must withstand 50×20, or 1000 amperes, without damage. Otherwise, the switch is likely to fail after a few operations.

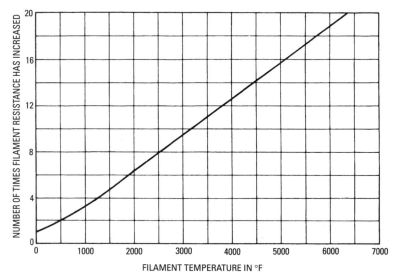

Figure 8-5 Increase of resistance by a tungsten filament at high temperatures, compared with its resistance at 70°F.

Aging Characteristics

As a tungsten lamp ages, its light output decreases for two reasons. Evaporation of the filament tends to cause the bulb to blacken. Also, evaporation makes the filament slowly decrease in diameter, which means that the *filament resistance increases*. Therefore, an old filament draws less current and operates at a lower temperature, which reduces its light output. In turn, the efficiency of the lamp (in lumens per watt) also decreases with age. The current drawn and the power consumed by the filament decrease at the same rate as the lamp ages. However, the efficiency decreases about four times as fast, and the light output decreases approximately five times as fast.

Electricians use a rule of thumb which states that when a lamp ages to a point that its light output has decreased 30 percent, it should be washed to remove any dust, dirt, or grease that may have accumulated on its outer surface. As a lamp ages, it may also reach a so-called smashing point, even if the filament does not burn out. This is defined as the point at which the light output of the lamp has fallen sufficiently that the cost of electricity per 1,000,000 lumen-hours is greater than the average cost of light produced up to that time, plus the initial cost of the lamp and electricity. The smashing

Electric Lighting 191

point is of practical importance only in large lighting systems where operating costs are comparatively high.

Lamp Bases and Bulbs

Various types of lamp bases are used in practice. The *screw-type* base shown in Figure 8-6 is the most common. The smallest size is called the *miniature* base, followed by the *candelabra, intermediate, medium,* and *mogul* bases. The medium base is used in ordinary home lighting systems. Other types of bases include the *bayonet, prefocus,* and *bipost* bases, as shown in Figure 8-7. Bayonet bases are made in a number of different sizes and have two projecting side pins, which seat the base in its socket. Automotive electricians use the *bayonet candelabra* size of base in automobile lamps.

Figure 8-6 Various screw bases used in incandescent lamps.

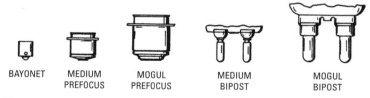

Figure 8-7 Bayonet, prefocus, and bipost bases.

The prefocus base has two side-flange projections and is made in medium and mogul sizes. It is actually a double base assembled at the factory so that when it is seated in its socket, the filament will have an exact position with respect to the optical system with which it is used. This feature is important for projector lamps. A bipost base has two metal pins sealed into the lower portion of the bulb. It also provides an exact position for the filament in an optical system.

Table 8-2 lists the most common miniature lamps and their characteristics. Outline drawings for each lamp are shown in Figure 8-8.

Table 8-2 Miniature Lamp Data

Lamp No.	Volts	Amps	Bead Color	Base	Bulb Type	Figure No.
PR2	2.4	0.50	Blue	Flange	B-3½	A
PR3	3.6	0.50	Green	Flange	B-3½	A
PR4	2.3	0.27	Yellow	Flange	B-3½	A
PR6	2.5	0.30	Brown	Flange	B-3½	A
PR12	5.95	0.50	White	Flange	B-3½	A
12	6.3	0.15		2-Pin	G-3½	H
13	3.8	0.30	Green	Screw	G-3½	B
14	2.5	0.30	Blue	Screw	G-3½	B
40	6.3	0.15	Brown	Screw	T-3¼	C
41	2.5	0.50	White	Screw	T-3¼	C
42	3.2	0.35*	Green	Screw	T-3¼	C
43	2.5	0.50	White	Bayonet	T-3¼	D
44	6.3	0.25	Blue	Bayonet	T-3¼	D
45	3.2	0.35†	Green‡	Bayonet	T-3¼	D
46	6.3	0.25	Blue	Screw	T-3¼‡	C
47	6.3	0.15	Brown	Bayonet	T-3¼	D
48	2.0	0.06	Pink	Screw	T3¼	C
49	2.0	0.06	Pink	Bayonet	T-3¼	D
50	6.3	0.20	White	Screw	G-3½	B
51	6.3	0.20	White	Bayonet	G-3½	E
55	6.3	0.40	White	Bayonet	G-4½	F
57	14.0	0.24	White	Bayonet	G-4½	F
112	1.1	0.22	Pink	Screw	TL-3	G
222	2.2	0.25	White	Screw	TL-3	G
233	2.3	0.27	Purple	Screw	G-3½	B
291	2.9	0.17	White	Screw	T-3¼	C
292	2.9	0.17	White	Screw	T-3¼	C
1490	3.2	0.16	White	Bayonet	T-3¼	D
1891	14.0	0.23	Pink	Bayonet	T-3¼	D
1892	14.0	0.12	White	Screw	T-3¼	C

*Some brands are 0.50 A.
†Some brands are 0.50 A and white bead.
‡Frosted.

Electric Lighting   193

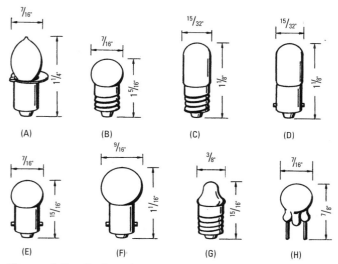

Figure 8-8   Outline drawings for miniature lamps.

Miniature lamps are widely used in electrical equipment such as instrument panels, flashlights, and on–off indicators. It is important for the electrician to observe the bead color code when replacing a miniature lamp. In other words, the color code indicates the voltage and current rating of the lamp.

Bulbs of larger lamps are made in various types and are described by letters, as shown in Figure 8-9. The bulbs may be clear, colored, inside-frosted, or coated with diffusing or reflecting materials. In addition to its letter description, a bulb is identified by a number that indicates its diameter in eighths of an inch. For example, a T-24 bulb tells the electrician that the lamp has a tubular shape and is 3 inches in diameter.

Lamps are specified in overall length from the top of the bulb to the bottom of the base. The *light-center length* is measured from the center of the filament to the bottom of the screw base, to the top of the base pins or flanges of a bayonet or prefocused base, to the shoulder of the post on a mogul bipost base, or to the bottom of the bulb (base end) of a medium bipost base.

Inside frosting is often chosen in selecting a bulb because it gives a moderate *diffusion* of the light, thereby reducing glare and shadow intensity. Since the outer surface of the bulb is smooth, it can be easily cleaned. Inside frosting results in a slight loss of light—usually

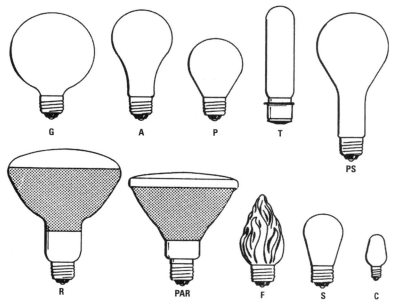

Figure 8-9 Bulb type designations.

between 1 percent and 2 percent. Some bulbs are made of white glass (milky appearance) or are white coated; these bulbs provide greater diffusion of the light but also lose from 15 percent to 20 percent of the light output from the filament. Electricians sometimes prefer high-vacuum lamps to gas-filled lamps for use in refrigeration rooms, because gas-filled lamps produce more heat. If meat is hung too close to a gas-filled lamp, it may be warmed enough to cause early spoiling.

Vapor-Discharge Lamps

Vapor-discharge lamps produce light by means of the flow of electricity through gases. Among the more important types are the *mercury, sodium, neon, neon glow, fluorescent,* and *ultraviolet germicidal* lamps. Figure 8-10 shows some familiar vapor-discharge lamps and bases. Vapor-discharge lamps are generally classified as *hot-filament, low-voltage* types; *high-voltage, cold-cathode* types; and *mercury-pool arc, low-voltage* types. The mercury lamp is a hot-cathode, low-voltage type that operates in the range from 100 to 200 volts. It produces light by conduction of electricity through

Electric Lighting 195

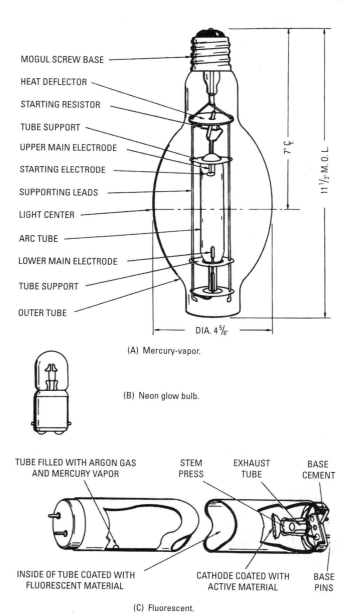

(A) Mercury-vapor.

(B) Neon glow bulb.

(C) Fluorescent.

Figure 8-10 Typical vapor-discharge lamps.

mercury vapor and has a blue-violet light output. This type is used in photographic work and for high-intensity illumination for use in industrial plants.

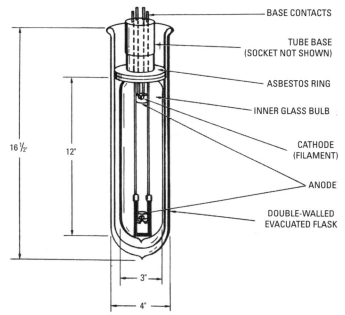

Figure 8-11 Construction of a sodium lamp.

The sodium type of lamp, shown in Figure 8-11, is a hot-cathode, low-voltage type that operates in the range from 100 to 200 volts. It produces light by conduction of electricity through sodium vapor and has a yellow light output. Sodium lamps are used for lighting bridges and highways. The neon lamp has a cold cathode and operates in the voltage range from 3000 to 15,000 volts. It is formed from glass tubes that contain neon gas and may be quite long. Neon gas produces orange-red light; this type of lamp is used in commercial signs and for display lighting.

A neon glow lamp has a cold cathode that is treated with special materials that provide conduction of electricity in the voltage range from 60 to 200 volts. These lamps have a power consumption from 0.5 to 3 watts and produce an orange-red glow. They are used chiefly as indicators, in electrical test equipment, and as night-lights or marker lights. Germicidal lamps use hot cathodes with mercury vapor in special glass envelopes that can pass ultraviolet light. They are used to sterilize the air in their vicinity.

Ordinary fluorescent lamps used in factories, shops, and homes are made with glass tubes that are coated on their inner surface with chemical powders that fluoresce when struck by ultraviolet light. When this type of lamp is filled with neon gas, it is called a hot-cathode, low-voltage neon lamp. Commercial signs and displays frequently use the high-voltage mercury, long-tube type of lamp, which operates in the range from 3000 to 15,000 volts. These lamps are similar to high-voltage neon lamps except that mercury vapor is added to produce blue light. The neon gas in these lamps is used only to provide easy starting. If this type of lamp is coated on the inner surface with fluorescent powders, it is called a high-voltage fluorescent lamp.

All vapor-discharge lamps operate on the basis of arc discharge and will be destroyed by excessive current flow unless connected in series with a resistor or reactor called a *ballast*. Table 8-3 lists various types of neon and argon lamps. Table 8-4 lists the values of resistance that are connected in series with each type. Vapor-lamp ballasts may consist of a series resistor or a reactor such as an inductor, a capacitor, or a transformer that has comparatively high leakage reactance. A ballast resistor drops from $1/3$ to $2/3$ of the line voltage and therefore wastes considerable power.

Power loss can be minimized by using an inductor as a ballast. Although there is some power loss in the winding resistance, most of the voltage drop occurs across the inductive reactance and consumes no true power. The power-factor angle is about $45°$, which means that reactive power surges back and forth in the line and increases the line losses. If a capacitor is used as a ballast, the power-factor angle is also about $45°$, but the current leads instead of lagging as with an inductor. Electricians sometimes use a capacitor ballast with one lamp and an inductor ballast with another lamp so that the total power-factor angle is practically zero.

Starters are used with fluorescent lamps, as shown in Figure 8-12. The starter consists of a neon-bulb thermostat. When the 117-volt AC input is applied, the neon gas conducts electricity and glows. This produces heat, which causes the thermostat contacts to close, and then the starting filaments in the fluorescent lamp heat up. The neon glow stops because the thermostat contacts are closed in the neon bulb. This cools the thermostat, and its contacts open. Opening of the filament circuit causes a high-voltage inductive kickback from the ballast reactor, which starts the arc discharge in the fluorescent lamp. The voltage drop across the lamp (arc drop) is too small to

Table 8-3 Gas-Filled Lamps

Number	Hours of Average Useful Life*	Type of Gas	Max. Length in Inches	Base	Amps	Volts	Watts[†]
AR-1	3000	Argon	$3^{1}/_{2}$	Medium screw	0.018	110–125	2
AR-3	1000	Argon	$1^{5}/_{8}$	Cand. screw	0.0035	110–125	$^{1}/_{4}$
AR-4	1000	Argon	$1^{1}/_{2}$	Double-contact bayonet	0.0035	110–125	$^{1}/_{4}$
NE-2	Over 25,000	Neon	$1^{1}/_{16}$[‡]	Unbased	0.003	110–125	$^{1}/_{25}$
NE-2A	Over 25,000	Neon	$^{27}/_{32}$[‡]	Unbased	0.003	110–125	$^{1}/_{25}$
NE-17	5000	Neon	$1^{1}/_{2}$	Double-contact bayonet[§]	0.002	110–125	$^{1}/_{4}$
NE-30	10,000	Neon	$2^{1}/_{4}$	Medium screw[§]	0.012	110–125	1
NE-32	10,000	Neon	$2^{1}/_{16}$	Double-contact bayonet[§]	0.012	110–125	1
NE-34	8000	Neon	$3^{1}/_{2}$	Medium screw	0.018	110–125	2
NE-40	8000	Neon	$3^{1}/_{2}$	Medium screw[§]	0.030	110–125	3
NE-45	Over 7500	Neon	$1^{5}/_{8}$	Cand. screw	0.002	110–125	$^{1}/_{4}$
NE-48	Over 7500	Neon	$1^{1}/_{2}$	Double-contact bayonet	0.002	110–125	$^{1}/_{4}$
NE-51	Over 15,000	Neon	$1^{3}/_{16}$	Miniature bayonet	0.0003	110–125	$^{1}/_{25}$
NE-56	10,000	Neon	$2^{1}/_{4}$	Medium screw[§]	0.005	220–250	1
NE-57	5000	Neon	$1^{5}/_{8}$	Cand. screw[§]	0.002	110–125	$^{1}/_{4}$
NE-58	Over 7500	Neon	$1^{5}/_{8}$	Cand. screw	0.002	220–250	$^{1}/_{2}$

* Life on DC is approximately 60% of AC value.
[†] For 110–125 V operation.
[‡] The dimension is for glass only.
[§] On DC circuits, the base should be negative.

Table 8-4 External Resistances Needed for Gas-Filled Lamps

Type	110–125 V	220–330 V	330–375 V	375–450 V	450–600 V
AR-1	Included in base	10,000	18,000	24,000	30,000
AR-3	Included in base	68,000	91,000	150,000	160,000
AR-4	15,000	82,000	100,000	160,000	180,000
NE-2	200,000	750,000	1,000,000	1,200,000	1,600,000
NE-2A	200,000	750,000	1,000,000	1,200,000	1,600,000
NE-17	30,000	110,000	150,000	180,000	240,000
NE-30	Included in Base	10,000	20,000	24,000	36,000
NE-32	7500	18,000	27,000	33,000	43,000
NE-34	Included in base	9100	13,000	16,000	22,000
NE-40	Included in base	6200	8200	11,000	16,000
NE-45	Included in base	82,000	120,000	150,000	200,000
NE-48	30,000	110,000	150,000	180,000	240,000
NE-51	200,000	750,000	1,000,000	1,200,000	1,600,000
NE-56	Included in base				
NE-57	Included in base	82,000	120,000	150,000	200,000
NE-58	Included in base				

fire the neon bulb. Therefore, the thermostat contacts remain open. Capacitor C in Figure 8-12 minimizes sparking at the thermostat contacts.

Figure 8-13 shows how two fluorescent lamps are connected to reduce the stroboscopic effect (strobe flicker). Capacitor C causes the upper lamp to draw a leading current and thereby stagger the peaks of light output from the pair of lamps. Strobe flicker appears as a broken series of images when a moving object is viewed under a fluorescent lamp. Note the starting switch in Figure 8-13. This switch is closed momentarily and then released. As the switch opens, kickback voltages from the ballast reactors fire up the fluorescent lamps. The arc discharges are then self-sustaining until the AC input is switched off.

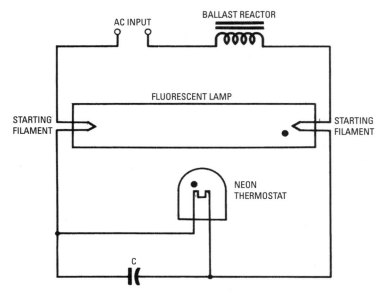

Figure 8-12 The neon thermostat operates as a starter.

A mercury-vapor lamp (Figure 8-14) consists of two electrodes at opposite ends of a glass tube containing vaporized mercury. These electrodes are made of coiled tungsten wire coated with chemical oxides for emission of electrons at a comparatively low temperature. The glass tube (arc tube) contains some argon gas to start the arc discharge before the mercury heats up and vaporizes. When voltage is applied to the lamp, an electric field is set up between the starting electrode and the upper main electrode. This causes electron emission from the surface of the main electrode. These moving electrons ionize the gas and start the arc discharge.

The sodium lamp shown in Figure 8-11 is larger and longer than a mercury-vapor lamp. However, its power consumption is less—about 200 watts. It contains sodium, which vaporizes after the lamp heats up. Some neon gas is placed in the lamp to provide easy starting and to heat the sodium. A thermostatic starting timer is used, which applies power to the filaments at each end for $1/2$ minute or longer. Then, the thermostat opens and full voltage is applied across the lamp, which starts the arc discharge. A dozen or more sodium lamps are often operated in series from a special transformer that supplies a constant current of 6.6 amperes.

Electric Lighting **201**

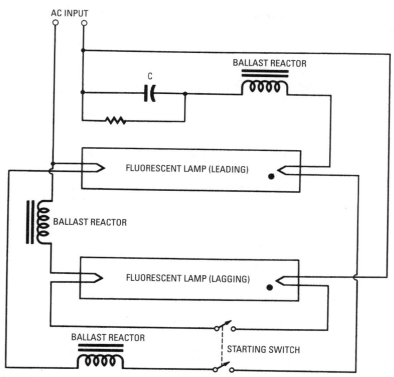

Figure 8-13 A capacitor in the upper ballast circuit reduces strobe flicker.

Metal Halide Lamps

Metal halide lamps are similar to mercury vapor lamps but use metal halide additives inside the arc tube along with the mercury and argon. These additives enable the lamp to produce more visible light per watt, along with improved color rendering.

Wattages range from 32 to 2000, offering a wide range of indoor and outdoor applications. The efficacy of metal halide lamps ranges from 50 to 115 lumens per watt (about double that of mercury vapor). The advantages of metal halide lamps include the following:

- High efficacy
- Good color rendering
- Wide range of wattages

202 Chapter 8

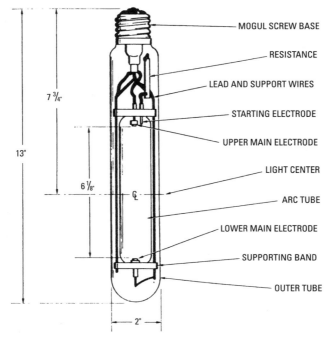

Figure 8-14 Construction of a mercury-vapor lamp.

They also have some operating limitations. The rated life of metal halide lamps is shorter than other HID sources (low-wattage lamps last less than 7500 hours, while high-wattage lamps last an average of 15,000 to 20,000 hours). Also, the color of light varies and may shift over the life of the lamp. Because of good color rendition and a high lumen output, these lamps are good for sports arenas and stadiums. Indoor uses include large auditoriums and convention halls. These lamps are sometimes used for general outdoor lighting, such as parking facilities, but a high-pressure sodium system is more popular for such uses.

High-Pressure Sodium Lamps

The high-pressure sodium (HPS) lamp is widely used for outdoor and industrial applications. Its higher efficacy makes it a better choice than metal halide for many applications, especially when good color rendering is not a priority. HPS lamps differ from mercury and metal halide lamps in that they do not contain starting electrodes; the ballast circuit includes a high-voltage electronic starter. The arc tube is made of a ceramic material, which can withstand temperatures over 2300°F. It is filled with xenon to help start the arc, as well as a sodium–mercury gas mixture.

The efficacy of the lamp is high—as much as 140 lumens per watt. For example, a 400-watt high-pressure sodium lamp produces 50,000 initial lumens. The same wattage metal halide lamp produces 40,000 initial lumens, and the 400-watt mercury vapor lamp produces only 21,000 initial lumens.

Sodium, the major element used, produces a gold color that is characteristic of HPS lamps. Although HPS lamps are not generally recommended for applications where color rendering is critical, HPS color rendering properties are being improved. Some HPS lamps are now available in deluxe and white colors that provide higher color temperature and improved color rendition. The efficacy of low-wattage white HPS lamps is lower than that of metal halide lamps (lumens per watt of low-wattage metal halide is 75–85, while white HPS is 50–60 lumens per watt).

Low-Pressure Sodium Lamps

Although low-pressure sodium (LPS) lamps are similar to fluorescent systems (because they are low-pressure systems), they are commonly included in the HID family. LPS lamps are the most efficacious light sources (that is, they put out the most light per watt of electricity), but they produce the poorest quality light of all the lamp types. LPS lamps are *monochromatic* light sources, putting out only one wavelength of light. When illuminated by LPS lamps, all colors appear as black, white, or shades of gray. LPS lamps are available in wattages ranging from 18 to 180.

LPS lamp use has been generally limited to outdoor applications such as security or street lighting and low-wattage indoor applications where color quality is not important (stairwells and similar areas). However, because the color rendering of LPS lamps is so poor, many municipalities do not allow them for roadway lighting. Because the LPS lamps are long tubes (like thin fluorescent lamps), they are less effective in directing and controlling a light beam than *point sources* such as high-pressure sodium and metal halide. Therefore, lower mounting heights will provide better results with LPS lamps. To compare an LPS installation with other alternatives, calculate the installation efficacy as the average maintained foot-candles divided by the input watts per square foot of illuminated area.

The input wattage of an LPS system increases over time and maintains consistent light output over the lamp life.

One hazard: Low-pressure sodium lamps can explode if the sodium comes in contact with water. You should always dispose of these lamps according to the manufacturer's instructions.

Sun Lamps

Sun (sunlight) lamps, such as the one shown in Figure 8-15, provide a combination of visible radiation and ultraviolet light. A tungsten filament is used, which is connected in parallel with electrodes for a mercury-vapor arc. Some argon gas is also placed in the bulb. When voltage is applied to the lamp, the filament is heated to a sufficiently high temperature, and it emits electrons. The moving electrons ionize the argon gas, and an arc discharge is started. In turn, the mercury quickly vaporizes, and the resulting mercury arc draws a heavy current; the voltage drop across the filament decreases, and the filament temperature drops several hundred degrees.

Sun lamps are powered by transformers with high leakage reactance so that the current is limited. The transformer is constructed to serve as a ballast for the lamp. Electricians must make certain that correct voltage is supplied to the transformer because the lamp is likely to operate unsatisfactorily if the supply voltage is 5 percent too high or too low. The most popular sun lamp for home use is rated at 275 watts. More-powerful ultraviolet lamps are used in hospitals. Since ultraviolet radiation can damage the retina of the eye, protective goggles must be worn.

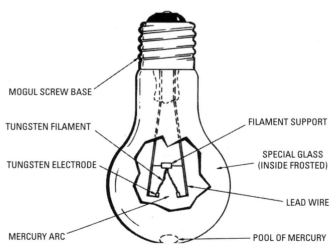

Figure 8-15 Construction of a typical sun lamp.

The action of a transformer used to power a sun lamp can be compared to the action of an ideal transformer with an inductor connected in series with the primary circuit. In other words, leakage reactance occurs when all of the magnetic flux produced by the primary does not cut the secondary turns. When only a fraction of the primary flux passes through the secondary, it is the same as if the inactive part of the primary winding were connected externally in series with the primary of an ideal transformer. This inactive part of the primary winding develops the leakage reactance that limits current output from the secondary.

If we check the secondary voltage in the absence of a load, the primary draws a very small amount of current. In turn, there is only a very small voltage drop across the leakage reactance. On the other hand, if we check the secondary voltage under heavy load, the primary draws a large amount of current. In turn, there is a large voltage drop across the leakage reactance. This is just another way of saying that the secondary voltage decreases rapidly with increasing loads when a transformer has high leakage reactance. This is why the amount of current that can be drawn from the secondary is limited by the value of the leakage reactance.

Ozone-Producing Lamps

Ozone is a form of oxygen that is formed in air by radiation of suitable wavelengths. Ozone-producing lamps are mercury-vapor lamps that produce radiation at a wavelength of 1849 angstroms. A typical lamp is rated at 4 watts and operates at 105 volts. This type of lamp is used as a deodorizer; ozone is a very active form of oxygen that reacts with common ill-smelling vapors and changes them into inoffensive chemical compounds. Ozone has a sharp and distinctive odor that is not unpleasant. Electricians who work in the vicinity of sparking electrical machinery are familiar with the odor of ozone.

Flash Tubes

Flash tubes use straight or coiled sections of glass or quartz tubing, as shown in Figure 8-16, and are filled with a gas such as xenon. When a voltage surge is applied to the tube, it produces a short flash of white light useful for high-speed photography. Operating voltages range from 800 to several thousand volts. Figure 8-17 shows a flash-tube circuit. The system is battery-powered, with a vibrator to chop the DC supply into an AC voltage to energize the transformer. In turn, a high AC voltage is provided by the secondary winding. The rectifier is a one-way valve for electric current, which charges the flash capacitor in the polarity shown in the diagram.

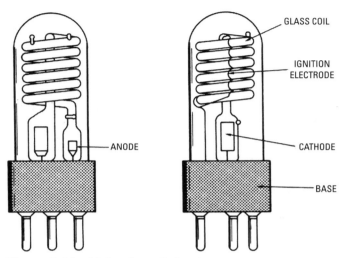

Figure 8-16 High-voltage flash tube.

Electrical energy is stored in the flash capacitor shown in Figure 8-17 but does not produce an arc discharge until a high-voltage pulse is applied to the ignition electrode. Resistors R_1 and R_2 operate as a voltage divider, and the ignition capacitor is charged to a lower voltage than the flash capacitor. To trigger the flash tube, the camera contact is closed. In turn, a surge of current flows through the primary of the ignition coil. The stepped-up voltage from the secondary is applied to the ignition electrode, and the xenon gas ionizes, permitting the flash capacitor to produce a brief arc discharge in the flash tube.

A flash tube produces a brilliant light of up to 90,000,000 peak lumens that lasts from a thousandth to a millionth of a second. The life of a typical flash tube extends for several thousand operations. Simpler flash tubes with less light output are used in amateur equipment, as shown in Figure 8-18. With suitable high-speed switching facilities, flash tubes can be operated to produce a rapid series of light pulses of up to 30,000 per second. This method is used to freeze the motion of high-speed objects stroboscopically, thus providing a series of photos that show the details of the motion.

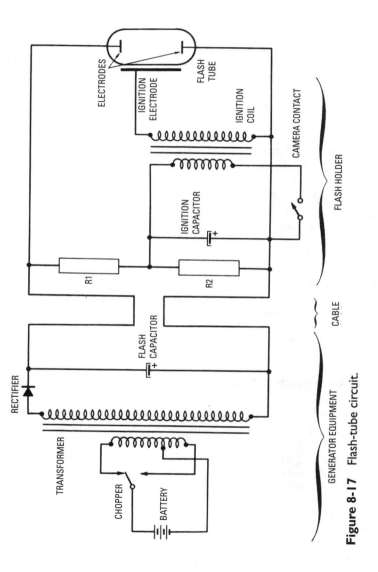

Figure 8-17 Flash-tube circuit.

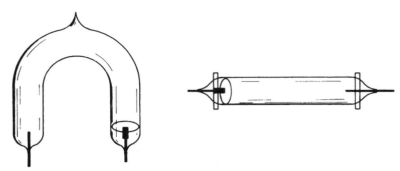

Figure 8-18 Flash tubes used in amateur equipment.

Infrared Lamps

The infrared range is below the visible range of radiant energy, as indicated in Table 8-1. Infrared lamps are used chiefly for heating and drying. They are often called heat lamps and are rated from 125 to 1000 watts. Heat lamps usually operate at 115 volts. Tungsten filaments are used at a comparatively low temperature, which produces much less light than an ordinary incandescent lamp. Most of the radiant energy is in the infrared region of wavelengths. Therefore, electricians describe this type of lamp as an infrared heating device. They are often used in radiation ovens. The lamps surround or partially surround the product that is to be dried or baked. Radiation ovens are best adapted for surface heating.

Electroluminescent Panels

Figure 8-19 illustrates the construction of electroluminescent panels. When a fluorescent material is embedded in the insulating dielectric between the plates of a capacitor, light can be produced by the electrostatic field between the plates of the capacitor. If one of the plates is made of a transparent conducting substance, the light will be radiated into surrounding space. This type of light source is called an EL plate or panel.

EL panels are not lamps as ordinarily defined by electricians. However, they are similar to lamps in that they provide lighted surfaces that are useful in many practical applications. An EL panel operates in the range from 115 to 600 volts. Typical applications are in faces of electric clocks and cover panels for electric switches or outlets. They are also used in aircraft signs, such as *Fasten Seat Belts*. This type of EL panel remains illuminated only while voltage

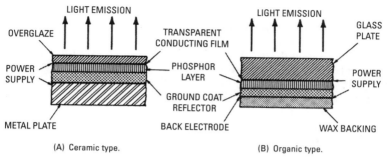

Figure 8-19 Electroluminescent panel construction.

is applied to the capacitor plates. AC is commonly used to energize ordinary EL panels.

Another type of EL panel is commonly called an image-retaining panel and is used in industrial plants. It employs a sulfide phosphor in a ceramic plate which is energized by DC voltage. This type of panel does not radiate light, however, until visible light, X rays, or electron beams strike the panel. At any point on the panel, the radiated light has an intensity that depends on the intensity of the visible light, X rays, or electron beams. Therefore, an image is formed which is retained as long as the DC voltage is applied. After the DC voltage is switched off, the image disappears. This type of panel is useful in X-ray testing of manufactured products because the image can be inspected at leisure after the X-ray beam is turned off.

Summary

A unit of illumination is called a foot-candle. If the entire inner surface of a sphere is illuminated by an amount of 1 foot-candle, the total light output from the source is 1 lumen. The intensity of illumination falls off as the square of the distance from the source.

Only metal with a high melting point is used as a lamp filament. Tungsten melts at approximately 6120°F, and a typical tungsten lamp operates at 4800°F. The average lamp life is approximately 1000 hours because the tungsten filament gradually evaporates at this temperature. Bulb blackening is caused by slow tungsten evaporation. Most filaments are coiled in the form of a long helical spring, which reduces the tendency of the gas to cool the filament.

Vapor-discharge lamps produce light by means of the flow of electricity through gases. The general types of lamps are mercury, sodium, neon, neon glow, fluorescent, and ultraviolet germicidal.

Vapor-discharge lamps are generally classed as hot-filament, low-voltage; high-voltage, cold-cathode; and mercury-pool arc.

Infrared lamps, sometimes called heating lamps, are used for heating and drying and are often used in radiation ovens. Tungsten filaments are used at a comparatively low temperature, which produces much less light than an ordinary incandescent lamp.

Sun lamps are powered by transformers so that the current is limited. The transformer serves as a ballast, which stabilizes the voltage. Tungsten filaments are used and are connected in parallel with electrodes for a mercury-vapor arc.

Test Questions

1. What is the difference between various colors of light in terms of wavelength?
2. To what color is the human eye most sensitive?
3. State an application for an infrared lamp; for an ordinary tungsten lamp; for an ultraviolet lamp.
4. What is an instrument used to measure illumination called?
5. Why is tungsten a good material to use for lamp filaments?
6. Define a lumen.
7. Explain the inverse-square law.
8. Describe bulb blackening.
9. Why does a sudden surge of current flow when a tungsten lamp is switched on?
10. What do electricians mean by the smashing point of a tungsten lamp?
11. How are the various types of lamp bulbs identified? How are their diameters identified?
12. Describe a vapor-discharge lamp.
13. How does a fluorescent lamp operate?
14. Why is a ballast required with a vapor-discharge lamp?
15. Explain the operation of a sodium lamp.
16. How is a sun lamp constructed?
17. Describe an ozone-producing lamp.
18. How is a flash tube constructed?
19. Explain the circuit used to operate a flash tube.

20. Describe an infrared lamp.
21. What is an electroluminescent panel?
22. State an application for an electroluminescent panel.
23. Explain an image-retaining panel.
24. State an application for an image-retaining panel.

Chapter 9
Lighting Calculations

Electricians are often concerned with lighting calculations to determine how much illumination is required in a given installation and to determine the necessary installation details. Two basic methods are used to calculate illumination requirements:

- Point-by-point method
- Lumen method

The final calculations are made in terms of *watts per square foot*.

Point-by-Point Method

This method permits the calculation of the illumination at any given point by means of simple arithmetic. We consider the candlepower distribution of the light source and its position with respect to the given point. It is not necessarily the best method, although electricians prefer it for certain lighting problems. The point-by-point method is an application of the inverse square law. For the reader's convenience, an illustration of the inverse square law is repeated in Figure 9-1. It shows that the intensity of illumination produced by a point source of light decreases as the square of the distance from the source. From a *candlepower distribution curve* of a reflector, the foot-candles at any given point may be found from the

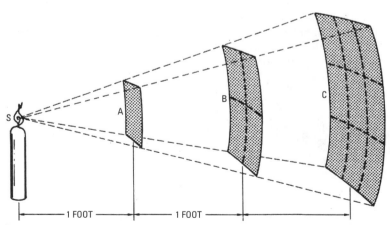

Figure 9-1 Illustrating the inverse square law.

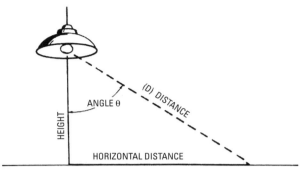

Figure 9-2 Illustrating point-by-point method of lighting calculation.

following formulas:

$$\text{Foot-candles (perpendicular to the beam)} = \frac{\text{CP(candlepower)}}{D^2(\text{distance in feet})}$$

$$\text{Foot-candles (on horizontal plane)} = \frac{\text{CP}}{D^2} \times \text{cosine of angle } \theta$$

The meaning of *perpendicular to the beam* is shown in Figure 9-1. The meaning of *angle* θ is shown in Figure 9-2. The number of degrees in an angle is measured with a *protractor*, as shown in Figure 9-3. Thus, the angle AOB is a $35°$ angle. With reference to Figure 9-2, cosine θ is the height divided by the distance D of the right-angled triangle. Table 9-1 lists the cosines of angles from $0°$ to $90°$. Therefore, when we wish to find the number of foot-candles on the horizontal plane in Figure 9-2, we measure the angle θ. Then, we look in Table 9-1 and find the cosine of this angle. We must also measure the distance D and the height in Figure 9-2.

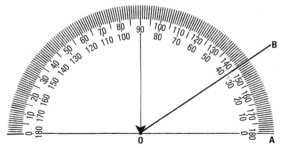

Figure 9-3 How a protractor is used to measure an angle.

Table 9-1 Sines, Cosines, and Tangents of Angles from 0° to 90°

Angle	Sin	Cos	Tan	Angle	Sin	Cos	Tan
0°	0.0000	1.0000	0.0000	46°	0.7193	0.6947	1.0355
1°	0.0175	0.9998	0.0175	47°	0.7314	0.6820	1.0724
2°	0.0349	0.9994	0.0349	48°	0.7431	0.6691	1.1106
3°	0.0523	0.9986	0.0524	49°	0.7547	0.6561	1.1504
4°	0.0698	0.9976	0.0699	50°	0.7660	0.6428	1.1918
5°	0.0872	0.9962	0.0875	51°	0.7771	0.6293	1.2349
6°	0.1045	0.9945	0.1051	52°	0.7880	0.6157	1.2799
7°	0.1219	0.9925	0.1228	53°	0.7986	0.6018	1.3270
8°	0.1392	0.9903	0.1405	54°	0.8090	0.5878	1.3764
9°	0.1564	0.9877	0.1584	55°	0.8192	0.5736	1.4281
10°	0.1736	0.9848	0.1763	56°	0.8290	0.5592	1.4826
11°	0.1908	0.9816	0.1944	57°	0.8387	0.5446	1.5399
12°	0.2079	0.9781	0.2126	58°	0.8480	0.5299	1.6003
13°	0.2250	0.9744	0.2309	59°	0.8572	0.5150	1.6643
14°	0.2419	0.9703	0.2493	60°	0.8660	0.5000	1.7321
15°	0.2588	0.9659	0.2679	61°	0.8746	0.4848	1.8040
16°	0.2756	0.9613	0.2867	62°	0.8829	0.4695	1.8807
17°	0.2924	0.9563	0.3057	63°	0.8910	0.4540	1.9626
18°	0.3090	0.9511	0.3249	64°	0.8988	0.4384	2.0503
19°	0.3256	0.9455	0.3443	65°	0.9063	0.4226	2.1445
20°	0.3420	0.9397	0.3640	66°	0.9135	0.4067	2.2460
21°	0.3584	0.9336	0.3839	67°	0.9205	0.3907	2.3559
22°	0.3746	0.9272	0.4040	68°	0.9272	0.3746	2.4751
23°	0.3907	0.9205	0.4245	69°	0.9336	0.3584	2.6051
24°	0.4067	0.9135	0.4452	70°	0.9397	0.3420	2.7475
25°	0.4226	0.9063	0.4663	71°	0.9455	0.3256	2.9042
26°	0.4384	0.8988	0.4877	72°	0.9511	0.3090	3.0777
27°	0.4540	0.8910	0.5095	73°	0.9563	0.2924	3.2709
28°	0.4695	0.8829	0.5317	74°	0.9613	0.2756	3.4874
29°	0.4848	0.8746	0.5543	75°	0.9659	0.2588	3.7321
30°	0.5000	0.8660	0.5774	76°	0.9703	0.2419	4.0108
31°	0.5150	0.8572	0.6009	77°	0.9744	0.2250	4.3315
32°	0.5299	0.8480	0.6249	78°	0.9781	0.2079	4.7046
33°	0.5446	0.8387	0.6494	79°	0.9816	0.1908	5.1446
34°	0.5592	0.8290	0.6745	80°	0.9848	0.1736	5.6713
35°	0.5736	0.8192	0.7002	81°	0.9877	0.1564	6.3138

(continued)

Table 9-1 *(continued)*

Angle	Sin	Cos	Tan	Angle	Sin	Cos	Tan
36°	0.5878	0.8090	0.7265	82°	0.9903	0.1392	7.1154
37°	0.6018	0.7986	0.7536	83°	0.9925	0.1219	8.1443
38°	0.6157	0.7880	0.7813	84°	0.9945	0.1045	9.5144
39°	0.6293	0.7771	0.8098	85°	0.9962	0.0872	11.4300
40°	0.6428	0.7660	0.8391	86°	0.9976	0.0698	14.3010
41°	0.6561	0.7547	0.8693	87°	0.9986	0.0523	19.0810
42°	0.6691	0.7431	0.9004	88°	0.9994	0.0349	28.6360
43°	0.6820	0.7314	0.9325	89°	0.9998	0.0175	57.2900
44°	0.6947	0.7193	0.9657	90°	1.0000	0.0000	
45°	0.7071	0.7071	1.0000				

We multiply the candlepower of the light source by the cosine of angle θ and divide this product by the square of distance D. The answer is the number of foot-candles on the horizontal plane.

Electricians are sometimes concerned with the sine or the tangent of an angle. Values of sines and tangents are also given in Table 9-1. Note that the sine of angle θ in Figure 9-2 is equal to distance D divided by the horizontal distance. The tangent of angle θ is equal to the horizontal distance divided by the height. Let us observe the vertical plane, perpendicular plane, and horizontal plane shown in Figure 9-4. We know that the illumination on the perpendicular plane is found as follows:

$$\text{Foot-candles (at point } d \text{ on the perp. plane)} = \frac{CP}{D^2}$$

We also know that the illumination on the horizontal plane is found as follows:

$$\text{Foot-candles (at point } d \text{ on the horiz. plane)} = \frac{CP}{D^2} \times \cosine \theta$$

Next, the illumination on the vertical plane is found as follows:

$$\text{Foot-candles (at point } d \text{ on the vert. plane)} = \frac{CP}{D^2} \times \sin \theta$$

The ratio of illumination on the vertical plane to illumination on the horizontal plane can be found from the two foregoing formulas, or the ratio can be found as follows:

$$\frac{\text{Foot-candles at } d \text{ on vert. plane}}{\text{Foot-candles at } d \text{ on horiz. plane}} = \frac{CP}{D^2} \times \tangent \theta$$

Lighting Calculations 217

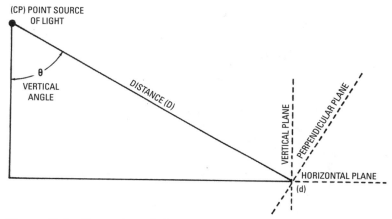

Figure 9-4 Illustration of the vertical, perpendicular, and horizontal planes.

When we use a CP value in an illumination formula, we must consider the candlepower distribution curve of the light source. Figure 9-5 shows distribution curves for a tungsten lamp suspended vertically from a ceiling and for a tungsten lamp with a reflector. It is customary to draw a distribution curve to show the candlepower 10 feet from the light source and at all possible angles from the source. For example, curve A in Figure 9-5A shows that a tungsten lamp has zero candlepower in the direction of its base. However, the lamp has 60 candlepower in a direction at right angles to its bulb. Again, the lamp has about 32 candlepower in a direction vertically below the bulb.

With reference to Figure 9-5B, we observe that a reflector tends to concentrate the light in an area below the bulb. The lamp and reflector have 155 candlepower in a direction vertically below the bulb. However, there is zero candlepower in a direction at right angles to the bulb. At an angle of 45°, the lamp and reflector have about 205 candlepower. Distribution curves are very useful for comparison purposes. For example, curves A and B show that the addition of a reflector to the lamp makes the candlepower 155/32, or almost 5 times as great in a direction vertically below the bulb. This ratio holds true at 10 feet, at 20 feet, or at any distance.

Lumen Method

The lumen method of lighting calculation is based on the *average* level of illumination desired over a given area. Since one foot-candle

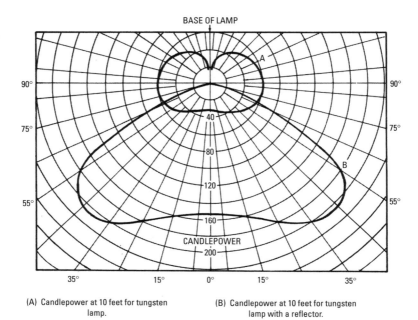

(A) Candlepower at 10 feet for tungsten lamp.

(B) Candlepower at 10 feet for tungsten lamp with a reflector.

Figure 9-5 Distribution curve.

is equal to one lumen per square foot, the total number of lumens that must be supplied to an area is found by multiplying this area by the required level of illumination in foot-candles. To find the total number of lumens that must be produced by the lamps in order to supply the required number of lumens to the area, we must take light losses into account. These losses are due to absorption of light, both in the fixture and by the walls of the room, as well as the depreciation of the system due to gradual collection of dust, smoke, and so on. The lumen method is used with the aid of suitable tables.

Reference is made to tables to determine the required level of illumination for various interiors. The type of lighting unit is selected. The mounting height and spacings are chosen with respect to the general dimensions of the room. In turn, the *room factor* or *room index* is based on the relation of these values. The *coefficient of utilization* is then found from the foregoing data, the efficiency and type of distribution of the lighting equipment, and the reflection factors of the walls and ceiling. Table 9-2 lists the percentages of light reflected from typical walls and ceilings. The coefficient of

Table 9-2 Percent of Light Reflected from Typical Walls and Ceilings

Surface	Class	Color	Percent of Light Reflected
Paint		White	81
Paint	Light	Ivory	79
Paint		Cream	74
Paint		Buff	63
Paint	Medium	Light green	63
Paint		Light gray	58
Paint		Tan	48
Paint		Dark gray	26
Paint		Olive green	17
Wood	Dark	Light oak	32
Wood		Dark oak	13
Wood		Mahogany	8
Cement		Natural	25
Brick		Red	13

Note: Each paint manufacturer's reflection values differ for similar colors, but the above table gives average figures on the colors and their average reflecting qualities.

utilization is the percentage of the source light that actually reaches the given area. A depreciation factor must be estimated, and then the total number of lumens required from each lamp is found as follows:

$$\text{Lumens required from each lamp} = \frac{\text{foot-candles desired} \times \text{area per lamp in sq. ft.}}{\text{coefficient of utilization} \times \text{depreciation factor}}$$

or, when computing for lamps of various sizes, the equation becomes:

$$\text{Foot-candles} = \frac{\text{lamp lumens} \times \text{coeff. of util.} \times \text{depreciation factor}}{\text{area per lamp sq. ft.}}$$

In practice, it is necessary to select the nearest lamp size because lamps are rated in steps. It is helpful to use tables that summarize the results for the more common types of lighting units when using the

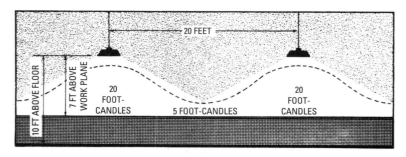

(A) Improper spacing.

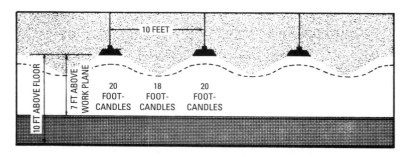

(B) Normal spacing.

Figure 9-6 Illustrating light-fixture spacing.

lumen method. This simplified method can be outlined for *interior lighting installation*:

1. Determine a suitable spacing of the lighting units (see Figure 9-6).
2. Obtain the room factor (room index). This is explained hereafter.
3. Determine the foot-candles required, as explained hereafter.
4. Determine the wattage necessary for the lamp or lamps, as explained hereafter.
5. Calculate the necessary wire size, as explained in Chapter 2.

The correct spacing of lighting units to obtain reasonably uniform illumination throughout a room depends on several facts. Strictly speaking, the spacing required for uniform illumination depends on the height of the light source. Basically, the ceiling height

Lighting Calculations **221**

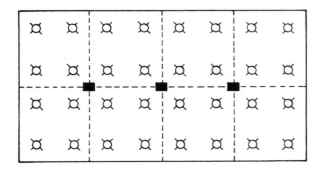

(A) Four units per bay.

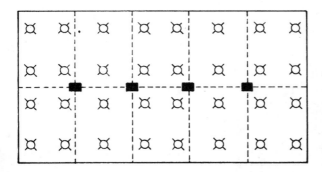

(B) Four-two system layout.

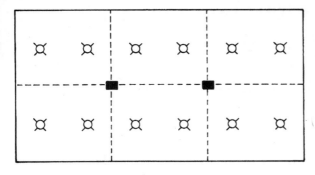

(C) Two units per bay.

Figure 9-7 Apportioning lighting units among partitions.

222 Chapter 9

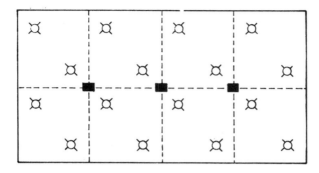

(D) Two units per bay staggered.

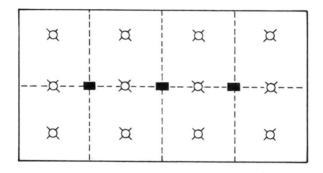

(E) Inter-spaced layout.

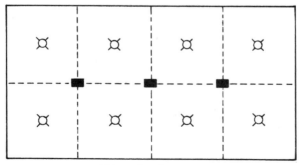

(F) One unit per bay.

Figure 9-7 *(continued)*.

limits the maximum permissible spacing. Note that the spacing of lighting units does not depend so much on the size or type of lamp used but on the distribution curve of the reflector, as exemplified in Figure 9-5.

Rooms or halls may be divided by columns or beams into a definite number of sections or bays. Electricians usually try to make a symmetrical layout in accordance with the bays, partitions, and similar architectural features. Thus, the lighting units may be spaced as shown in Figure 9-7. Arrangements in which each bay is treated as a separate unit have an additional advantage in that no change in lighting units will be required in case the partitions are later rearranged.

The room factor is an appraisal: first, from the standpoint of its general proportions; second, from the reflection factors of its walls and ceiling in order to determine what percentage of light is lost in traveling from the lamp to the illuminated area; third, from the type of lighting equipment to be used. In general, large rooms of average height use light more efficiently than small rooms of the same height because less of the total light from the source is absorbed by the walls. In using the simplified method, it is necessary only to note whether the room width is approximately equal to two, three, or four times the ceiling height.

Next, we note the color of the walls and ceiling. As listed in Table 9-2, the illumination in a room is dependent on the amount of light reflected from the walls and ceiling. We use three classifications—light, medium, and dark. With the foregoing data, we can find the room factor from Table 9-3. We will use the room factor in combination with other data to find the lamp wattage required. At this point, let us briefly consider various types of lighting units.

Table 9-3 Light Distribution of Various Types of Lighting Systems

Classification	Approximate Light Distribution	
	% Upward	% Downward
Direct	0–10	90–100
Semidirect	10–40	60–90
General diffusing	40–60	40–60
Semi-indirect	60–90	10–40
Indirect	90–100	0–10

Lighting Equipment

Lighting fixtures have changed considerably over the years, with most of the changes affecting artistic form rather than function. New diffusing and reflecting materials have been developed, and there is a continuing trend toward built-in lighting systems. Thus, the electrician has a much wider range of lighting equipment to choose from than in the past. However, basic lighting calculations are changed very little by these developments. Lighting systems are grouped into four main types: direct, semidirect, semi-indirect, and indirect.

Direct lighting is defined as any system in which practically all of the light on the illuminated area is essentially downward and comes directly from the lighting units. Direct lighting systems range from spotlight and concentrating types of equipment, through the many types of bowl and dome reflectors, to extended light source areas such as large glass panels and skylights. Figure 9-8 illustrates the general classifications of lighting fixtures. Open-type reflectors are most efficient, although it is difficult to provide high levels of illumination without glare unless considerable care is taken in locating and shielding this type of lighting unit.

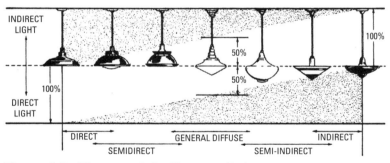

Figure 9-8 The general classification of lighting units.

Table 9-3 lists the light distributions of the various lighting systems illustrated in Figure 9-8. To reduce the glare problem of open-type reflectors, louvered downlights with concentrating reflector or lens control may be used. This confines the light narrowly to the illuminated area with a minimum of light in the direction of the eyes. To obtain good distribution, careful location of equipment is required; harsh shadows and glaring reflections from shiny or polished surfaces should be avoided. Direct lighting from large-area sources with good diffusion has some of the advantages of indirect

Table 9-4 Common Units of Brightness and Their Relations

1 candle per square inch	= 452 foot-lamberts
	= 0.487 lamberts
	= 487 millilamberts
1 foot-lambert	= 1 lumen per square foot reflected or emitted
	= 0.00221 candles per square inch
	= 1.076 millilamberts
1 lambert	= 1 lumen per square centimeter reflected or emitted
	= 1000 millilamberts
	= 929 foot-lamberts
	= 2.054 candles per square inch
1 millilambert	= 0.929 foot-lamberts
	= 0.002054 candles per square inch

lighting in that harsh shadows and both direct and reflected glare are minimized.

Semidirect lighting is provided by systems in which the predominant light on the illuminated area is obtained directly from the lighting units, with a considerable amount of light obtained by reflection from the ceiling. This is the type of lighting provided by opal or prismatic glass globes that surround the lamps. Lighting units of this type direct their light out at all angles and are likely to be too bright for offices, schools, and similar installations unless the electrician uses oversize globes. However, it is sometimes practical to use parchment shades with the smaller globes to reduce brightness toward the eyes and at the same time redirect the light more efficiently to the illuminated area.

Semi-indirect lighting is defined as any system in which some light (such as 5 percent to 25 percent) is directed downward, with more than half of the light directed upward and reflected from the ceiling. The surface brightness of the bowl should not be greater than 500 foot-lamberts. The foot-lambert is equal to the incident foot-candles multiplied by the reflection factor, assuming a diffusing illuminated area. (See Table 9-4.) The reflection factor is the ratio of light reflected by an illuminated area to the incident light. Opaque lighting units that have baffles or shielded openings to redirect a small portion of the light to their undersurfaces for decorative effects are classed as indirect lighting units.

Indirect lighting has a soft and subdued appearance due to low brightness and absence of sharp shadows. Practically all of the light is diffusely reflected from large ceiling areas. Installations may consist of suspended or portable lighting units or built-in concealed sources in the form of coves, ceiling coffers, column urns, or wall boxes. A fair uniformity of ceiling brightness is required. Practical problems may occur in the case of long, low-ceilinged rooms, where large expanses of ceiling area are in the line of vision, if the installation produces more than 25 foot-candles. These installation problems are less serious if the ceiling area is broken by crossbeams or if ceiling valances are used to break up the flat area.

Level of Illumination

During past years there has been a trend toward higher levels of illumination in nearly all applications. Six classifications are recognized by the electrician, as follows:

1. Close and prolonged tasks with small details, low contrast, and high speed or operation. A level of at least 100 foot-candles is recommended, which may be obtained by supplementary lighting.
2. When speed of operation is not a factor, the lighting level may be reduced to a minimum of 50 foot-candles. The required level may be obtained by supplementary lighting.
3. For conventional industrial and commercial applications, a range of 20 to 50 foot-candles is adequate. Economy may be realized in some installations by use of supplementary lighting.
4. Familiar tasks of comparatively brief duration, recreational activities, and so on, require lighting levels from 10 to 20 foot-candles. General lighting systems are commonly used in these applications.
5. With large objects, slow movement, and good contrast, 5 to 10 foot-candles from a general lighting system is adequate.
6. Passageways with light traffic and no hazards may have lighting levels of less than 5 foot-candles. The quality of illumination is not of great importance.

Lamp Wattage

After the electrician has selected a suitable illumination level, the necessary lamp wattage is determined. The tables at the back of the book are used for this purpose. We locate the spacing between lighting units in the second column and check the minimum mounting height given in the first column to make sure that the mounting height to be used is greater than the minimum height listed for the given spacing. Then, in the column headed *Room Factor*, we select a permissible room factor (A, B, C, D, or E) among those listed, in accordance with the room factors specified in Table 9-5.

Following horizontally to the right of this line in the design data table, we locate the column that contains the foot-candle value recommended for the installation. At the head of the column will be found the size of Mazda lamp required. Although installation details vary in practice, these tables serve as a useful guide in almost any situation. When in doubt because of unusual ceiling or sidewall surfaces, or because of unusual architectural features, it is good practice to use the next larger size of lamp.

Wire Capacity

The basic facts concerning wire current-carrying capacity were explained in Chapter 2. The need for adequate wire capacity cannot be too strongly emphasized. Operation of lighting systems with poor wiring installations defeats the goal of better lighting. Overload is not only a nuisance from the standpoint of fuse trouble or circuit breaker tripping but also potentially dangerous. Also, line-voltage fluctuation is inevitable and will cause unsatisfactory operation of electrical devices in overloaded circuits.

The *National Electrical Code* is the basic guide for safe wiring. Note, however, that the *Code* merely specifies wiring practice with regard to fire hazards, with little consideration to economy of operation. For example, the size of wire used in lighting installation may conform strictly to the *Code* and yet, because of the length of the circuit, have an excessive voltage drop. Lamps may burn dimly, illumination can vary considerably with the load, and there may be excessive power consumption in line losses.

On either new or remodeling jobs, where the wattage has been determined, wiring specifications should be based on wattage for the next larger size of lamps. It costs about $1/3$ more to double the capacity of an installation. In branch circuits for general illumination, the reasonable *voltage drop* between panelboard and outlet would be a maximum of 2 volts.

In some localities, wiring installations must also meet specifications set down in city, county, or state codes. Public utilities may have certain rules also, based on considerations of the best service that can be provided to the greatest number of customers. Any installation should be inspected by a qualified official to make certain that it is in accordance with *NEC* and local codes. In cities, such inspection is usually required by ordinances. Even if inspection is not mandatory in a given locality, it is good practice to request an

Table 9-5 Room Factors

Proportions of Room	Color of Ceiling and Upper Sidewalls	Direct Lighting		Semidirect Lighting	Semiindirect Lighting	Indirect Lighting
		Distributing	Concentrating			
Width approximately four or more times ceiling height	Light	A	A	C	C	C
	Medium	A	A	C	D	D
	Dark	A	A	D	D	E
Width approximately twice ceiling height	Light	B	A	C	C	D
	Medium	B	B	D	D	D
	Dark	B	B	D	E	
Width approximately equal to ceiling height	Light	C	B	D	D	D
	Medium	C	B	E	E	E
	Dark	C	B	E		

inspection by a fire underwriter. Installations in federal and state buildings generally require inspection by authorized federal or state personnel.

Load and Length of Run

For 15-amp circuits, the initial load per circuit should not exceed 1000 watts, with No. 12 minimum wire size used where the length of run does not exceed 50 feet. No. 10 wire should be used for runs between 50 and 100 feet, and No. 8 wire for runs between 100 and 150 feet. For heavy-duty lamp circuits, with No. 8 wire and a 3000-watt load, runs may extend up to 50 feet; for runs from 50 to 100 feet, No. 6 wire should be used. No. 4 wire should be used for runs between 100 and 150 feet. It is recommended that panelboards be located so that the length of run does not exceed 100 feet whenever it is practical to do so. Table 9-6 lists required sizes for a maximum line drop of 2 volts.

Panelboards

A panelboard is an electrical panel containing switches, fuses, or circuit breakers housed in a metal cabinet. A panelboard may also be called a service or distribution panel. One spare circuit should be provided for each five circuits used in the initial installation. Concealed branch-circuit conduits should be large enough to add one additional circuit for every five or fewer circuits that it contains.

Service and Feeders

A line running from the source of supply to various branch lines is called a *feeder*. A feeder in a farm installation may be a *building feeder* from the main load panel to a branch-circuit panel in the same building or as a *service* to another building. Again, a feeder might be a *service feeder* running from a yard pole to a house or other building. The latter are often called *service drops,* but they operate as feeders. The *service* carries power into a building and to the load center where the power is then distributed to interior circuits.

The current-carrying capacity of service wiring and feeders should be such that there will be less than a 2-volt drop under full load. Provision should be made for a 50 percent increase in current-carrying capacity over the initial installation. Convenience outlets should not be connected to branch circuits that supply fixture outlets because the load that the user might place on a convenience outlet is somewhat unpredictable. A minimum wire size of No. 12 should be used for convenience outlets, and No. 10 should be used if the run exceeds 100 feet.

Table 9-6 Wire Size Required (Computed for Maximum of 2-Volt Drop on Two-Wire 120-Volt Circuit)

Load per Circuit	Current 120-Volt Circuit	Length of Run (Panel Box to Load Center)—Feet																	
Watts	Amps	30	40	50	60	70	80	90	100	110	120	130	140	150	160	170	180	190	200
500	4.2	14	14	14	14	14	14	12	12	12	12	12	12	10	10	10	10	10	10
600	5.0	14	14	14	14	14	12	12	12	12	10	10	10	10	10	10	10	8	8
700	5.8	14	14	14	14	12	12	12	10	10	10	10	10	10	8	8	8	8	8
800	6.7	14	14	14	12	12	12	10	10	10	10	8	8	8	8	8	8	8	8
900	7.5	14	14	12	12	12	10	10	10	10	8	8	8	8	8	8	8	8	6
1000	8.3	14	14	12	12	10	10	10	10	8	8	8	8	8	8	8	6	6	6
1200	10.0	14	12	12	10	10	10	10	8	8	8	8	8	6	6	6	6	6	6
1400	11.7	14	12	10	10	10	8	8	8	8	8	6	6	6	6	6	6	6	6
1600	13.3	12	12	10	10	8	8	8	8	6	6	6	6	6	4	4	4	4	4
1800	15.0	12	10	10	10	8	8	8	6	6	6	6	6	4	4	4	4	4	4
2000	16.7	12	10	10	8	8	8	6	6	6	6	6	6	4	4	4	4	4	4
2200	18.3	12	10	10	8	8	8	6	6	6	6	6	4	4	4	4	4	4	4
2400	20.0	10	10	8	8	8	6	6	6	6	6	4	4	4	4	4	4	2	2
2600	21.7	10	10	8	8	6	6	6	6	4	4	4	4	4	4	4	4	2	2
2800	23.3	10	8	8	8	6	6	6	6	4	4	4	4	4	4	4	2	2	2
3000	25.0	10	8	8	6	6	6	6	6	4	4	4	4	4	4	2	2	2	2
3500	29.2	10	8	8	6	6	6	4	4	4	4	2	2	2	2	2	2	2	2
4000	33.3	8	8	6	6	6	4	4	4	4	2	2	2	2	2	2	1	1	1
4500	37.5	8	6	6	6	4	4	4	2	2	2	2	2	2	1	1	1	1	1

Figure 9-9 Typical receptacles. A duplex receptacle (right) is used in most applications; a single outlet may be found for certain uses, such as with a high-wattage appliance.

Electricians should have a *National Electrical Code* book at hand and check requirements for outlets in various installations. Duplex receptacles are specified in most situations (see Figure 9-9). In office space, there should be one convenience outlet circuit for each 800 feet of floor area and at least one duplex outlet for each 20 linear feet of wall. In manufacturing spaces, there should be one convenience outlet for each 1200 square feet of floor space, with at least one duplex outlet in each bay. In stores, there should be at least one convenience outlet in each supporting column or at least one floor outlet for each 400 square feet or fraction thereof in floor space. For windows, at least one outlet for each 5 linear feet of plate glass, with an additional floor outlet for each 50 square feet of platform area, should be provided. Provision for signs should be made by installing a 1-inch conduit from the distribution panel to the front face of the building for each individual store space (see Chapter 10 for house wiring).

Watts per Square Foot

In determining illumination on a watts-per-square-foot basis, the floor area is computed from the outside dimensions of a building, the area, and the number of floors, not including open porches, garages connected with dwellings, or unfinished spaces in basements or attics. Although the watts-per-square-foot basis is not exact, a fairly accurate illumination level may be found. One watt per square foot may produce from 3 to 10 foot-candles, depending on room size, ceiling and wall color, type of lighting units, and method of lighting used. Therefore, any watts-per-square-foot load estimates should be based on required foot-candles and also by the installation considerations.

One watt per square foot will provide about 5 foot-candles in dead storage areas, locker rooms, inactive spaces, and so on. If a storage space is used as an active work area, 2 watts or more per square foot should be provided. *Two watts per square foot* will provide an illumination of 10 to 15 foot-candles in industrial areas for familiar tasks of comparatively brief duration. In commercial areas, 2 watts per square foot will provide from 8 to 12 foot-candles when used with standard reflecting equipment.

Four watts per square foot will provide about 20 foot-candles when somewhat greater illumination is required. *Six watts per square foot* provides about 30 foot-candles and is adequate for many conventional industrial and commercial applications. If indirect lighting is used, a high allowance should be made. *Eight watts per square foot* will provide from 30 to 35 foot-candles when somewhat greater illumination is required. Illumination requirements in excess of 35 foot-candles are often met by use of supplementary lighting.

Summary

Basically, two methods are used to calculate illumination requirements: the point-by-point method and the lumen method. The point-by-point method permits calculation by means of simple arithmetic. The lumen method of lighting is calculated on the basis of the average level of illumination desired over a given area.

Lighting equipment is grouped into four main types: direct, semidirect, semi-indirect, and indirect. Direct lighting is defined as any system where practically all the light on the illuminated area is essentially downward and comes directly from the lighting fixture.

Semidirect lighting is provided by a system in which the light on the illuminated area is obtained directly from the lighting unit, with a certain amount of light obtained by reflection from the ceiling.

Semi-indirect lighting is defined as any system in which 5 percent to 25 percent of the light is directed downward, with more than half of the light directed upward and reflected from the ceiling.

Indirect lighting has a soft and subdued appearance due to low brightness and the absence of sharp shadows. Practically all of the light is reflected from a large ceiling area.

A line running from the source of supply to various branch lines is called a feeder. A feeder could be a service feeder, which is a line running from a yard pole to a house where the power is then distributed to interior circuits.

Test Questions

1. What are the two basic methods used to calculate illumination requirements?
2. Describe the point-by-point method.

3. How do we find the cosine of an angle?
4. What do electricians mean by the perpendicular plane? Horizontal plane? Vertical plane?
5. Explain how a protractor is used.
6. Discuss a candlepower distribution curve.
7. Describe the lumen method of lighting calculation.
8. How many foot-candles produce one lumen per square foot?
9. About how much light is reflected by medium-class paint?
10. Why must we observe a minimum spacing between lighting units in many installations?
11. Define the coefficient of utilization.
12. Give an outline of procedure for interior lighting installations.
13. What is the difference between direct, semidirect, semi-indirect, and indirect luminaires?
14. Define a foot-lambert.
15. What is the recommended lighting level for close and prolonged tasks? For conventional industrial and commercial applications? For passageways?
16. What is meant by a room factor? How do we find a room factor?
17. State the maximum voltage drop that should be permitted between a panelboard and an outlet.
18. How many spare circuits should an electrician provide in a panelboard?
19. Describe service and feeder lines.
20. What is the maximum voltage drop that should be permitted in a service or feeder line?
21. What percentage increase in current-carrying capacity should be provided for future requirements in service and feeder installations?
22. Explain the watts-per-square-foot method of determining illumination requirements.
23. Why is the watts-per-square-foot method less exact than other methods?
24. About how many foot-candles will 1 watt per square foot provide in dead storage areas?
25. Will 8 watts per square foot provide 8 times as many foot-candles as 1 watt per square foot?

Chapter 10
Basic House Wiring

All wiring must conform to the *National Electrical Code* (NEC), as well as to any codes of the particular locality. Many methods of wiring have been included in the *NEC* for residential use. Open conductors and knob-and-tube wiring are found only in older homes, and the electrician can use these only for extensions of existing circuits. The most common type of wiring used for homes is nonmetallic sheathed cable, as discussed in the previous chapter. Armored cable, also known as BX and more properly called Type AC, is also found in extensive use, as are various types of conduit, including electrical metallic tubing (EMT). Approved, but less widely used, are various raceways, wireways, and busways. These are illustrated in Figure 10-1.

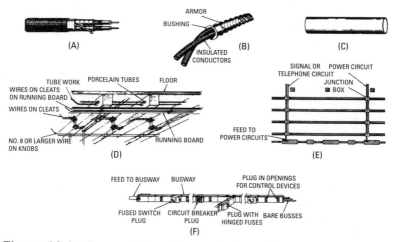

Figure 10-1 Some of the wiring methods used in residential work: (A) Type NM, nonmetallic sheathed cable, which is the most common; (B) Type AC, armored cable (BX), which is also widely used; (C) various types of conduit, such as the thin wall (EMT); (D) knob-and-tube wiring, which was once used in homes but not for new residences. Special situations may call for underground raceways (E) or busways (F).

The wiring of a new house is usually a simple job for an experienced electrician. The most difficult part is determining and planning the circuitry. Once that is accomplished, the actual installation usually progresses smoothly.

When the electrician is called upon to add wiring or to improve wiring in old houses, the job can become quite difficult. The older the home, the harder it is to predict what type of framing is used inside the walls. Various and often totally unexpected obstructions can make running the wires from one location to the next most complicated.

In the next few chapters, we concentrate less on the theory of electricity and more on the practical aspects of the electrician's task. It is important that even such seemingly minor or simple tasks as wire stripping be done in the proper manner. The electrician should always remember that careless or slipshod work can result in serious consequences for the homeowner. Faulty installation can result not only in the loss of an electrician's license but also in the loss of someone's life.

As always, be sure that any work performed is in conformance with the latest mandates of all codes, both national and local.

Service Connections

A *service* is that portion of the supply conductors that extends from the street main (or duct or line transformer) to the entrance or service panel inside the house. There are various methods of making a service entrance into a building, and they may be classified as conduit or underground services.

Figure 10-2 shows a typical overhead service drop, with Figure 10-3 illustrating the minimum clearances over certain areas, as required by the *NEC*.

Many utility companies now install meters outside residences for easier reading, and so where a meter pole is not used, rigid conduit is installed from the service cap to an exterior meter. The service wires are then connected through more conduit through a bushing to the entrance panel, from which power is distributed to the house through the various circuits (see Figure 10-4).

With reference to Figure 10-5, rigid conduit is used from the cutout switch to a point at least 8 feet above the ground. The wires enter the conduit through a fitting called a *service cap* in order to protect the wires at the entrance point and prevent water from entering the conduit. Note in Figure 10-5 that the meter is installed inside the home. A Pierce wire-holder insulator may be used, as shown in Figure 10-6.

To install a conduit service entrance, a hole is drilled through the wall to pass the conduit. The conduit is then bent so that the end passing through the wall extends $3/8$ inch inside the main switch

Basic House Wiring 237

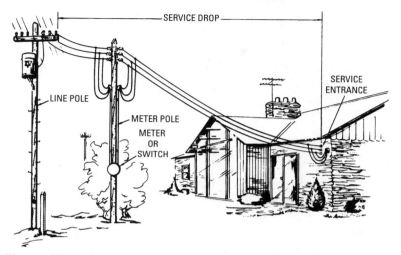

Figure 10-2 A service drop to a residence using a separate meter pole.

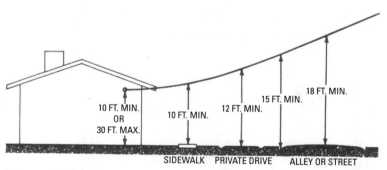

Figure 10-3 The *NEC* calls for minimum clearances above certain aspects of the landscape. Note here that the service must be not less than 10 feet and not more than 30 feet from ground level where it enters the residence.

cabinet. Instead of bending the conduit, an approved L (condulet) fitting is often used, as shown in Figure 10-7.

The end of the conduit is secured to the panel box by a locknut and bushing. The locknut is screwed onto the conduit before it enters the panel. The bushing protects the wires where they leave the pipe

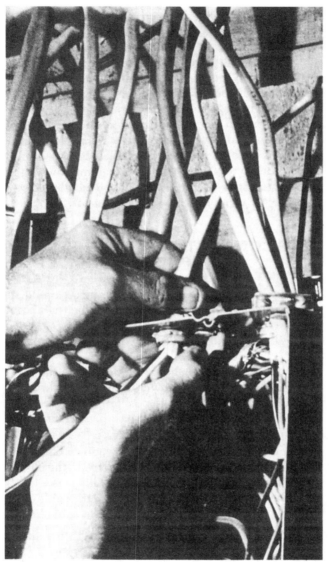

Figure 10-4 The service or entrance panel distributes power to the house through the various circuits.

Basic House Wiring

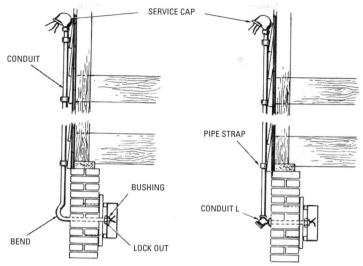

Figure 10-5 Service entrance cap and installation. In many localities, utility companies insist that the meter be placed outside the house.

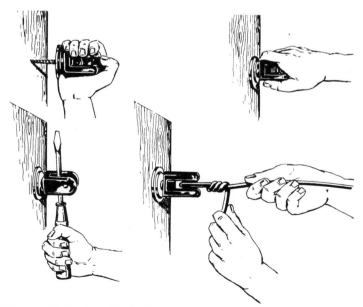

Figure 10-6 Installing a Pierce wire-holder insulator.

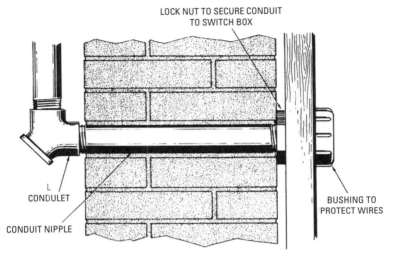

Figure 10-7 Fitting used in securing rigid conduit to switch box through foundation or other wall.

and should be tightened with a pair of pliers. The locknut is then tightened against the wall of the cabinet to hold the conduit securely in the switch box.

That portion of the conduit that is on the outside of the building is held in place by pipe straps, which in turn are fastened with screws. The L condulet must be of the weatherproof type. Figure 10-8 shows one type of L condulet. This fitting is made weatherproof by placing a rubber gasket between the body of the fitting and the cover.

Residential Underground Service

There is a considerable trend to underground service in densely populated areas. An underground service lateral (wire or cable extending in a horizontal direction) is installed, owned, and maintained by the public utility. It is run to the residential termination facility, which is usually the kWh meter or meter enclosure, as depicted in Figure 10-9. Meters are ordinarily located within 36 inches of wall nearest to the street or easement where the public utility's distribution facilities are located. The meter is mounted from 48 to 75 inches above final grade level. If there is likelihood of damage, meters must be adequately protected. Larger conduit is now required than formerly, and public utilities are frequently specifying minimum 2-inch inside diameter conduit for the service entrance. Note that aluminum

Basic House Wiring 241

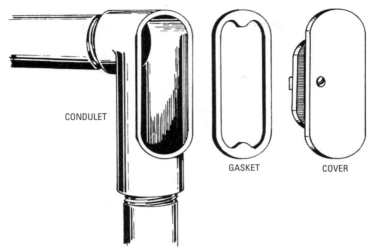

Figure 10-8 Waterproof conduit L.

conduit cannot be installed below ground level. The service conductors are usually No. 2 aluminum wire, and the service is customarily three-wire 120/240 volt single-phase 60 Hz. The third wire is a grounded neutral conductor.

Making Connections to the Entrance Panel

As mentioned earlier, the service or entrance panel distributes the power to the house through circuits. Each circuit has its own overcurrent protective device, be it fuse or circuit breaker. Electric ranges and other high-wattage appliances will ordinarily require two breakers joined together, usually as one double-pole unit.

All work within the panel should be done with the power disconnected. If there is already power to the panel, throw the dual breakers labeled "Main" to the off position, but be careful, since there may still be power to the range or other circuits with an unusual or older type of panel.

Plan your circuitry as discussed here and in Chapter 12 and determine where each cable will be connected to the panel, where it will enter the panel, and the path it will take (see Figure 10-10).

Remove the knockouts for the breaker and the cable connection to the outside of the panel. Before cutting the cable, make a trial run to make sure you have enough wire to fit (see Figure 10-11).

242 Chapter 10

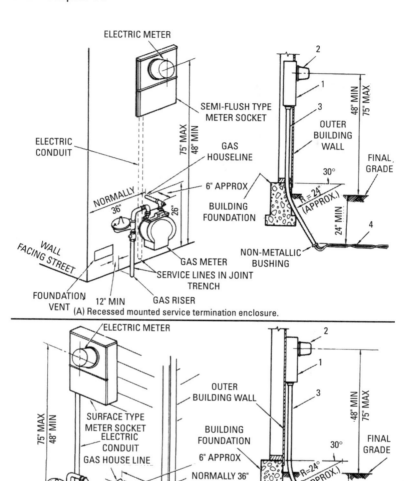

Figure 10-9 Details of several underground service entrance installations.

Basic House Wiring 243

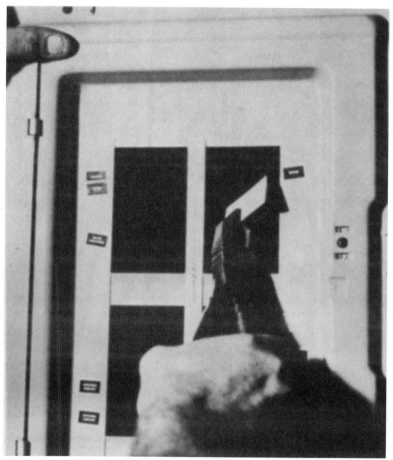

Figure 10-10 Plan each circuit ahead of time and determine where it will be connected to the entrance panel. Tags next to breakers identify the main route of the circuit and help the homeowner identify circuits in an emergency.

Select a breaker to fit the type of panel and size of wire used (15 ampere for No. 14 wire, 20 ampere for No. 12). Cut the cable and strip it as you would any other device with terminal screws. Secure the cable to the panel with standard connectors. Attach the black wire to the breaker (see Figure 10-12) and the white wire to

Figure 10-11 Make sure you allow enough cable inside the panel before you make the cut.

the common grounding bar. In most panels, the bare or green wire will also be attached to the grounding bus bar. If there is a separate bar, connect your ground wire to that. Push the breaker into place as shown in Figure 10-13.

Circuitry

Circuitry, or *circuiting,* is a term used by practical electricians for planning the wiring to connect outlets, switches, fixtures, and so on, to the source of electricity—in this case, the service entrance panel. Although the architect's or builder's plans will show locations for these, the actual wiring plan is usually left for the electrician.

Basic House Wiring 245

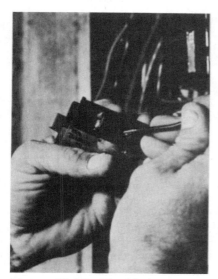

Figure 10-12 The black wire is attached to the dark screw of the circuit breaker. White (neutral) and bare, or green, insulated ground wires are ordinarily attached to the common grounding bar.

Figure 10-13 After all wires are in place and securely attached, snap the circuit breaker into its slot inside the panel.

Home circuitry can vary widely, depending on the size and age of the home, the availability of other types of energy, and even the region in which you live. A small, older home in a temperate climate with a natural gas line may have only a few electrical circuits (although it could probably use more). A large, new home in the Sunbelt, where there is no natural gas, can have many more circuits. Not only would such a house have more rooms, it would need circuits for heating, cooking, clothes drying, and especially air-conditioning. It would also need a much bigger service (more housepower) than the former, since the number of circuits and the capacity of each would require a higher capacity entrance panel.

Symbols that are used in electrical installation work are shown in Table 10-1.

Individual Appliance Circuits

These are provided for such high-wattage appliances as ranges, central air conditioners, and electrical dryers. These usually operate at 240 volts and are served by two lines working together. Other individual appliances use only one line—for example, oil burners, water heaters, certain workshop tools, and clothes washers. Individual appliance circuits, whether they operate on 120 (one line) or 240 volts (two lines), either terminate in a single wall outlet (such as an electric clothes dryer) or are wired directly into the appliance itself (as in the case of an electric range). Modern homes should also have a single outlet for the refrigerator because of current demands for frost-free features.

Regular Appliance Circuits

These are used primarily in the kitchen but can also be found in the pantry, breakfast room, laundry, and dining room. These circuits are designed for higher-energy-consuming appliances such as toasters, electric coffeepots, and similar kitchen appliances. They are wired with No. 12 cable and 20-ampere fuses.

General-Purpose Circuits

These are found in most rooms of the house. They are used primarily for lighting and low-wattage appliances, such as clocks, stereos, and TVs. Such circuits generally use No. 14 cable and are fused at 15 amperes.

The rule of thumb is that there should be at least one 15-ampere general-purpose circuit for every 375 square feet of floor space, or one 20-ampere circuit for every 500 square feet. The *NEC* also requires at least two 20-ampere general appliance circuits for the

Table 10-1 Standard Electrical Symbols Used in Architectural Drawings

Ceiling	Wall	General Outlets	Switch Outlets
○	–○	Outlet	S Single-pole switch
Ⓑ	–Ⓑ	Blanked outlet	S_2 Double-pole switch
Ⓓ		Drop cord	S_3 Three-way switch
Ⓔ	–Ⓔ	Electrical outlet—for use only when circle used alone might be confused with columns, plumbing symbols, etc.	S_4 Four-way switch S_D Automatic door switch S_E Electrolier switch S_K Key-operated switch
	----	Branch circuit— concealed in floor	[FS] Automatic fire-alarm device
	-----	Branch circuit— exposed	[W] Watchman's station
	→→	Home run to panelboard. Indicate number of circuits by number of arrows. *Note:* Any circuit without further designation indicates a two-wire circuit. For a greater number of wires, indicate as follows:	[|W|] Watchman's central station [H] Horn [N] Nurses's signal plug [M] Maid's signal plug [R] Radio outlet [|SC|] Signal central station □ Interconnection box

(continued)

Table 10-1 (*continued*)

Ceiling	Wall	General Outlets	Switch Outlets
		—///— (three wires) —//—//— (four wires, etc.)	ıIıIıIıI Battery ———— Auxiliary system circuits
	▬	Feeders. *Note:* Use heavy lines and designate by number corresponding to listing in feeder schedule.	*Note:* Any line without further designation indicates a two-wire system. For a greater number of wires, designate with numerals in manner similar to—
	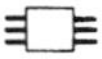 ⊒⊏	Under-floor duct and junction box—triple system. *Note:* For double or single systems, eliminate one or two lines. This symbol is equally adaptable to auxiliary system layouts.	12-N, 18 W-$^{3}/_{4}''$ C, or designate by number corresponding to listing in schedule.
Ⓖ		Generator	☐a, b, c
Ⓜ		Motor	Special auxiliary outlets.
Ⓘ		Instrument	Subscript letters refer to notes on plans or detailed description in specifications.
		Auxiliary Systems	
	▣	Pushbutton	Ⓣ Bell-ringing transformer
	▱	Buzzer	Ⓓ Electric door opener

(*continued*)

Basic House Wiring 249

Table 10-1 (*continued*)

Ceiling	Wall	General Outlets	Switch Outlets
▢○		Bell	[F]○
			Fire alarm bell
◇		Annunciator	[F]
			Fire alarm station
◄		Outside telephone	◼
			City fire alarm station
◁		Interconnecting telephone	[FA]
			Fire alarm central station
▷◁		Telephone switchboard	
Ⓕ	─Ⓕ	Fan outlet	S_P Switch and pilot lamp
Ⓙ	─Ⓙ	Junction box	S_{CB} Circuit breaker
Ⓛ	─Ⓛ	Lamp holder	S_{WCB} Weatherproof circuit breaker
Ⓛ$_{PS}$	─Ⓛ$_{PS}$	Lamp holder with pull switch	S_{MC} Momentary contact switch
Ⓢ	─Ⓢ	Pull switch	S_{RC} Remote control switch
Ⓥ	─Ⓥ	Outlet for vapor-discharge lamp	S_{WP} Weatherproof switch
Ⓧ	─Ⓧ	Exit light outlet	S_F Fused switch
Ⓒ	─Ⓒ	Clock outlet (specify voltage)	S_{WF} Weatherproof fused switch
		Convenience Outlets	
	⊖	Duplex convenience outlet	
	 ⊖$_{WP}$		
		Convenience outlet other than duplex. 1 = single, 3 = triplex, etc. Weatherproof convenience outlet	**Special Outlets** ○$_{a,b,c,\text{etc.}}$ 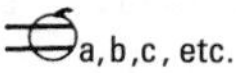 ⊖$_{a,b,c,\text{etc.}}$ $S_{a,b,c,\text{etc.}}$

(*continued*)

Table 10-1 *(continued)*

Ceiling	Wall	General Outlets	Switch Outlets
	⊖R	Range outlet	Any standard symbol as given above with the addition of a lowercase subscript letter may be used to designate some special variation of standard equipment of particular interest in a specific set of architectural plans.
	⊖S	Switch and convenience outlet	
		Radio and convenience outlet	
	⊖-R	Special-purpose outlet (description in specification)	
	●	Floor outlet	
⊙			When used, they must be listed in the Key of Symbols on each drawing and, if necessary, further described in the specifications.
		Panels, Circuits, and Miscellaneous	
	■	Lighting panel	
	▨	Power panel	
	—	Branch circuit—concealed in ceiling or wall	

kitchen/dining area and one for the laundry room, independent of lighting fixtures.

New Work in Old Houses

The electrician is often asked to add new outlets, new circuits, or even a whole new wiring system. In very old houses, a new service may be required to provide adequate housepower. When doing this type of work, the electrician must be able to compute the capacity of each circuit. Most homes have 120 volts delivered by the power company. To determine how many watts the circuit can deliver, multiply the amperage of each circuit by 120:

15-amp circuit × 120 = 1800 watts

20-amp circuit × 120 = 2400 watts

To calculate probable consumption on each circuit, use Table 10-2. (Use actual wattage if available.)

Improving Circuits

When working on home circuits, it is wise to also make prudent repairs and improvements—or to suggest them to the homeowner. If a switch is now in an inconvenient place, this is the time to move it. Or perhaps you can add an extra switch so that the light can be turned on at two—or three—places. Stairways and long hallways are excellent candidates for three- and four-way switches.

If possible, it is a good idea to have more than one circuit serve a room. This is difficult to achieve in existing rooms, but you can and should try to plan that way for new work. Perhaps you can combine a new circuit with an extension of an existing circuit. Run the new cable along one side of the room, and then tap into an existing circuit for another wall. That way, if a fuse blows on one circuit, there will still be some light in the room so that the homeowner can find his way around until the problem is corrected.

Placing Outlets

In general, space outlets according to the following guidelines (equally apart as far as possible). Even if you vary from this, do not leave more than 12 feet between each outlet. Twelve-foot spacing allows lamps with their usual 6-foot cords to be placed where needed and still reach an outlet (see Figure 10-14).

- General-purpose outlets should be placed within 6 feet of lamps, television sets, and other small appliances.
- There should be one outlet at least every 12 feet on every wall.
- Outlets for general use should be about 12 inches off the floor.
- Switches should be 4 feet from the floor and within 6 inches of doors and archways.
- Kitchen receptacles are placed about 12 inches above countertops, or 4 feet off the floor.
- There should be a kitchen outlet at least every 4 feet in the work area.

Table 10-2 Power Consumed by Appliances (Average)

	Watts
Air conditioner, room type	800–1500
Blanket, electric	175
Broiler, rotisserie	1400
Clock, electric	2
Coffeemaker	600
Dishwasher	1800
Dryer, clothes	4500
Fan, portable	175
Freezer	400
Fryer, deep-fat	1320
Frying pan	1000
Garbage disposer	900
Heater, portable	1200
Heater, wall-type permanent	1600
Heat lamp (infrared)	250
Heating pad	75
Hot plate (per burner)	825
Iron, hand	1000
Iron, motorized	1650
Mixer, food	150
Motor, per hp	1000
Oven, built-in	4000
Radio	75
Razor	10
Range (all burners and oven on)	8000–16,000
Range, separate	5000
Refrigerator	250
Roaster	1380
Sewing machine	75
Stereo	300
Sun lamp (ultraviolet)	275
Television	250
Toaster	1100
Vacuum cleaner	400
Waffle iron	800
Washer, automatic	700
Washer, electric, manual	400
Water heater, standard 80 gal	4500
Personal computer	200
PC monitor	200
Printer	100

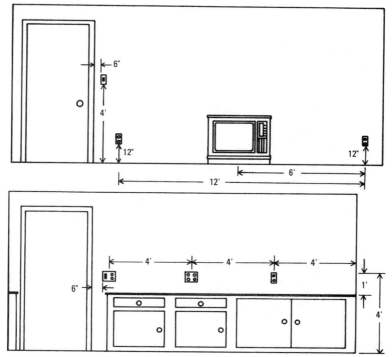

Figure 10-14 Top: maximum distances for placement of general outlets and switches. Bottom: *NEC* requirements for placement of outlets and switches in the kitchen.

- Outdoor wiring and bathroom outlets must be channeled through a ground-fault circuit interrupter (GFCI).
- Consider fluorescent fixtures for softer, less-energy-consuming light.
- All branch circuits serving 125-volt, 15- or 20-amp receptacles in the bedrooms of dwelling units must be protected with arc fault circuit interrupters (AFCIs).

When wiring a family room, remember that there will probably be a television set, perhaps with a rotary antenna, requiring two outlets at one spot. There may also be a stereo, room air conditioner, or other power users in the same area, so space outlets close together or consider multiple outlets in one spot, which requires two or more ganged boxes. Be generous with your outlets when putting in new work. It is simpler, easier, and less expensive to do it now than at a later date.

Use similar foresight in all your planning. Don't skimp on outlets just because you cannot see a need for them in the immediate future. Twenty years ago, no one could predict a home with multiple TVs and stereos, videotape recorders, individual hair dryers and hair setters, microwave ovens, computers, and other equipment.

Placing Switches

Wall switches are another often-neglected planning area. Just because there are no wall or ceiling lights does not mean that there is no need for switches. Even when all lamps are the plug-in type, switches near room entrances are always a good idea for any room. These switches can control one or more outlets in the room and prevent fumbling in the dark. In a bedroom, for example, connect a doorway switch to an outlet near a dresser or vanity. That way, a person can flick on the light as soon as he or she enters the room instead of stumbling about trying to find the lamp. In many parts of the United States, codes require certain outlets to be connected to a wall switch by a door. Use the same reasoning for outdoor lights. An indoor switch can turn on patio or driveway lights and prevent one from falling over the garbage or down the steps.

Roughing-In

Both electricians and plumbers do their work in two stages. After the framing of a new house (or addition) is completed, the basic electrical work is done. It is called roughing-in the new work and consists of putting up the required boxes and stringing the cable. Cabling to switches, receptacles, and so on, is brought to the points where the new switches, receptacles, and the like will be located and left hanging (with enough length to ensure connection) to await finishing. After the wallboard and other finishing materials are attached to the framing, the switches, receptacles, lights, and other electrical fixtures are installed.

You should work the same way. As soon as you have the framing of your room in place, the first step in your wiring job is to install the boxes. After they are put in place, the cable is then strung between the boxes. (Actually, you can put up one box at a time, adding the cable as you go. It's the same process, either way.)

Complete all new wiring before hooking up to the current source.

Install the boxes, using the right-sized boxes to avoid *Code* violation (see Table 10-3). Standard $2 \times 3 \times 2\frac{1}{2}$-inch switch, Gem,

Table 10-3 Maximum Number of Conductors per Box

Type	Size	Capacity (cu. in.)	No. 14	No. 12	No. 10	No. 8
Rectangular switch or gem	3 × 2 × 2¼	10.5	5	4	4	3
	3 × 2 × 2½	12.5	6	5	5	4
	3 × 2 × 2¾	14.0	7	6	5	4
	3 × 2 × 3½	18.0	9	8	7	6
Square	4 × 1¼	18.0	9	8	7	6
	4 × 1½	21.0	10	9	8	7
	4 × 2⅛	30.3	15	13	12	10
	4^{11}/$_{16}$ × 1¼	25.5	12	11	10	8
	4^{11}/$_{16}$ × 1½	29.5	14	13	11	
Round or octagonal	4 × 1¼	12.5	6	5	5	4
	4 × 1½	15.5	7	6	6	5
	4 × 2⅛	21.5	10	9	8	7

or device boxes could present a problem if you install a series of wall receptacles in which outlets in the middle of the run contain cable both entering and leaving. To get around this, either use larger boxes, square boxes, or forgo the simpler interior cable clamps and substitute exterior connectors.

Boxes should be installed so that the outside edges are flush with the finishing material. Presumably, you know what materials you will be using, but make sure that you know the thickness of any paneling, wallboard, or combinations. (For ³⁄₈-inch drywall plus ¼-inch paneling, for example, the box should stick out ⅝ inch from the front of the framing.) The *Code* does allow boxes to be as deep as ½ inch behind the surface of noncombustible materials, such as gypsum wallboard, brick, or concrete block. This could present a problem lining up the receptacle and cover plate, however, so have the outside edges flush with the finishing material in all cases.

Mounting Boxes

There are a great many different types of electrical boxes, but there are basically only two ways of attaching them to walls and ceilings. The easiest method, used almost universally in new construction, is to nail the box directly to the framing of the new work before any finishing materials are attached to walls or ceilings.

The most convenient boxes are equipped with mounting brackets welded to the box itself. Simply nail through the bracket into the front or sides of the studs or joist bottoms with 1-inch roofing nails (see Figure 10-15). Other boxes are nailed with 8d (8-penny) nails into the sides of the studs through projections in the top or bottom or through holes predrilled in the boxes themselves (see Figure 10-16). Some boxes, usually plastic ones, come with nails already attached through in-line projections.

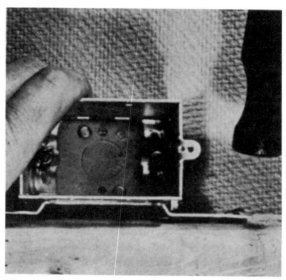

Figure 10-15 If the electrical box has built-in mounting bars, nail the box to the framing with roofing nails.

Occasionally a box must be located away from the framing members. This is often true of ceiling fixtures, and sometimes wall fixtures, when exact placement is more desirable than it is for a switch or outlet. In new work, use wood cleats, metal mounting straps, or adjustable bar hangers (see Figure 10-17), which are nailed into the studs or joists on each end. The box can then be slid and locked in place at the optimum location.

Old Work

When you are working with existing walls or ceiling, box mounting, like everything else in old work, is a little more difficult. When the proper location of the box is determined, a hole is cut into the

Basic House Wiring 257

Figure 10-16 When the box has no mounting device, 8d nails are nailed through predrilled holes at the top and bottom of the box.

wallboard or paneling to accept the new box. Make a paper or cardboard template of the box by laying it face down and tracing around it. (Some box manufacturers supply a template with the box.) Trace around the template onto the wall to mark the rough opening. If only one or two boxes are involved, it may be easier just to hold the box itself to the wall and trace around it.

If the walls consist of gypsum wallboard or paneling, drill holes about $1/2$ inch in diameter at the corners of the box opening and cut out the opening with a keyhole saw. When the walls are made of real plaster, chisel away some of the plaster near the center of the box first (see Figure 10-18). If there is metal or gypsum lath behind the plaster, proceed as previously described for regular walls, but use a fine-toothed blade, such as a hacksaw blade, to avoid damaging

258 Chapter 10

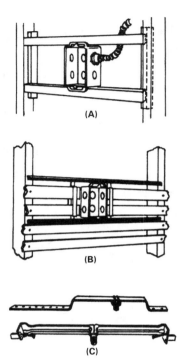

Figure 10-17 Especially with switches and ceiling fixtures, a box must be located in an exact position between framing members. Use either (A) wood cleats, (B) metal mounting strips, or (C) bar hangers.

the plaster. In homes built prior to World War II, you will probably find wooden lath behind the plaster. If so, chip away a little more until you expose a couple of pieces of wood lath. Then, adjust the box location, if necessary, so that the top and bottom of the boxes will fall in the middle of the lath strips. The lath strips are about $1^{1}/_{2}$ inches wide. Cut out the opening as before with a fine-toothed blade. Then, chip away about $3/_{8}$ inch more plaster above and below the opening to allow direct mounting of the box to the wooden lath with No. 5 wood screws.

Special Mounting Devices
For all other walls, special mounting devices will be needed. There are several types, many of which are attached to the boxes themselves. Some have clamplike devices that hug the back of the wallboard when the attached screws are turned. Boxes without mounting devices can be attached to wallboards or thin paneling with Madison hangers, which are slipped between the box and wall on both sides

Basic House Wiring 259

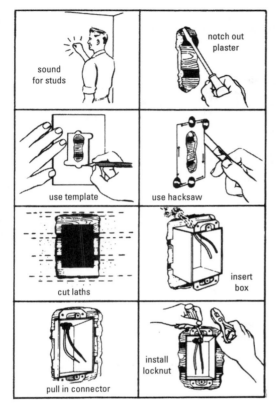

Figure 10-18 Step-by-step directions for installing a box in a wall made of plaster and wood lath.

and then bent back over the insides of the boxes. (The longer length at top and bottom keeps the clips from falling out.) On thick paneling, boxes can be screwed directly to the wood.

Ceiling boxes in old work should be mounted from above where possible, as in an attic, using adjustable bar hangers.

Running Sheathed Cable

Type NM cable is easier and less expensive to use than other types of cable and is perfectly safe for any permanently dry interior use. There are several ways of installing type NM cable in new work, but the best way for most jobs is to drill $5/8$-inch holes through the centers of the joists and studs of the framing when it runs at right angles and to clamp it along the studs or joists for parallel runs (see Figure 10-19).

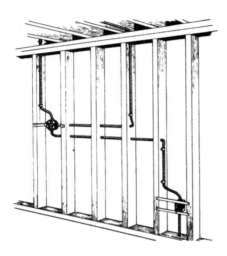

Figure 10-19 Running cable in new work. This is not too difficult because everything is exposed.

When you run the cable through drilled holes in the framing, no straps or staples are needed for support. Cable running lengthwise along joists or studs must be secured every 4 feet or less and within 12 inches of any metal box. Use small straps or special staples designed for use with this type of cable. Do not use staples that are designed for use with type AC cable (see Figure 10-20).

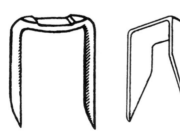

Figure 10-20 The staples at the left are used for attaching nonmetallic sheathed cable to framing. The staples at the right are to be used only for attaching armored cable (type AC).

If you use staples, be very careful. Don't drive them in so hard that they crush or otherwise damage the insulation. Tap them in only until the cable is secured.

Metal Plates

The *NEC* also requires that metal plates be used in the front of any framing where the holes are located less than 2 inches from the front edge of the finishing material. If you drill a hole in the center of a 2 × 4-inch stud (which is actually 1½ × 3½ inches), its center will be only 1¾ inches from the front of the stud. When you add the finishing material, though, this places the hole just about 2 inches away, perhaps a little more.

The purpose of this is to protect the wiring in case someone drives a nail through later on. It is good practice to use these plates whenever drilling through a stud. They should be at least $^{1}/_{16}$ × $^{3}/_{4}$ inches (see Figure 10-21). Nail them to the front of the stud wherever the cable passes through. (There is no need for plates in a joist, which is always a 2 × 6 or larger, placing the hole at least 2¾ inches from the bottom of the joist.)

Figure 10-21 Whenever the cable is less than 2 inches from the front of the wall finishing material (not the stud), the wiring should be protected from accidental damage by hammering the special plates shown to the front of the stud.

Running Boards

Another method of stringing cable at right angles to joists in exposed attic floors or basement ceilings is to use running boards (see Figure 10-22). Pieces of 1 × 3-inch furring lumber are nailed to the joists, and the cable is attached to that in the same way as it is for parallel runs. This may not be a good technique for basements if the homeowner wishes to finish the ceiling, for it necessitates rerouting the cable.

If the attic is accessible by means of stairs or a permanent ladder (including the pull-down type), the cable must be protected with wood guard strips on each side, at least as high as the wiring. Where

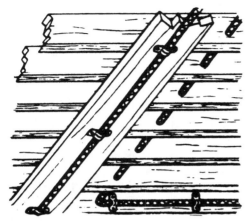

Figure 10-22 Attaching cable to a running board (left) is quick and easy. If joints are to be covered with finishing material, it is best to run the cable through drilled holes as shown at right.

there is only a crawl space opening to the attic, the *NEC* requires guard strips within 6 feet of all sides of the opening. Use 1 × 2 furring for guard strips when using typical No. 12 or No. 14 residential cable. You can also string cable along the rafters, but it must be at least 7 feet above any joists over a living area.

Connecting Cable to Boxes

Attach both types NM and UF cable to interior boxes by clamping the wires. When attaching cable to metal boxes, remove the knockouts on the box where the cable will enter by twisting them out with a screwdriver or by tapping with a hammer. If the box has built-in clamps, loosen the setscrew that holds down the clamp, insert the cable to the proper length, and turn the screw in again until the cable is held tightly (see Figure 10-23). Most clamps go over two knockout holes, so insert both incoming and outgoing cables beforehand if you use both of these knockouts.

For boxes without built-in clamps, purchase either metal or plastic clamp-type connectors. Metal connectors, which are the more common of the two, come in two parts. Remove the locknut, and then slip the clamp part over the cable before inserting into the box (see Figure 10-24A). Turn down the setscrews to hold the cable in place, and insert the connector into the box knockout. Hand-tighten the locknut over the clamp portion, and then secure firmly

Figure 10-23 Some boxes have built-in cable clamps.

by tapping a screwdriver blade against the points of the locknut (see Figure 10-24B). With plastic connectors, firmly push through the knockout hole, and then insert the plastic wedge into the slot provided.

See the next chapter for instructions on making connections using other types of cable.

Fishing Cable in Old Work

Unlike new work, old construction may present the electrician with a difficult task when attempting to string cable through existing framing. Experience and trial and error are often the only teacher, but some of the basic techniques can be explained to provide guidance.

Some jobs are relatively easy. If you are merely adding another outlet or two to an existing circuit, and the floor is accessible from

Figure 10-24 Cable connectors must be used when no clamps are built into the box. First, slip the exterior bushing over the cable, and then insert into the box through the knockout (left). Turn the inner locknut ring onto the connector by hand. Tighten by tapping the points of the ring with a screwdriver.

the basement, the cable can be strung without too much difficulty down into the basement and back up again (see Figure 10-25).

When the outlets are in an interior partition, drill $5/8$-inch holes from the basement straight up through the subfloor and bottom plate. On an outer wall, use a long extension bit and bore diagonally through the plate into the wall cavity. Needless to say, make sure

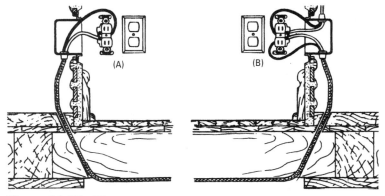

Figure 10-25 The easiest way to add a new outlet (A) to an existing circuit is to run cable from the outlet at the former end of the circuit (B), drilling up through the walls of each and stringing cable through the basement.

that you have lined up the holes with the outlets. Also, deenergize all circuits in that area, since you may inadvertently drill into some of the existing wiring.

Remove the existing box and device, chiseling it out if necessary, and install the new boxes. Run the cable through the basement as previously described. You should now be able to push the cable up from the basement into the wall cavity and the boxes. If you must cut the cable before you do this, leave plenty of slack. You can always cut off any excess later.

With the boxes out, you should be able to grab the cable through the box holes and connect it to the boxes. If you have difficulty, see the following section.

If you have an open attic, you can approach the job from above in much the same way. If an attic floor is in the way, you will have to lift a few floorboards to gain access to the wall plates. Then, simply drill holes and push the cable down to the two box openings.

Those are the easy ways, but if they are not feasible, you can remove the baseboard between the two outlets (in the same room or adjacent rooms) and cut a groove in the plaster or wallboard a few inches above the floor. Check local codes on this first, though. The groove must be deep enough to accommodate the cable without crushing it. It is a good idea to use armored cable or conduit for such a run so that it will withstand crushing and is less likely to be damaged by errant nails when the baseboard is replaced.

The reverse is true when a cable run includes an existing doorway. Type NM cable *should* be used instead of type AC or conduit because of the sharp bends around the doorframe. It may be necessary to match and groove parts of the framing (see Figure 10-26), so use the metal plates previously discussed. The same holds true if you *must* use type NM along the baseboard.

Two-Person Technique

For all but the shortest runs, some use of a fish tape will probably be required. Normally, the fish wire is worked into the new box opening and back to the power source. Since it is steel and springy, you can work fish wire through more easily than you could the cable itself.

In some cases, such as going around corners, two fish wires may be needed. One comes from the old opening, the other from the new. An assistant is often essential. Work both fish tapes in until they meet, and then turn them slowly until the hooks engage. Then, carefully pull one of them back through the opening, keeping it taut enough so that the hooks stay together. Make sure that the tape being pulled through is long enough to reach between the openings.

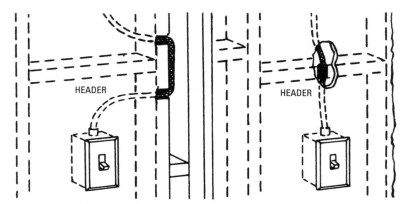

Figure 10-26 When headers or braces get in the way of the new cable, a way must be found to get around them (left). If this is impossible, notch the framing, insert the cable, and cover with a metal plate (right).

When you have finally brought the fish wire from one opening to the other, attach the cable by baring a few inches of the wire and bending them around the fish tape hook. To be safe, tape them all together as well (see Figure 10-27). Then, pull the fish tape and the cable back through the wall to the other box opening. (Make sure the cable is long enough to make the run, with enough left over for connections at both ends.) Detach the fish wire, and you can make the connections.

Figure 10-27 To secure cable to a fish wire, bare a few inches of the wire, and then bend around the fish hook (top). Tape wire and fish hook together (bottom).

Bypassing One Floor of a Two-Story Home

The best advice here is to avoid having to run cable from a basement to the second story, or from the attic to the first floor, whenever possible. There may be times, however, when you simply cannot

Basic House Wiring 267

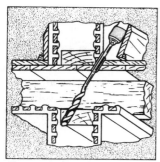

If you can get into attic or upper room, remove the upstairs baseboard. Drill diagonal hole downward as shown.

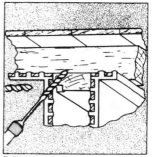

Drill diagonal hole upward from opposite room. Then drill horizontally till holes meet. This procedure requires patching plaster.

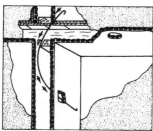

Push 12-foot fish wire, hooked at two ends, through hole on 2nd floor. Pull one end out at outlet on 1st floor.

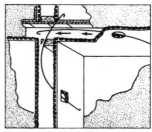

Next, push 20–25-foot fish wire, hooked at both ends through ceiling outlet (arrows). Work the fish wire until you touch the first wire.

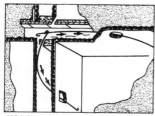

Withdraw either wire (arrows) until it hooks the other wire; then withdraw second wire until both hooks hook together.

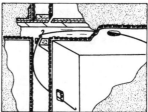

Finally, pull shorter wire through switch outlet. When hook from long wire appears, attach cable and pull through wall and ceiling.

Figure 10-28 Step-by-step directions for fishing a cable from one floor of the house to the next.

make the connection unless you go through an intervening floor. (This may happen in 1½-story and split-level homes.)

Start by removing the baseboard on the second floor where the cable will pass through. Using a long extension bit, drill diagonally from the junction of floor and wall into the top plate of the floor below as shown in Figure 10-28. Chisel a hole in the wall above the bottom plate (plates are the horizontal beams at the tops and bottoms of walls) of the second-floor wall, keeping it large enough to run the cable through but not so high that it will show when the baseboard is replaced.

You have now completed a passageway so that fish tapes can be used to bring the cable either down from the attic or up from the basement, as the case may be. One person runs the tape up or down, and the other seizes it (by hand or with another tape) and brings it around the second-floor bottom plate and through the hole in the first-floor top plate.

Completing House Wiring: Splicing

When roughing-in is finished, other workmen put up the gypsum wallboard, paneling, and so on. The electrician then returns to complete his or her work. This involves installing outlet receptacles, switches, lighting fixtures, and similar tasks.

At one time, splicing was a primary and time-consuming part of the electrician's job. Modern solderless connectors, such as wire nuts, have made splicing only an occasional job, such as when there will be excessive pull on the wires. In fact, it is against the *National Electrical Code* to make a splice anywhere but in an accessible box. In most cases, solderless connectors do the job faster and easier.

If and when a splice is needed, it must be done in the proper manner. A good splice should be as strong as the wire itself. If it is not, it should not be used.

To make a proper splice, bare each wire for about 3 inches, tapering the insulation back about 20° as shown in Figure 10-29. All wires should be clean and shiny. Cross the wires (black to black and white to white) about 1 inch beyond the insulation on each end, and then twist the wires six to eight times tightly around each other.

Place a soldering gun or iron on the wires, with the end of the rosin-core solder nearby, and heat until the wires are hot enough to melt the solder (see Figure 10-30). Don't apply so much heat that it melts the insulation. The solder is soft and doesn't require excessive heat. Slide the solder along the wires so that it flows into every crevice, completely coating each wire. Let the soldered joint cool naturally, without disturbing the molten solder; otherwise, a

Basic House Wiring   269

BARE WIRES   START SPLICE   FINISH SPLICE

Figure 10-29 When soldering wires, bare each one about 3 inches, tapering insulation about 20°. Twist each wire around the other six to eight times.

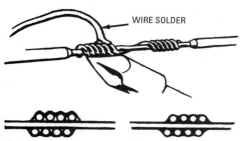

Figure 10-30 Apply soldering rod or gun to wires, not to the solder. Heat until the solder melts and starts to flow onto the hot wires. A properly soldered splice should be solid throughout, as shown in cross section (bottom right), not superficial (as at bottom left, where wires were too cold).

crystallized joint will result. Crystallized joints are mechanically weak and electrically unreliable. You can recognize such a joint by its dull, often rough appearance. When the solder cools, cover the splice from one end to the other with plastic electrical tape, stretching the tape tightly (see Figure 10-31). Keep taping back and forth

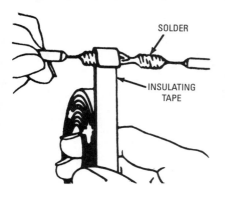

Figure 10-31 After a splice is soldered, wrap it with black plastic tape as shown. Keep adding layers of tape until the splice is at least as thick as the regular wire insulation.

over the splice until it is about as thick as the insulted wire, or perhaps a little thicker. Tape about 1/2 inch back onto the insulation at both ends.

Solderless Connectors

Figure 10-32A shows the solderless type of connector called a wire nut. It is screwed over the bare ends of the wires to make the connection. A wire nut has an insulating shell, so the connection does not need to be taped. Wire nuts are available in several different sizes. Figure 10-32C shows the spring-type solderless connector. After it is screwed over the bare ends of the wires, the protruding lever is broken off. The completed connection must then be taped. Figure 10-32B shows a type of solderless connector used on runs to buildings and feed lines for power. If a tap line exerts strain on an existing line, the type A connector is used. On the other hand, if there is little or no strain exerted, as in the service connection, the type B connector is used.

Solderless connectors can be used in all cases for copper wire joints. However, certain solderless connectors may not be permitted

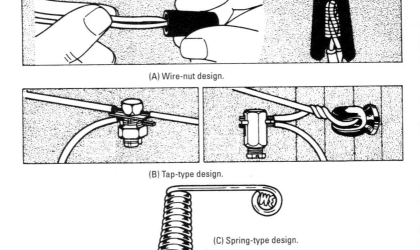

(A) Wire-nut design.

(B) Tap-type design.

(C) Spring-type design.

Figure 10-32 Typical solderless connectors.

for aluminum wire joints. If a connector is suitable for use with aluminum wire, a statement to this effect will be printed on the connector package. Large-scale connectors will be individually marked, as follows: A CU marking means that the connector can be used only with copper wires, an AL marking means that the connector can be used only with aluminum wires, and a CU/AL marking means that the connector may be used with either copper or aluminum wires. However, a copper wire cannot be joined to an aluminum wire in the same connector. If, in addition, the connector has a divider that prevents the copper and aluminum wires from touching each other, it can be used to join a copper wire to an aluminum wire. An unmarked large-size solderless connector can be used only with copper wires.

Installing Outlet Receptacles

Receptacles have two sets of screws: copper-colored for black, or hot, wires, and nickel-colored for white, or neutral, wires. Modern devices also have a green screw for the bare, or green, ground wire. Loosen the screws to which the respective wires will be attached.

Remove enough insulation so that the wire can be bent around the screw at least three-quarters of the way.

You should always loosen the terminal screw as far as you can (without releasing it from the switch or receptacle). That way, the wire can be bent around the body of the screw, instead of trying to slip it over the head. A loop that is wide enough to go over the head will be too large for a tight connection.

Some receptacles may be back-wired. These are designed so that you can insert the wire through an aperture that automatically grips for a solid connection (see Figure 10-33). To *detach* wires from the push-in terminals, insert a small screwdriver blade into the special release aperture next to each terminal aperture where the wire is inserted, pulling on the wire at the same time.

Attach the ground wire, if any, to the grounding screw (see Figure 10-34). If there is no ground screw, use a clip such as the one shown in Figure 10-35 to form a ground with the box. When the wiring continues to another box, pigtail with the jumper wire.

Push the receptacle carefully back inside the box, making sure that no connections have come loose in the process. To avoid the possibility of a short circuit, make sure that no wire ends are sticking out and all wire nuts (if used) are secure. Attach the mounting screws through the slots at bottom and top, straightening the receptacle if

272 Chapter 10

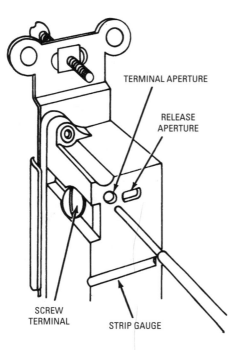

Figure 10-33 A back-wired outlet receptacle. Stripped wire is simply pushed into the aperture.

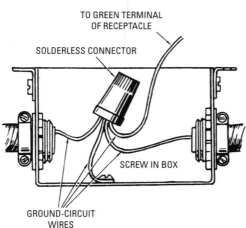

Figure 10-34 Grounding a box with an attached screw for the grounding wire.

Basic House Wiring 273

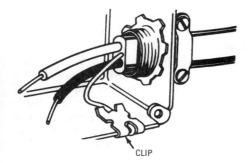

Figure 10-35 When there is no ground screw, a clip is used for the grounding wire. When wires continue to another box, ground wires are pigtailed with a jumper wire.

necessary by adjusting the screws in the mounting slots, moving to right or left as needed. Attach the cover plate.

Single-Pole Switches

Uncomplicated single-pole switches (see Figure 10-36) are installed in boxes in a manner similar to outlet receptacle installation. The one important distinction is that *only* the colored wires (though never green) are attached to the terminals. The white wires are attached together with wire nuts.

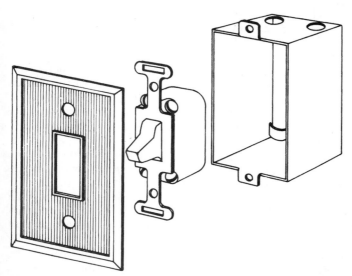

Figure 10-36 A single-pole switch with box and cover plate.

See Chapter 11 for details on more complex switch circuiting.

Ceiling Box and Fixture Mounting

As depicted in Figure 10-37, fixtures are attached to outlet boxes in various ways. For example, an outlet box with a central stud requires only an adapter and a nipple to attach the fixture. Ceiling drop fixtures are usually mounted on a stud, using two nipples and a hickey to join them. If the box does not provide a stud, it will have threaded ears to which a strap can be attached for securing the fixture. A strap can also be installed in a box with a stud, using a threaded nipple and locknut. A fluorescent fixture may be mounted on the ceiling with a stud, nipple, and strap. An extension nipple can be used in case the stud happens to be too short for a certain stud and nipple assembly.

Most lighting fixtures are equipped with prestripped stranded wire. If not, strip about $1^{1}/_{2}$ inches of stranded wire. Wrap the stranded wire around the solid wire and bend the end of the solid wire down over itself. Attach the wire nuts.

Summary

House wiring must conform to the *National Electrical Code* regulations, as well as to all local regulations. Many methods for wiring are approved by the *Code* and include conduit wiring, surface metal raceway, armored cable, underfloor raceway, nonmetallic sheathed cable, electrical metallic tubing, cast-in-place raceways, and wireways and busways.

A service connection is that portion of the supply conductor that extends from the street main, duct, or transformer to the service switch, switches, or switchboard of the building supply. The service entrance enters the conduit through a fitting called the service cap, which protects the wires at the entrance point and prevents water from entering the conduit.

The power is distributed throughout the house from the entrance panel to numerous circuits. The number of circuits will vary according to the amount of electricity consumed in the home. The three main types of circuits are individual appliance circuits, such as those used by a range; regular appliance circuits, used primarily in the kitchen; and general-purpose circuits for use in most areas of the home.

When the electrician plans a circuit, he or she must follow the recommendations of the *NEC* for number and placement of outlets, position and type of switches, and other mandates. Roughing-in is

Basic House Wiring 275

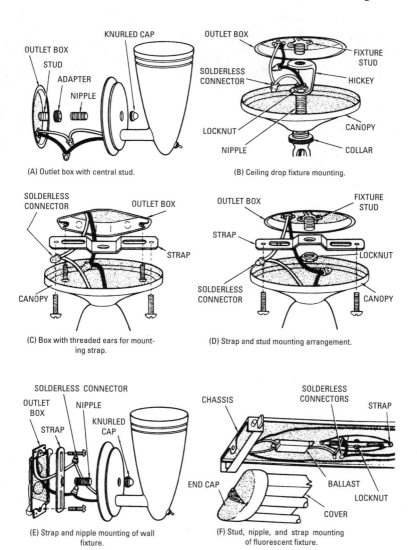

Figure 10-37 Various ways to mount lighting fixtures. A combination of studs, nipples, straps, hickeys, and so on, is sometimes necessary.

done first and consists of stringing cable and installing boxes in their proper places. Outlet receptacles, switches, and other finishing work are done after the walls and ceilings have been completed by other workers.

In old work, cable is fished from one location to the next. This often requires a fish wire and helper. Although there is still an occasional need to splice wires, wire nuts are used for almost all house wiring. Splicing must be done with care and according to codes.

Test Questions

1. Do all localities have the same electrical code?
2. What is the first step to take in wiring a finished house?
3. What is the definition of a service?
4. Name the two basic methods of installing a service entrance.
5. Describe an underground service entrance.
6. How is a Pierce wire holder installed?
7. What are the features of a weatherproof condulet?
8. Explain what is meant by a pull box.
9. Define the term circuiting as used by practical electricians.
10. Why are black wires and white wires used in an installation?
11. How are circuit breakers installed?
12. Name and describe the three main types of circuits.
13. How much wattage can a 120-volt, 15-ampere circuit handle? A 20-ampere circuit?
14. What is the maximum distance allowed between outlets?
15. Where should switches be located in a room?
16. What is meant by roughing-in?
17. How do we compute the capacity of a box?
18. What does conductor mean as applied to box capacity?
19. Describe how to mount a box in new work.
20. Describe how to mount a box in plaster and lath.
21. Describe how to mount a box in gypsum wallboard.
22. Describe the various methods of running type NM cable.
23. How is type NM cable attached to boxes?
24. Describe some of the methods used to fish cable.

25. What problems might be encountered in running cable around doors, and how are they solved?
26. What are the proper procedures for making a splice?
27. How are wires attached to outlet receptacles?
28. What happens to the white wires in a single-pole switch?
29. Name three attaching devices used for installing lighting fixtures.

Chapter 11
Wiring with Armored Cable and Conduit

Armored cable (see Figure 11-1) can be used in dry locations unless local codes impose restrictions. Ordinary armored cable contains two wires, one with black insulation and the other with white insulation. In a three-wire cable, the third has red insulation. As in other types of conductors, armored cable is available in a wide range of wire sizes. While an installation with armored cable does not have the advantage of a conduit installation insofar as withdrawing old wire and inserting new wires, armored cable nevertheless has certain other advantages.

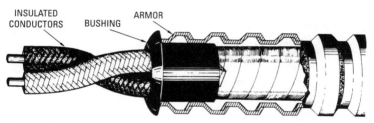

Figure 11-1 Cutaway view of armored cable.

An armored cable is flexible and can be fished between floors and between partition walls. It is also less expensive than conduit. On the other hand, armored cable is more expensive than nonmetallic sheathed cable. Where subject to moisture, special types of armored cable are required by codes.

Cutting Armored Cable

Armored cable is supplied in rolls, much the same as other electrical cables. Almost all types have an internal bonding strip of copper or aluminum in contact with the armor to provide good grounding. Types ACV, AC, and ACT may be used in dry locations, except as noted in applicable codes. In order to remove the metal casing from any type of armored cable, a fine-toothed hacksaw (24 teeth to the inch) should be used, as shown in Figure 11-2. A special armored cable tool may be used, if desired. Service-entrance cable is available in two types—ASE, with interlocked armor protection, and USE, which is underground service-entrance cable for direct burial in the ground. Although constructional details vary, the method of removing the metal casing is basically the same.

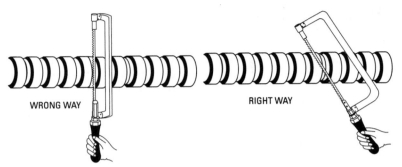

Figure 11-2 The right and the wrong way to cut armored cable.

Armor is cut diagonally across the sheath. In the commonly used type AC, for example, the armor is spiraled so that the cut is made at 90° to the *wires*, but diagonal to the *armor* (see Figure 11-2). Do not cut entirely through the sheath, but make the cut deep enough so that the sheath will break when the cable is bent. Otherwise, you may damage the insulation or cut the wires. The saw may be used as shown in Figure 11-3, supporting the cable on your knee, sawhorse, or any convenient object.

Installing Armored Cable

Before the cable is installed, it should be examined at each end to see if the insulation might be punctured by any part of the sheath. This is an important precaution, because grounds and short circuits are common in careless installations. To protect the insulation on the conductor, a bushing is inserted as seen in Figure 11-4. Outlet and switch boxes should be located and installed before the cable is installed. In turn, the electrician can size up the holes that will be necessary to be bored in order to run the cable. Holes should be bored through the floor beams at right angles to the run so that the cable can be pulled through easily. Figure 11-5 shows a run of armored cable.

After the holes have been bored for a run between two outlets, the cable may be pulled through. Then, the holes may be bored for a run between the next pair of outlets, and so on. In other words, it is not necessary to wait until all the holes have been bored for the complete installation before starting the cable-pulling job. The cut end of the cable is first bushed, as shown in Figure 11-6A. This

Wiring with Armored Cable and Conduit 281

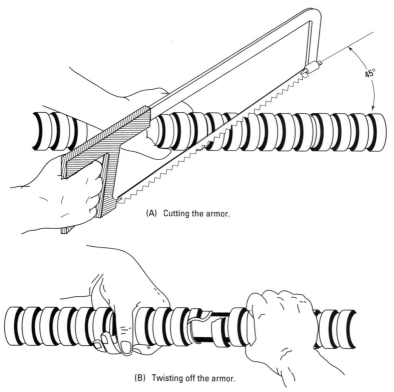

Figure 11-3 Removing the metal casing.

is an important part of the job because the edges of the cut will have sharp edges projecting inward, as seen in Figure 11-6B. When properly bushed, the wires will be completely protected from these sharp edges.

The cable is then pulled through the holes that have been bored and fastened with a clamp or connector to the outlet box. Knockouts in boxes are removed with a screwdriver or hammer; this leaves an access hole for the cable. If a knockout has been removed and then is not used, it should be closed with a knockout closer. One type of connector is shown in Figure 11-7A, and the method of connecting it to the cable and box is seen in Figures 11-7B through 7E.

After fastening the cable to the outlet box, it is pulled fairly tight and is cut off at the proper point to connect up with the other box. When cutting off the cable, make sure to cut it at the right length so

(C) Removing the armor.

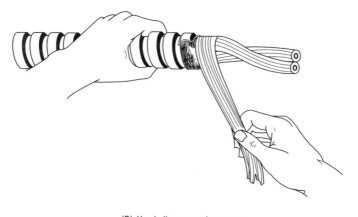

(D) Unwinding protective cover.

Figure 11-3 (*continued*)

that it will project into the box about 6 inches, or enough to make a workmanlike joint. In case a cable is cut too short, it is necessary to remove it and run another cable of correct length, because joints are not allowed between boxes. When a cable runs parallel with joists or studs, it is fastened by pipe clamps.

In wiring an old house, the cable must be fished from one outlet to the next and then fastened into the outlet boxes. It is good practice to fasten the armored cable to timbers with pipe straps wherever possible. Approved straps or staples are used at $4^{1}/_{2}$-foot intervals. At various points in an installation, the cable must be bent. All bends should be made as gradually as practical to avoid damaging and opening the armor. A cable should be securely fastened at all bends; this is particularly important when installing cable in the

Wiring with Armored Cable and Conduit **283**

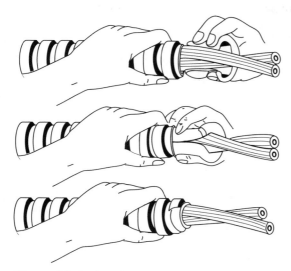

Figure 11-4 Installing protective bushing.

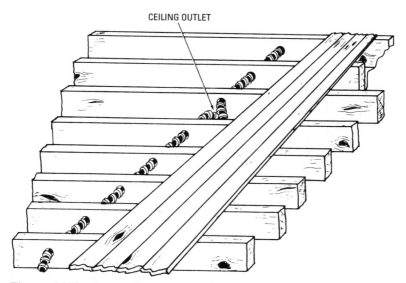

Figure 11-5 A run of armored cable.

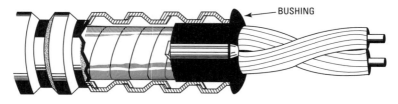

(A) Bushing inserted.

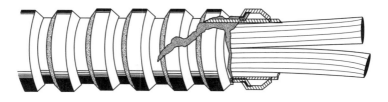

(B) Ragged edges that are protected from the conductor with the bushing.

Figure 11-6　How the bushing protects the conductor.

vicinity of machinery. Figure 11-8 shows the right and wrong ways to bend armored cable.

Conduit Fittings

Conduit fittings are the accessories necessary for the completion of a conduit system, such as boxes, bushings, and access fittings. A conduit box is a metal box adapted for connection to conduit and is used to simplify wiring procedure or for mounting electrical devices. An access fitting is used to permit one to work with conductors elsewhere than at an outlet in a concealed or enclosed wiring system. Conduit differs from electrical metallic tubing in that conduit is comparatively heavy and uses threaded fittings.

A conduit fitting differs from an ordinary pipe fitting in that it has an opening with a removable cover, as shown in Figure 11-9. This opening permits pulling wire through. A conduit fitting is used accordingly as a *pull box* and as a fitting to join two lengths of conduit. Various types of covers are used with conduit fittings.

The conduit elbow shown in Figure 11-10 is a 90° type. Other elbows are also used in particular installations, as explained subsequently. Before an electrician can determine the types of conduit fittings that will be required in an installation, he or she must prepare

Wiring with Armored Cable and Conduit 285

(A) Connector assembly.

(B) Installing connector bushing.

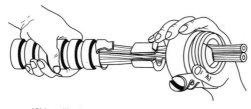

(C) Installing connector.

(D) Locking connector in position.

Figure 11-7 Armored cable connector for junction box.

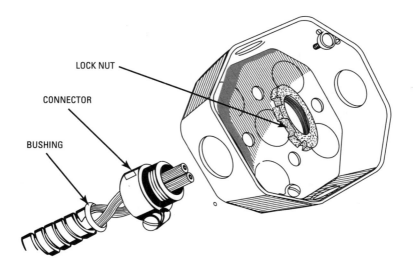

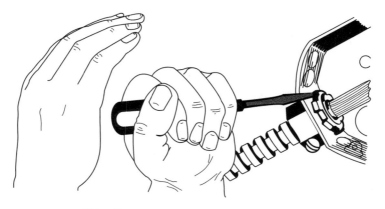

(E) Installing cable to junction box.

Figure 11-7 (*continued*)

a wiring diagram. This is a sketch showing each wire in an installation or part of an installation, with all the connections between the lamps and devices to the source of electricity. Note that *black* insulated wire is used on one side of a run, and *white* insulated wire is used on the other side of the run. White insulation is used for return conductors, called the *grounded* or *neutral* wire.

Wiring with Armored Cable and Conduit **287**

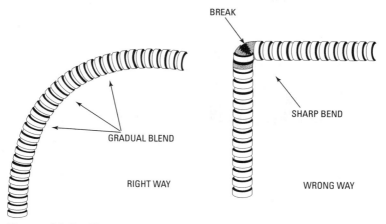

Figure 11-8 The right and the wrong way to bend armored cable.

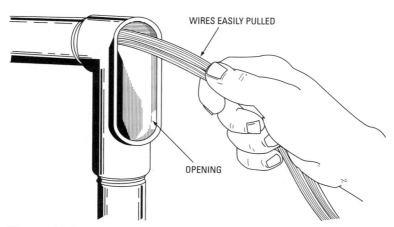

Figure 11-9 Conduit elbow with removable cover. Cover is replaced after wires are pulled through.

Conduit fittings are made in a large number of varieties, only a few of which can be illustrated in this chapter.

Conduit fittings should be carefully chosen so that the conduit will be bent as little as possible. When selecting the size of conduit and fittings, applicable codes should be checked to make sure that the installation will pass inspection. The size of conduit depends on the size and number of wires used. Table 11-1 gives a general

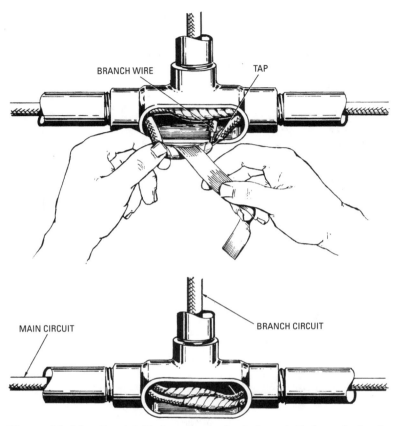

Figure 11-10 Conduit T, showing method of tapping a branch circuit.

idea of the minimum size of conduit that will be used in various situations.

In some installations, one small wire may be run with a larger wire; in such a case, a suitable size of conduit must be used. Table 11-2 gives a general idea of the size of conduit that will be used.

Conduit elbows, usually called *ells*, have 90° bends and are measured in terms of their *offset* and *radius*, as shown in Figure 11-11A. Ells are purchased ready-made because it is difficult to bend conduit to a short radius. However, if small conduit is bent in a fairly large radius, an ell can be made on the job. Various types of bench-mounted

Table 11-1 Minimum Conduit Sizes

Size of Wire	Number of Wires in One Conduit								
	1	2	3	4	5	6	7	8	9
	Minimum Size of Conduit in Inches								
No. 14	1/2	1/2	1/2	3/4	3/4	1	1	1	1
12	1/2	1/2	3/4	3/4	3/4	1	1	1	1 1/4
10	1/2	3/4	3/4	1	1	1	1 1/4	1 1/4	1 1/4
8	1/2	3/4	1	1	1	1 1/4	1 1/4	1 1/4	1 1/4
6	1/2	1	1 1/4	1 1/4	1 1/2	1 1/2	2	2	2
5	3/4	1 1/4	1 1/4	1 1/4	1 1/2	2	2	2	2
4	3/4	1 1/4	1 1/4	1 1/2	2	2	2	2	2 1/2
3	3/4	1 1/4	1 1/4	1 1/2	2	2	2	2 1/2	2 1/2
2	3/4	1 1/4	1 1/2	1 1/2	2	2	2 1/2	2 1/2	2 1/2
1	3/4	1 1/2	1 1/2	2	2	2 1/2	2 1/2	3	3
0	1	1 1/2	2	2	2 1/2	2 1/2	3	3	3
00	1	2	2	2 1/2	2 1/2	3	3	3	3 1/2
000	1	2	2	2 1/2	3	3	3	3 1/2	3 1/2
0000	1 1/4	2	2 1/2	2 1/2	3	3	3 1/2	3 1/2	4

Table 11-2 Three-Conductor Convertible System

	Size of Wires			Size Conduit Electrical Trade Size
Two	14	and one	10	3/4 inch
Two	12	and one	8	3/4 inch
Two	10	and one	6	1 inch
Two	8	and one	4	1 inch
Two	6	and one	2	1 1/4 inch
Two	5	and one	1	1 1/4 inch
Two	4	and one	0	1 1/2 inch
Two	3	and one	00	1 1/2 inch
Two	2	and one	000	1 1/2 inch
Two	1	and one	0000	2 inches

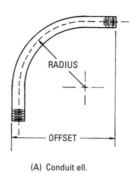

(A) Conduit ell.

(B) Conduit bending jaws or hickey.

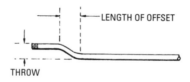

(C) An offset bend in rigid tubing.

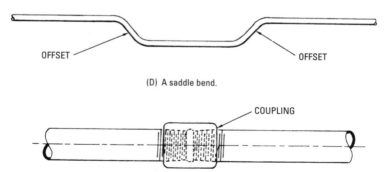

(D) A saddle bend.

(E) Conduit coupling.

Figure 11-11   Various conduit bends.

tools are used for bending conduit. Portable tools are preferred whenever they are practical. A portable bending tool is called a *hickey*. It consists of jaws (see Figure 11-11B) mounted on a long handle to provide good leverage.

The conduit to be bent is placed on the floor or ground, and the jaws of the hickey are placed around the conduit at a suitable point. Holding the conduit in place with his foot, the electrician then presses against the handle of the hickey and makes a small bend. Then, the bend is sized up and the position of the hickey is changed if necessary to complete the bend properly. Note that the inside radius of a bend should not be less than six times the diameter of the conduit. Otherwise, the conduit is likely to flatten, and it will be difficult or impossible to draw the wires through.

A hickey is used only to bend rigid-wall conduit; other tools are used to bend thin-wall conduit. An *offset* bend appears as shown in Figure 11-11C. Two offsets produce a *saddle* bend, as shown in Figure 11-11D. Other bends are also used, but the offset and the saddle bends are the basic types. Threaded *couplings* are used to connect lengths of conduit together, as shown in Figure 11-11E. Practice is required to make accurate and neat bends with a hickey. When the length of throw and offset is comparatively small, two hickeys separated by a suitable distance may be used.

Conduit is *reamed* at both ends at the factory to remove sharp corners that could damage insulation on wires. When conduit is cut on the job, it should be reamed. A conical tool with cutting edges, called a *reamer*, is used for this purpose. Conduit is usually cut with a hacksaw or tubing cutter. The conduit should be held in a vise in order to make a workmanlike cut; the point of cutting should be carefully measured to avoid waste due to lengths that are a little too short for the installation. After the conduit is cut, it must be threaded using a stock and die as in plumbing work, though the threads for electrical conduits do not taper, as those of plumbing pipes do. An electrician uses a vise for cutting and threading conduit; the vise also serves for bending.

Conduit must always be grounded because an ungrounded run of conduit is both a shock hazard and a fire hazard. Conduit grounds use connections made with ground straps. A grounding conductor is run from the conduit to a water pipe on the supply side of the water meter. Grounding requirements are strictly regulated by electrical codes, which should be checked by the electrician to make sure that the installation will pass inspection.

Threadless conduit fittings are used with electrical metallic tubing and may be used with rigid conduit in exposed runs. EMT cannot be threaded because its walls are quite thin. A threadless fitting has split sleeves, which are tightened around the end of the conduit by locknuts. These fittings are made in many varieties and forms and are commonly called *threadless fittings*. For example, conduit fittings with 90° and 45° projections are also made in no-thread types.

EMT is lighter and easier to work with than rigid-wall conduit and costs less. It can be bent easily with a suitable bender. Much time can be saved by using EMT in an installation if electrical codes permit its use. Conduit must be well supported and held firmly in place by suitable devices. Pipe clamps may be used in many installations. Where parallel runs of two or more conduits are required, a *hanger* support is commonly used. A *plate hanger* consists of a metal plate with a lip bent at right angles. Large holes are drilled in the plate to pass the conduit, and small holes are drilled in the lip for wood screws.

A plate hanger may be bolted instead of screwed to a wall or ceiling, or it may be suspended by metal rods from a ceiling. Instead of a plate hanger, an electrician may use a *trapeze hanger* such as the one shown in Figure 11-12. Many types of trapeze hangers can be obtained from supply houses. A single conduit may be suspended by a *single-ring conduit hanger*. It consists of a ring-type support for the conduit, suspended from the ceiling by a metal rod, strap, or pipe.

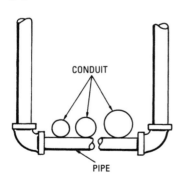

Figure 11-12 One form of trapeze hanger.

Any run of conduit must be electrically continuous so that the entire run is well grounded. If a sharp edge or other defect in a conduit run cuts through the insulation of a live wire, the installation will be defective. Conduit is wired *after* the installation is completed, but covers from elbows and other fittings are removed to facilitate fishing the wiring. The wires are pushed or pulled (drawn) through the conduit, depending on its length. When wires are drawn through a long run of conduit, the procedure is called *fishing*. A *fish tape* (occasionally called a *snake*) is used for this purpose. The fish tape is fed into the conduit at one end by hand pressure as far as possible.

If more pressure must be applied in feeding the fish tape, a pair of pliers will give a firm grip. However, the fish tape can be fed only a few inches at a time if considerable pressure is necessary. After the fish tape comes out of the far end of the conduit, it is hooked to the wires, which are then drawn back through the conduit. Fishing from both ends of a long conduit is sometimes necessary. After a fish tape has been pushed as far as possible into one end of the conduit, another is pushed in from the other end. The hooks in the fish tapes must then be caught. To do this, the electrician works with the short tape, while his helper shakes and works the long one. After the hooks are caught, the electrician can finish pulling the long fish tape through.

Wires must be firmly attached to a fish tape's hook, and it is good practice to tape the loops and turns. If a smooth tape covering is used, the attached wires are much less likely to jam in the conduit while being drawn. Drawing the wires requires one person to pull the fish tape and another person to feed the wires into the conduit. Wires must be prevented from kinking or crossing each other as they feed in. It is helpful to exert a steady pressure on the fish tape and to feed the wires smoothly instead of exerting a series of sudden jerks.

Wiring in Flexible Conduit

A flexible conduit consists of a continuous flexible steel tube that is made from convex and concave metal strips in a spiral winding upon each other so that the concave surfaces interlock (see Figure 11-13). Flexible conduit has considerable strength and is supplied in lengths from 500 to 200 feet. Elbow fittings are not required because the conduit can be bent to almost any radius. There are fissures between the strips, which provide some ventilation. This is an advantage in some applications and a disadvantage in others. Note that wires

Figure 11-13 Double-strip flexible conduit.

must be fished through flexible conduit. Many electricians call this type of conduit Greenfield.

The *National Electrical Code* forbids the use of flexible conduit in wet locations unless the conductors are approved for such locations. Flexible conduit is also prohibited in many hoistways, in storage-battery rooms, and in most hazardous locations. Various other restrictions apply to the installation of flexible conduit, by both national and local codes; therefore, the electrician should check before going ahead with a flexible-conduit installation. Flexible conduit *must* be used if a fluorescent light is not connected directly to a box.

Although flexible conduit is easy to work with, it is seldom desirable to install an entire wiring job with it. Electricians usually combine runs of flexible conduit with rigid conduit for extensions that are short and irregular. Occasional use of flexible conduit for short and irregular runs can save considerable pipe fitting. It can also be passed through joists and studs with comparatively little difficulty. Note that flexible conduit is also made in single-strip form, as shown in Figure 11-14. Single-strip conduit is good for use where considerable pulling and bending is required. Its flattened outer surface makes it easier to draw through bored holes, and its inside surface is smooth and even, which allows wires to slide easily during fishing procedures.

Figure 11-14 Single-strip flexible conduit.

On the other hand, double-strip flexible conduit has double armor for protection of the conductors, and it may be preferred for this reason. For quick and easy installation, various fittings are used. The fittings can be used with armored cable as well as with flexible conduit. Fittings can be classified as follows:

- Terminal bushings
- Couplings
- Connectors
- Adapters

Wiring with Armored Cable and Conduit 295

(A) Flexible-to-flexible squeeze-type coupler.

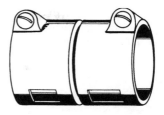

(B) Flexible-to-flexible tangent-screw coupler.

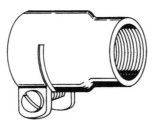

(C) Flexible-to-rigid squeeze-type coupler.

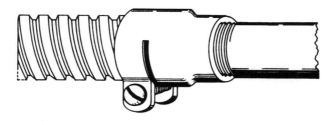

(D) Flexible-to-rigid coupler.

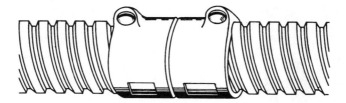

(E) Flexible-to-flexible coupler.

Figure 11-15 Various types of couplers.

Connectors

Connectors are fittings used for connecting the conduit to devices such as outlet boxes. Various types are classified as squeeze, setscrew, slip-in, duplex, and angle types. Examples are shown in Figure 11-16. *Squeeze connectors* give a firm grip entirely around

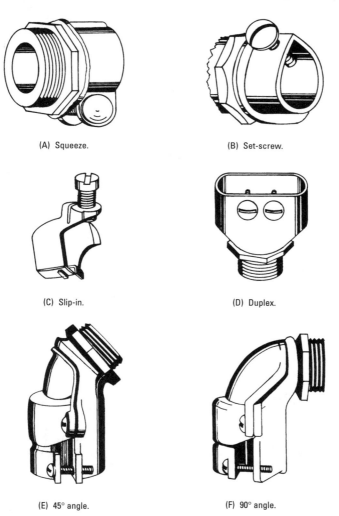

(A) Squeeze.　　　　　　　(B) Set-screw.

(C) Slip-in.　　　　　　　(D) Duplex.

(E) 45° angle.　　　　　　(F) 90° angle.

Figure 11-16 Illustrating various connectors.

the armor and are supplied with locknuts. *Setscrew connectors* employ a long tangential screw, which is forced in the groove of the armor between the armor and the side of the connector. *Slip connectors* are used to save time; no locknut is used, and the setscrew is simply tightened after the connector is pushed into the knockout.

Duplex connectors are used in installations where two cables are run into one knockout, as shown in Figure 11-17. The use of one duplex connector instead of two separate connectors in two different knockouts saves time and cable, and doubles the effective number of outlets. Duplex connectors are of tangential setscrew type. *Angle connectors* are used to install conduit at an angle to devices such as outlet boxes. Both 45° and 90° angles are available, as shown in Figure 11-16E and F. A few of these connectors will save time and make a better job in cramped locations.

An *adapter*, such as those shown in Figure 11-18, is a fitting used with rigid conduit. A short section of rigid conduit may use an adapter at one end and a coupling at the other, as shown in Figure 11-15D. A summary of the operations in attaching *flexible* conduit to an outlet box is shown in Figure 11-19.

Summary

Armored cable does not have the advantages of conduit installation, where old wires can be removed and new wires inserted, but armored cable is flexible and can be fished between floors and between partition walls. It is also less expensive than rigid conduit or flexible steel conduit. There are many buildings where the installation of armored cable is not permitted.

Conduit fittings are made in a large number of varieties. Fittings should be carefully chosen so that the conduit will be bent as little as possible. Size of conduit will depend on the number of wires that are used. Conduit elbows that are ready-made are used where it is difficult to bend the conduit to a short radius.

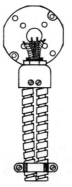

Figure 11-17 Use of a Duplex connection.

Various types of bench-mounted tools are used for bending conduit. Portable tools are preferred whenever they are practical. A portable bending tool is called a hickey. A hickey is used only to bend rigid-wall conduit; other tools are used to bend thin-wall tubing.

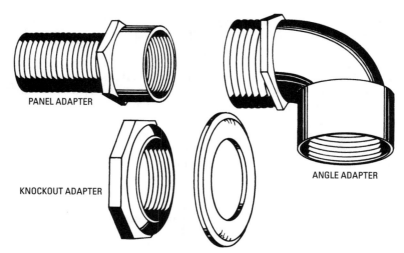

Figure 11-18 Flexible conduit Adaptors.

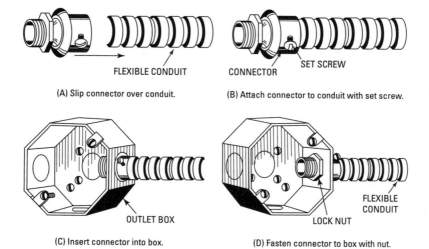

Figure 11-19 Connecting flexible conduit to outlet box.

When conduit is cut to length on the job, a tool called a reamer is used to ream out the sharp edges. These sharp edges can damage the insulation on the wires when pulled through the tubing.

Flexible conduit consists of a continuous flexible steel tube made from convex and concave metal strips in a spiral. Flexible conduit is also restricted in various installations where oil, gasoline, or other chemicals are used. Although flexible tubing is easy to work with, it is seldom desirable to install an entire wiring job with it. It is usually combined in short runs with rigid conduit or EMT.

Test Questions

1. What colors of insulation are used in a two-wire armored cable? In a three-wire cable?
2. Where would lead-covered armored cable be used?
3. Why is a bonding strip used in an armored cable?
4. How is armored cable cut for installation?
5. What is the purpose of an armored-cable bushing?
6. When is a knockout closer used?
7. Explain the shortest bend allowed in armored cable by the *National Electrical Code*.
8. Name several applications in which armored cable is forbidden.
9. How is armored cable passed around braces? Across doorways?
10. Describe the difference between armored cable and conduit.
11. What is an ell?
12. Describe an offset bend. A saddle bend.
13. How is a hickey used?
14. Where are threaded couplings used?
15. Explain how a threadless conduit fitting holds the end of a conduit.
16. What is the purpose of a hanger support?
17. Why is a run of conduit always grounded?
18. Describe the process of fishing a wire or wires through conduit.
19. What difficulties may be encountered in using wire snakes?
20. Why is EMT preferable to rigid-wall conduit in situations where its use is permissible?

21. Explain how to make a ground connection that will pass inspection.
22. Name the two basic types of flexible conduit.
23. What types of couplings are used with flexible conduit?
24. Describe two basic types of connectors.
25. Explain how flexible conduit can be attached to an outlet box.

Chapter 12
Home Circuiting, Multiple Switching, and Wiring Requirements

Circuiting details require careful consideration. Otherwise, after an installation is completed, it may be discovered that the lamp control facilities are less than desired. The chief lamp control arrangements may be listed as follows:

- A certain number of lamps are to be switched on or off from one location.
- A pair of lamps are to be individually controlled by a pair of switches.
- A pair of lamps are to be individually controlled by a single switch. In one arrangement, a main switch must be thrown to turn both lamps off.
- A pair of lamps are to be individually controlled from one location. In one arrangement, a main switch must be thrown to turn both lamps off.
- A pair of lamps are to be switched on or off from two locations.
- Three lamps and three switches are to be wired so that a switch will operate its own light for certain positions of the two other switches.

Figure 12-1 shows possible circuit arrangements. Three lamps are controlled from one position in A. Of course, any number of lamps may be used, provided the switch and conductors are suitably rated. In B, each lamp is controlled by its own switch; the switches may be in a single unit or separate units. Each lamp in Figure 12-1B may be a bank of lamps connected in parallel, provided the switches and conductors are suitably rated.

In Figure 12-1C a pair of lamps are individually controlled by one three-way switch. Only one lamp will be on at a time, and both lamps will be dark only if a main switch (not shown) is opened to remove current from the circuit. A three-way switch is basically a reversing switch. In Figure 12-1D, a pair of lamps are individually controlled by a three-way switch and a four-way switch at separate locations. Either of the lamps can be turned on from each location,

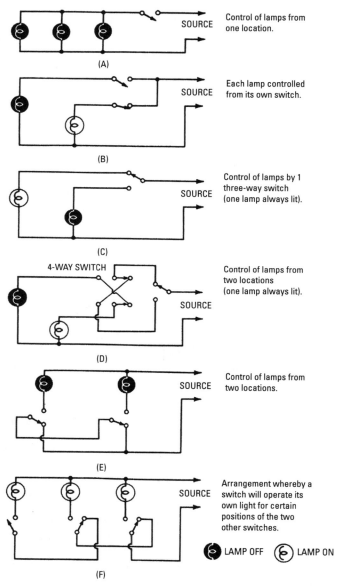

Figure 12-1 Possible circuit arrangements using various types of switches.

but one of the lamps will always be burning unless a main switch (not shown) is opened.

In Figure 12-1E, a pair of lamps are connected in parallel and can be turned on or off by three-way switches in two separate locations. A larger number of lamps can be connected in parallel if desired, provided that the switches and conductors are suitably rated. In Figure 12-1F, three lamps and three switches are installed in separate locations. The third light can always be turned on or off by means of the three-way switch. However, the second light cannot be turned on unless the third light is off. Similarly, the first light cannot be turned on unless both the second and the third lights are turned off. Circuits A, B, and E are in most common use.

Another useful circuit is shown in Figure 12-2. This is an arrangement for lamp control from two separate locations. In A, two three-way switches are used to turn the lamps on or off from either location regardless of the switch setting at the other location. With both switches off, as shown in Figure 12-2A, the lamps are dark. Next, if switch no. 2 is turned on, as shown in B, the lamps are turned on. If switch no. 2 is turned off, the lamps are turned off, as shown in Figure 12-2C. Then, if switch no. 1 is turned on, as shown in D, the lamps are turned on. If switch no. 1 is turned off, the lamps are turned off.

Note in Figure 12-2B that the lamps will be turned off if switch no. 1 is turned off. Similarly, the lamps in D will be turned off if switch no. 2 is turned off. In summary, when either switch is thrown, the lamps will be turned on or off as the case may be. However, the *direction* that a switch must be thrown to turn a lamp on (or off) depends on the setting of the other switch. In other words, the three-way switch does not have a position marked on or off.

Lamp Control from Electrolier Switch

The switch shown in Figure 12-3 is called a *two-circuit* switch. It is widely used in electroliers. An electrolier is a support for a lamp—for example, a pole lamp, a table lamp, a floor lamp, or a chandelier supported by an electrolier. A two-circuit switch permits the user to turn two groups of lamps on or off, either in individual groups or as a whole. For example, in the first position of the switch shown, both groups of lights are off. In the second position of the switch, the second group of lights is on and the first group is off. In the third position of the switch, both groups of lights are on. In the fourth position of the switch, the first group of lights is on and the second group is off. This switching circuit is not considered as standard but is only one of several arrangements.

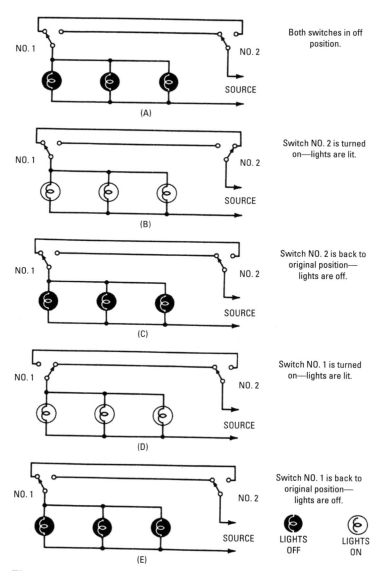

Figure 12-2 Possible circuit arrangements using three-way switches.

Home Circuiting, Multiple Switching, and Wiring Requirements **305**

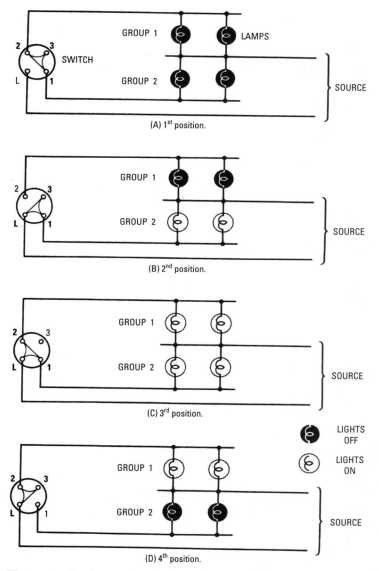

Figure 12-3 A two-circuit electrolier switch arrangement.

306 Chapter 12

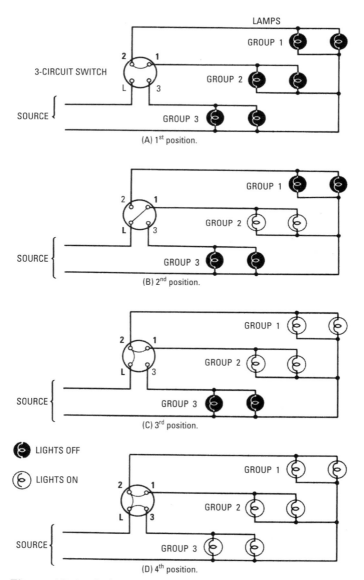

Figure 12-4 A three-circuit electrolier switch arrangement.

To control three groups of lights, a three-circuit electrolier switch is used, as shown in Figure 12-4. In this arrangement, the groups of lamps are not individually controllable. Instead, each group can be switched on progressively or all groups switched off simultaneously. With the switch in the first position, all lamps are turned off. In the second position of the switch, the second group of lamps is turned on. In the third position of the switch, both the first and second groups of lamps are turned on. In the fourth position of the switch, all three groups of lamps are turned on.

Control of Lamps from More Than One Location with Three-Way and Four-Way Switches

When a lamp (or parallel bank of lamps) is to be controlled by switches from three or more locations, three-way and four-way switches are used as shown in Figure 12-5. The lamps can be turned on or off with a switch at any location. Either three-way or four-way switches can be used at the ends of the switching circuit; however, four-way switches must be used at intermediate positions in the switching circuit. It is possible to use as many switching locations as desired by wiring in additional four-way switches.

Stairway Lamp Control Lighting

Three-way switches are widely used in stairway lamp control circuits. Double-pole master switches are used at the ends of the switching circuit. If a person is going upstairs, he or she operates the switch on one floor to light the lamp on the floor above and to turn out the light on the floor below. In going downstairs, the switching process is reversed. Before installing any switching system, it is important to check the applicable electrical codes.

Color-Coding of Switch Wiring

As noted previously, black wires are always connected to black wires, white wires are always connected to white wires, and so on. However, when three-way and four-way switches are installed, maintaining the color code requires that white wires from the switches be painted black at the ends, both at the switches and at the outlet. Figure 12-6 shows typical examples of this requirement. Note that the switch terminals marked A and B are light-colored terminals, to which red and white wires must be connected. On the other hand, terminal C is the dark-colored terminal, to which a black wire must be connected. (A light-colored terminal may be a

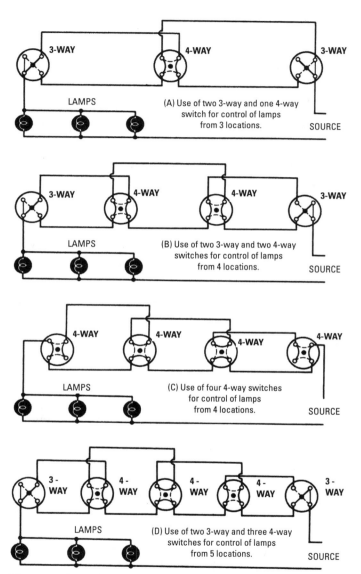

Figure 12-5 Switch arrangement using three-way and four-way switches.

Home Circuiting, Multiple Switching, and Wiring Requirements 309

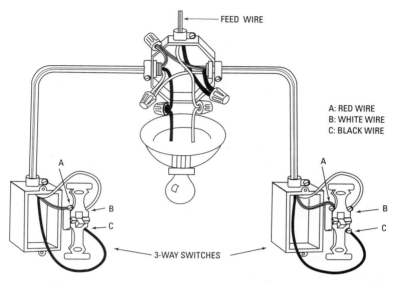

A: RED WIRE
B: WHITE WIRE
C: BLACK WIRE

(A) Three-way switches controlling an outlet between the switches.

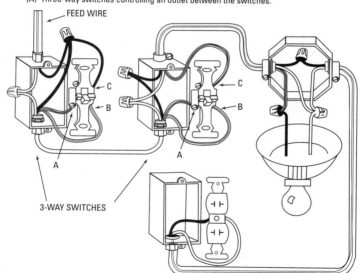

(B) Three-way switches controlling an outlet beyond the switches.

Figure 12-6 Color-coding of switch wiring.

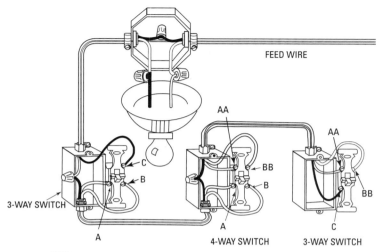

(C) Three- and four-way switches controlling an outlet from three locations.

Figure 12-6 (*continued*)

brass terminal in some switches.) It is evident from these examples that the color code in the completed installation would be incorrect unless the electrician painted the ends of the white wire black both at the switch and at the light outlet. (This may not be required, but it remains a good idea. A simple ring of black electrical tape is enough and takes only a moment.)

Functional Planning

In planning an electrical system, the first consideration is *safety*, and the next consideration is *function*. The functioning of a system should permit full and convenient use of both present and future electrical equipment. An effective and efficient home wiring system depends on the following:

- Sufficient circuits of adequately large wire to supply the various loads without uneconomical voltage drop
- A satisfactory number of outlets to permit convenient use of electrical equipment
- High-quality materials and good workmanship

Planning of Equipment

The size of a house is directly related to the extent of the electrical installation. Appliances are usually classified as fixed or portable. The owner must determine the type and number of lighting units and appliances to be used. Typical fixed appliances are water heaters, washer, ironers, dryers, ranges, refrigerators, air conditioners, dishwashers, and garbage disposers. If heating and air-conditioning units are to be installed, provision must be made in the wiring plan for blowers, furnace motors, and so on. Typical portable appliances are electric clocks, coffeemakers, fans, irons, heating blankets, mixers, radios, television receivers, roasters, sewing machines, shavers, sun lamps, toasters, and vacuum cleaners.

Lighting facilities must meet three basic requirements:

1. There must be sufficient light for all activities from the viewpoint of comfort and avoidance of eye fatigue.
2. Light must be provided so as to avoid objectionable shadows and controlled to eliminate glare.
3. Fixtures must be adapted to their purpose and appropriate to their surroundings. Figure 12-7 shows outlines of typical luminaires, or lighting fixtures. Note that the type of luminaire will partially determine how many will be required for a given area.

Built-in lighting has been briefly noted previously. It comprises luminous panels, columns, or similar structures on existing walls, or it may be part of the house architecturally. Figure 12-8 shows some examples of built-in lighting. Lamps range from general-service, tubular, and Lumiline types to specialized tungsten lamps. Lumiline is a thin tubular type of fluorescent lamp that starts instantly, as explained subsequently. Cove lighting is shown in Figure 12-9A. Entrance lighting should clearly illuminate the path, steps, house number, and bell switch (see Figure 12-9B and C). Weatherproof lanterns or built-in lighting units are often used.

If a house is located some distance from the sidewalk, an illuminated number may be installed at the driveway entrance. Small floodlights are sometimes placed in shrubbery to illuminate the front of the house. Entrance fixtures should be controlled by switches installed inside or just outside the front door; switches should be on the lock side—not on the hinge side. Night-lights make it easy to find a switch in the dark. Pilot lights serve as a useful reminder that a switch is turned on. Weatherproof convenience outlets should be

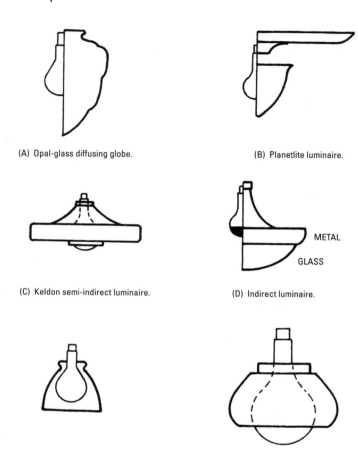

Figure 12-7 Various shapes of light fixtures.

installed outside of the house for electrical garden equipment and seasonal lighting. A vestibule between the hall and entrance is usually small and can be illuminated by a ceiling fixture with a switch adjacent to the entrance switch.

Lighting the Halls

Figure 12-10 shows a wiring plan for the first story of a two-story home. A hall may be illuminated by a suspended decorative lantern,

Home Circuiting, Multiple Switching, and Wiring Requirements 313

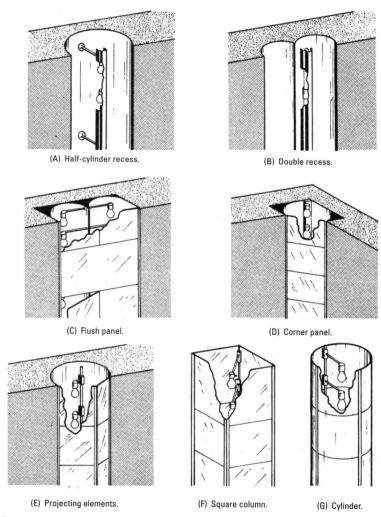

(A) Half-cylinder recess.
(B) Double recess.
(C) Flush panel.
(D) Corner panel.
(E) Projecting elements.
(F) Square column.
(G) Cylinder.

Figure 12-8 Built-in lighting.

close-fitting fixture, or a shaded candle-type fixture. Other possibilities are luminous panels around a mirror, torchères, a pole lamp, or a console table lamp. Note that light from a downstairs hall fixture should be supplemented by light from an upstairs fixture. These fixtures are usually controlled by separate switches, both upstairs and

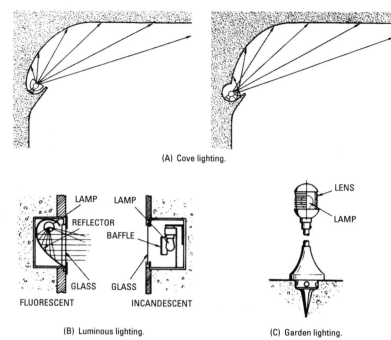

Figure 12-9 Examples of light projection.

down, so that they can be turned on or off from either floor. It is essential to install convenience outlets in halls for electrical appliances; these outlets may also be used for additional portable lamps.

Lighting the Living Room

A living room usually has the brightest illumination level, with the possible exception of a library or studio. Light sources for a living room are generally divided between fixed and portable units. Fixed units may include ceiling fixtures, wall brackets, cove lights, recess lights, or column lights. Portable units may include floor lamps, table lamps, small high-intensity lamps, or semiportable units such as pole lamps.

Although fewer ceiling fixtures are used in modern installations, they are not completely a thing of the past. Pendant fixtures such as chandeliers may be used in the more elaborate installations. Note that rooms with low ceilings require close-fitting and comparatively

Home Circuiting, Multiple Switching, and Wiring Requirements 315

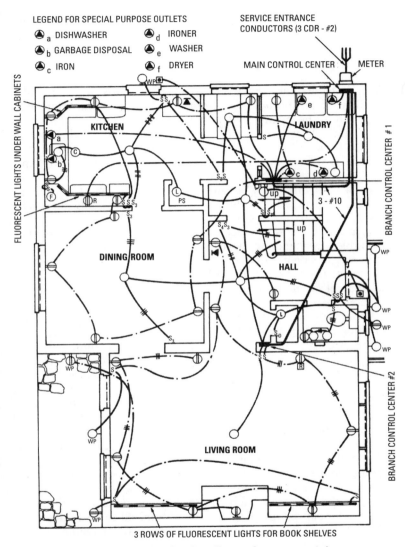

Figure 12-10 Wiring plan for first floor of a two-story house.

closely spaced ceiling fixtures, or built-in fixtures. Reasonably uniform general illumination is desirable in a living room. High-level local illumination for reading, sewing, or piano playing is obtained by various types of supplementary lighting units.

In planning the system, keep in mind that wall brackets and totally indirect wall urns have often-needed decorative value and also contribute to greater uniformity of general room illumination. Wall brackets must be located so that they do not interfere with tapestries or paintings. Three-way lamps (also called three-way lights) are often used for supplementary illumination at sofas. A three-way light bulb has two filaments, which may be operated together or separately. They generally provide 100-, 200-, or 300-watt levels from a switch with three "on" positions. The lamp is used in a mogul three-way socket, which provides one contact at the side of the base plus a button contact and a ring contact at the base.

Note that every wall space in a living room that is large enough for a table, easy chair, or television receiver should be provided with a duplex convenience outlet. In any case, a duplex outlet should be installed at 12-foot intervals, or less, along a living room wall. Also, each mantel should have at least one flush-mounted outlet on its top for an electric clock or decorative light.

A common fault in living room lighting plans is lack of functional switching facilities. Switch location and switching circuits should be planned for convenience in everyday use. Good planning requires careful attention to the architectural layout. For example, the plan shown in Figure 12-11 requires different switching facilities than the plan shown in Figure 12-10. Chandeliers, ceiling fixtures, or wall brackets are usually controlled by a switch at the main entrance of the living room and at each subsidiary entrance. Note that an outlet for an electric clock should not be connected to a switch.

Whenever possible, switching facilities should be discussed in detail with the homeowner and double-checked. To avoid overlooking various requirements, the planners should go over the rooms in the house step-by-step at twilight instead of in broad daylight. Then, by sizing up the plan from each entrance and from each seating location, functional requirements can be recognized to best advantage. All too often, electricians and homeowners plan switching installations in haste.

Lighting the Dining Room

Dining tables are often illuminated by a pendant fixture, such as the one shown in Figure 12-12. Wall brackets will provide pleasant background illumination if properly shaded. Today, electricians

Home Circuiting, Multiple Switching, and Wiring Requirements 317

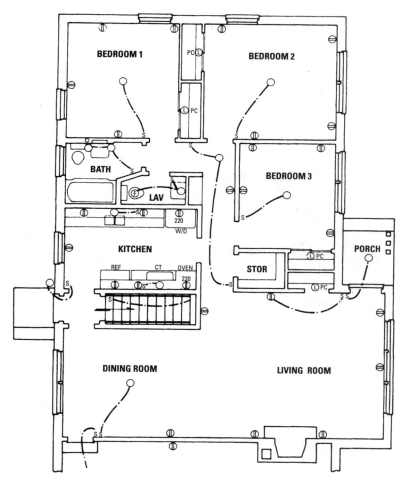

Figure 12-11 Floor plan of a house showing wiring diagram.

work with lower ceiling lights and with combined living room and dining areas, as shown in Figure 12-11. In turn, more careful planning is required to obtain good lighting proportions and balance. Colored lighting, such as amber or moonlight tones, may be used for general lighting, but this must be supplemented by white localized lighting at any table that may be used for study or close work.

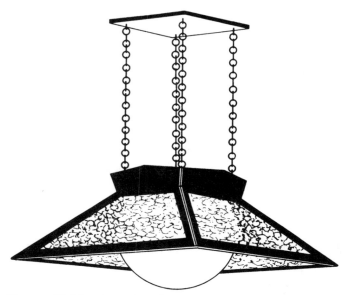

Figure 12-12 A pendant fixture.

Many interesting effects can be obtained by installations of white or colored Lumiline or larger fluorescent lamps installed in plaster, metal, or glass coves around a dining room or at opposite ends of a living room. Aside from provision for illumination, convenience outlets must be installed for electrical appliances on the table or at the buffet. Lack of foresight can result in the necessity for running extension cords over floors or under rugs where they are a continuing nuisance. Switching facilities for a dining room should follow a functional plan, as explained previously for living rooms.

Lighting the Kitchen

Kitchen lighting should be both cheery and utilitarian. A central ceiling fixture is almost always required. This is usually a single diffusing enclosed unit. Modern kitchens may have totally indirect fixtures. If the wall and ceiling have a light-colored surface, the problem of good kitchen lighting is more easily solved. The general rule is that plenty of light should be provided at any point that may be occupied. Therefore, local lighting is necessary over the range, sink, and associated work areas.

Home Circuiting, Multiple Switching, and Wiring Requirements 319

In most installations, the first requirement is a pendant fixture over the sink and range. A glass shade is essential to protect the cook from glare. There is a marked trend in modern kitchens to employ recessed units covered by diffusing glass panels. *Soffit lighting* can be defined as a tailored installation, which may be recessed into furring above a kitchen sink. Figure 12-13 shows how a ventilator can also be advantageously mounted in soffit space over kitchen cabinets. Lights can often be mounted advantageously beside the blower.

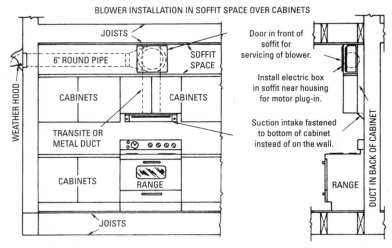

Figure 12-13 Soffit with blower mounted in place. Light mounted beside the blower.

In the foregoing arrangement, fluorescent lamps may be mounted in curved metallic reflectors, with the diffusing element provided by frosted glass covering the *lamp installation* beside the blower. Either Lumiline or ordinary fluorescent lamps are good choices for installation over the range, over the sink, or under the kitchen cabinets to illuminate the worktable or tables. In a butler's pantry, a small close-fitting ceiling fixture is adequate to light shelves or cupboards. In case a breakfast alcove is provided, a pendant-shaped fixture or indirect lighting unit will usually meet illumination requirements.

Obviously, an ample number of convenience outlets must be provided in a kitchen. Toasters, percolators, mixers, and roasters are always used at table height, and their outlets should be installed 42 inches above the floor. On the other hand, electric clocks and ventilating fans will usually be mounted higher, and their outlets will be installed from 6 to 8 feet above floor level in most kitchens. Dishwashers and refrigerators are generally connected to functionally located outlets. The range is connected to a specialized line as illustrated in Figure 12-14.

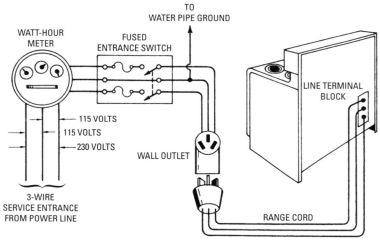

Figure 12-14 Typical installation of electric-range wiring.

As explained previously, a three-wire line is installed from the service entrance as shown in Figure 12-15. The neutral wire is grounded. Where substantial power is demanded, 230-volt operation provides greater economy than 115-volt operation, due to copper costs versus line drop. In other words, voltage drop is *directly* proportional to current, whereas power is proportional to *current squared*. Therefore, substantially greater efficiency can be obtained by supplying power at double voltage and half current.

Lighting the Bedroom

Bedroom illumination tends to be skimped in many installations. For example, low-wattage night-lights add considerably to functional utility. A night-light may consist merely of a small plug-in electroluminescent panel in an economy-type installation. Alternatively, a neon glow lamp may be installed near the bathroom or light switch. A bedroom should have a general lighting unit, such as a ceiling fixture, supplemented by local units for reading or possibly for

Home Circuiting, Multiple Switching, and Wiring Requirements **321**

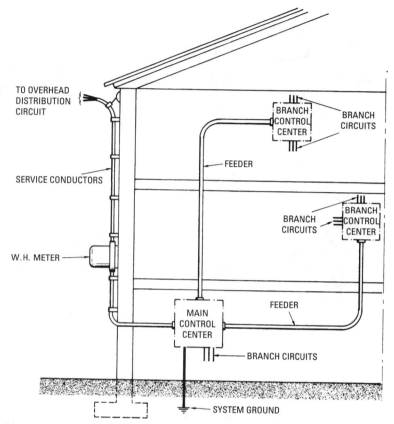

Figure 12-15 Diagram of three-wire line installation through the service entrance to various locations in the house.

sewing. Most types of boudoir units are inadequately illuminated, having only decorative lighting.

Wall brackets are not always indicated for bedroom illumination because they may limit furniture arrangements. Ceiling fixtures may be semidirect or totally indirect types. All bulbs must be shaded in a candle-type fixture. Typical lamp styles suggested by interior decorators for various rooms are shown in Figure 12-16. Note that there are three places in a bedroom that need special lighting: the vanity dresser, the bed, and the boudoir chair or chaise lounge. Boudoir

322 Chapter 12

(A) Living room.

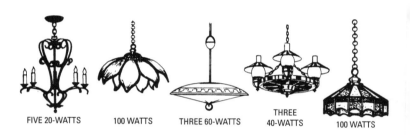

(B) Dining room.

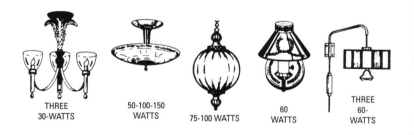

(C) Bedroom.

Figure 12-16 Lamp styles used in various rooms by interior decorators.

Home Circuiting, Multiple Switching, and Wiring Requirements 323

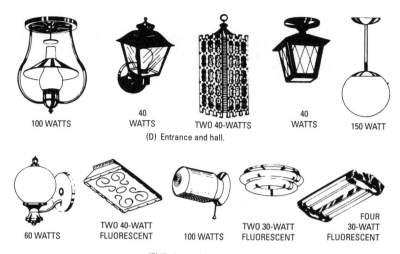

Figure 12-16 (*continued*)

lamps on the vanity should be tall, with light-colored shades to provide adequate lighting on both sides of the face. For reading in bed, a wall, table, or floor lamp can be used if designed so that glaring light does not occur at eye level.

A lamp with a white diffusing bowl under the shade is a good choice in most installations. Also, a floor lamp that provides a wide spread of diffused light downward beside the boudoir chair or chaise lounge will be suitable for mending, knitting, or reading. A small lamp placed under the bed in the baseboard will provide floor illumination. In turn, convenience outlets for the lamps must be installed in each wall. Closet illumination should not be overlooked. Each closet over 2 feet in depth should have a light installed inside on the wall or in the ceiling just above the door so that light is directed back on the clothes and shelves.

Lighting the Bathroom

Bathroom lighting should be planned so that the mirror can be used to advantage for shaving. Two brackets installed about $5^{1}/_{2}$ feet from floor level may be adequate. The advantage of two units is in minimizing shadows and providing uniform illumination from each side. Bulbs are usually shaded with diffusing glass. In the more elaborate installations, the mirror is framed along the top and the sides by built-in panels with lamps concealed behind frosted glass.

Most bathrooms require a ceiling fixture. A globe-type fixture is typical. Switch facilities should be provided just inside the door, with a separate switch for the mirror light. Note that showers and tubs may need lighting in some cases. These lights are waterproof types and are usually mounted flush with the ceiling. A night-light installed in the baseboard provides practical utility, just as in a bedroom. A convenience outlet installed at the right-hand side of the mirror is necessary for use of electric razors, curling irons, or other personal grooming items.

Lighting the Attic

An attic space usually requires a ceiling light and a convenience outlet. A standard dome reflector may be used, controlled by switches at the foot and at the head of the attic stairs. Additional wiring is required in case an attic fan or blower is installed. Cable television cables may be routed through an attic. Figure 12-17 shows a wiring plan for a small cable TV system.

Lighting the Basement

Basement areas may be divided into stairway, laundry, workshop, furnace room, and recreation room. A glareless light installed at the foot of the stairs and controlled by three-way switches from the head of the stairs and from the basement is usually required. A pilot light in the switch at the head of the stairs is useful as an indicator to show if the light has been left on.

A laundry room needs one or more ceiling lights, depending on the area to be illuminated. White diffusing globe luminaires or silvered bowl lamps are often installed. Local light is almost always required in the laundry and at various appliances. The general rule to be observed is provision of ample light at each work area. Convenience outlets are required at suitable locations for portable appliances.

Workbenches must be well illuminated and shadows avoided. This is particularly important in the operation of power tools. Diffusers installed over a workbench will direct light downward for close mechanical work. Local lighting units are also desirable in many cases. An ample number of well-located duplex outlets should be provided throughout the work area. A separate circuit is advised for any workroom where power tools will be used.

Furnace rooms do not require extensive illumination, and a ceiling reflector installed at the furnace is often adequate. The light

Home Circuiting, Multiple Switching, and Wiring Requirements 325

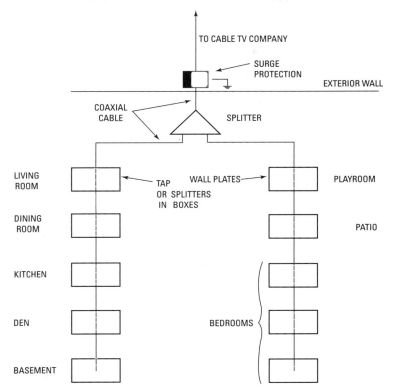

Figure 12-17 CATV system.

should clearly illuminate all controls and indicating scales. It is good practice to install at least one convenience outlet for use in maintenance. If a number of units are installed in a furnace room, it may be necessary to supplement a ceiling reflector with one or more local units. The rule to be observed is that of good utility under both normal and maintenance requirements.

Recreation rooms are often associated with the basement area. Low ceilings may have to be contended with, and recessed or close-fitting fixtures may be most suitable. Portable lamps are frequently necessary for card tables and the like. Bars require special lighting installations with more or less decorative effects. An ample number of convenience outlets should be provided for appliances, electric toys, and so on.

Lighting the Garage

Garages may be associated with basement or general living areas. In each case, the requirements must be carefully sized up by the electrician. Good functional illumination is the prime consideration. Unfortunately, garage lighting is often skimped, which not only imposes a hazard at night but also complicates any automobile maintenance work that might be required. Good lighting is essential at the rear door from a safety standpoint. This may be a simple bracket or porch ceiling light. An exterior light is often a decorative waterproof lantern or a reflector.

Lights should be controlled by switches installed inside the entrance to the garage and inside the most convenient door to the house. Most new garages have radio-controlled automatic door openers (see Figure 12-18). Although the wiring of door openers is a specialized job, it is not unduly difficult. Detailed instructions for installation will be provided with the equipment. Convenience outlets should be provided in a garage for use of lamps with extension cords and electrical tools.

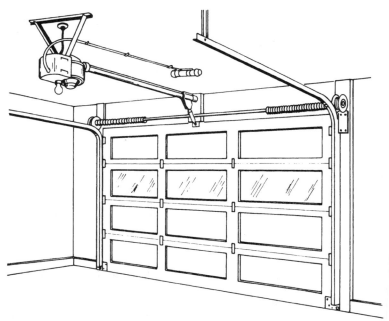

Figure 12-18 Typical overhead door with operator mechanism.

Fluorescent Lights

Modern fluorescent lamps are available in four types:

1. Hot-cathode, preheat starting
2. Hot-cathode, instant starting
3. Cold-cathode
4. RF lamp

Figure 12-19 shows the circuits that are typical for each lamp. When a manual starting switch is used, the preheating period depends on the time that the operator keeps the switch depressed. However, glow-type starters, such as the one shown in Figure 12-20, are automatic; they are preferable because the preheating period is closely controlled, with the result that the lamp is usually longer-lived.

A thermal-type starter employs heat from a resistance, instead of neon gas as in a glow-type starter. When the lamp is turned on, the bimetal strip touches a set of carbon contacts. In turn, current flows through the lamp cathodes for preheating. The resistance of the carbon also produces heat, which soon causes the bimetal strip to move away from the contacts, thus striking an arc inside the lamp. Thereafter, the bimetal strip is kept away from the contacts by current flowing through the resistor, which is in parallel with the lamp. If the current should be momentarily interrupted, the voltage across the resistor increases and the additional heat causes the bimetal strip to bend until it touches the restart contact. In turn, the cathodes are preheated again, but the resistor is now short-circuited; the bimetal strip soon moves away from the restart contacts, and the arc is struck inside the lamp.

In instant-starting lamps, hot cathodes may be used, and the arc is struck immediately by a high kickback voltage from the ballast reactor. The cathodes are specially manufactured to withstand the effects of starting voltages between 450 and 750 volts. Instant-starting lamps also use cold cathodes. This type of lamp has the advantage that its life is not shortened by brief operating periods with frequent operation of the switch. An RF lamp is similar to a low-pressure mercury-arc lamp, except that tubing is used which is coated with a phosphor that converts ultraviolet light into visible light. RF lamps are used to a greater extent in industry than in home lighting systems.

Ballast reactors are usually installed in the raceway behind the lamp. A ballast may be used alone or in combination with a step-up

328 Chapter 12

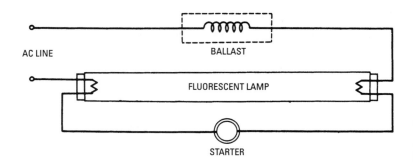

(A) Hot-cathode, preheat starting type.

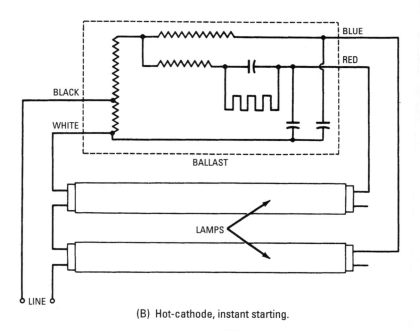

(B) Hot-cathode, instant starting.

NOTE: BALLAST REACTORS ARE SYMBOLIZED ⁀⁀⁀⁀⁀ OR ∿∿∿∿∿ IN VARIOUS ELECTRICAL DIAGRAMS.

Figure 12-19 Four types of fluorescent lamps.

Home Circuiting, Multiple Switching, and Wiring Requirements 329

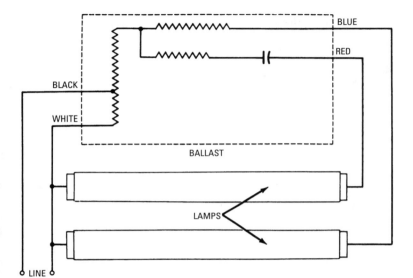

(C) Cold-cathode.

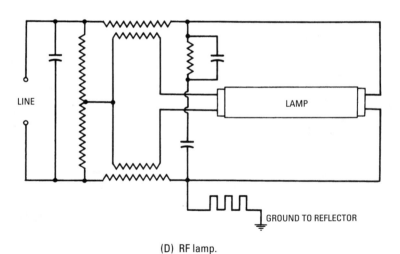

(D) RF lamp.

Figure 12-19 (*continued*)

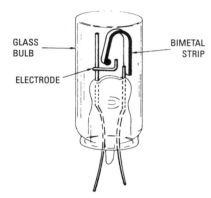

Figure 12-20 Principal parts of a glow-type starter.

transformer to obtain increased starting voltage. Ballasts are also used in combination with capacitors for power-factor correction so that the lamp circuit draws current in-phase with the applied voltage. Power-factor capacitors are seen in Figure 12-19. All of these lamp auxiliaries are usually provided as a single enclosed unit, which is mounted in the raceway.

Ground Circuit Fault Interrupters

Fuses do not blow and circuit breakers do not trip until the current demand exceeds the ampacity rating of the fuse or breaker. In turn, if there is electrical leakage to the frame of an appliance such as a lawnmower, a person can be fatally shocked by a very small fraction of 1 ampere. If three-prong plugs on three-wire cords (cable with a grounded conductor) are used, the chance of receiving a shock is quite small. However, ground connections in wiring systems can become defective. In addition, many existing wiring systems do not have grounding conductors, and no protection is provided against electrical leakage in appliances. For this reason, the *National Electrical Code* now requires the installation of ground fault circuit interrupters (GFCIs) in various locations. A GFCI is a special outlet device that includes a relay that trips on 5 milliamperes of leakage current. In turn, the user of an appliance is protected against defects that make the frame of the appliance hot. Although 5 milliamperes will shock the user, the relay trips in 0.025 second, so the shock is felt for only an instant.

Arc-Fault Circuit Interrupters

An arc-fault circuit interrupter (AFCI) protects persons and equipment from an arc fault by recognizing the characteristics unique to an arcing fault and deenergizing the circuit when an arc fault is detected. All branch circuits supplying 15- or 20-ampere, single-phase 125-volt outlets installed in dwelling-unit bedrooms must be AFCI-protected by a listed device that protects the entire branch circuit. This change extends AFCI protection to all 125-volt outlets in dwelling-unit bedrooms, whereas the 1999 *NEC* only required AFCI protection for all branch circuits that supply 15- or 20-ampere, single-phase 125-volt receptacle outlets in dwelling-unit bedrooms. The *Code* defines an outlet as a point on the wiring system at which current is taken to supply utilization equipment. This includes openings for receptacles, luminaires, or smoke detectors.

The traditional practice of separating the lighting from the receptacle circuits in dwelling-unit bedrooms will, since the addition of this requirement, require two AFCI circuit breakers. The 125-volt limitation to the requirement means that AFCI protection is not required for a 240-volt circuit, such as one for an electric heater.

It is important to note that *NEC* requirements change over time. An installation that meets *Code* this year may not match the requirements 10 years from now (see Figure 12-21). This is not to say that every existing installation has to be upgraded for every new edition of the *NEC*, although it would be a good idea to consider upgrading electrical systems from time to time.

Summary

When wiring a house, the light control system must be carefully considered. Otherwise, after an installation is completed, it may be discovered that the lamp control facilities are different from what was actually desired. When lighting is to be controlled by switches from three or more locations, three-way and four-way switches are used. The lighting can be turned on or off with a switch at any location. Either three-way or four-way switches can be used at the end of the switching circuit; however, four-way switches must be used at intermediate positions in the switching circuits.

Safety and function are the primary considerations in planning an electrical system. An effective and efficient home wiring system depends on sufficient circuits, satisfactory number of outlets, and high-quality material and workmanship.

Lighting facilities must meet three basic requirements—sufficient light for all activities, fixtures that are adapted to their purpose and

332 Chapter 12

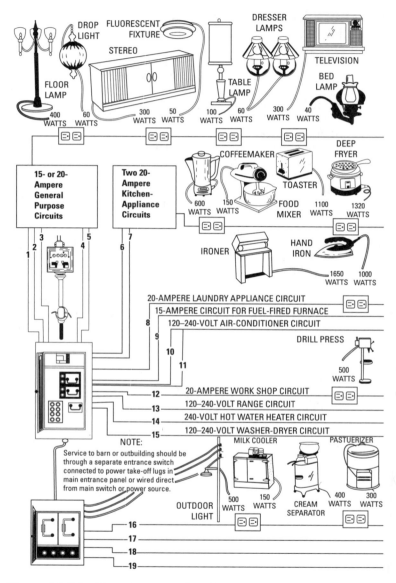

Figure 12-21 Typical circuitry found in existing farm home with outbuildings. Although similar circuits can be planned by the electrician, some of the installations shown may violate the *National Electrical Code* for new homes. For example, a refrigerator must now have its own circuit.

Home Circuiting, Multiple Switching, and Wiring Requirements 333

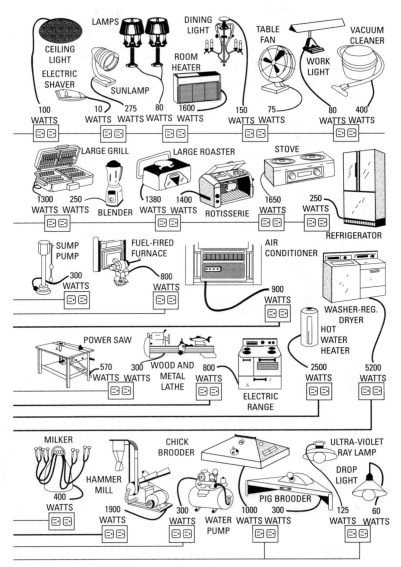

Figure 12-21 (*continued*)

appropriate to the surroundings, and the avoidance of objectionable shadows and glare.

Lighting in the living room generally has the brightest illumination level and is divided between fixed and portable fixtures. In more elaborate homes, ceiling fixtures or chandeliers are used, but in smaller homes, where lower ceilings are generally the design, wall fixtures, table lamps, and pole lighting are used. Many times, totally indirect wall lighting is used, giving a uniform room illumination. In this type of lighting, wall brackets must be located so that they do not interfere with tapestries or paintings.

In dining rooms, more careful planning is required to obtain good lighting proportions and balance. Colored lighting, such as amber or moonlight tones, may be used for general lighting but must be supplemented by white localized lighting at tables used for study or close work. Convenience outlets must be installed for electrical appliances used at the table or the buffet. Lack of foresight can result in the necessity for running extension cords over floors or under rugs, which can be a general nuisance.

Kitchen lighting should be both cheery and utilitarian. A central ceiling light is almost always required, which is usually a single diffusing enclosed unit. There is a trend in modern kitchens to employ recessed units covered by diffusing glass panels that also serve as the kitchen ceiling.

Bedroom and bathroom fixtures should provide adequate lighting at dressing mirrors. There are three places in the bedroom that need special lighting: the vanity dresser, the bed, and the boudoir chair or chaise lounge. Bathroom lighting should be planned so that the mirror can be used to advantage for shaving. The shower and tub may need lighting in some cases.

The basement may be divided into multiple areas, such as laundry, workshop, furnace room, and recreation room. The laundry and furnace room may need one or more ceiling fixtures, depending on the area to be illuminated. White diffusing globes or silvered bowl lamps are often used. Low ceilings may exist in the recreation room, and recessed or close-fitting fixtures may be most suitable.

Test Questions

1. What is a three-way switch?
2. Discuss a simple application for a three-way switch.
3. What is a four-way switch?

Home Circuiting, Multiple Switching, and Wiring Requirements 335

4. How would a four-way switch be used in a circuit with a three-way switch?
5. Explain how switches are used to control a lamp or a group of lamps from two separate locations.
6. What is an electrolier?
7. Describe a two-circuit switch.
8. How is a three-circuit electrolier switch wired to control three groups of lamps?
9. How would a circuit be wired to control a lamp bank from five separate locations?
10. Explain what is meant by central-point lamp control.
11. What are the three chief requirements for a good home lighting system?
12. Name several types of luminaires that are suitable for home lighting installations.
13. How should entrance lighting be planned? Where should switches be installed?
14. Discuss the requirements for good hall lighting.
15. Explain how an electrician plans a living room installation when the ceiling is low.
16. What are the requirements for installation of convenience outlets?
17. Why should colored lighting in a dining room be supplemented by white localized lighting?
18. Describe the requirements for good kitchen lighting.
19. What does an electrician mean by soffit lighting?
20. Why does an electric range require separate wiring?
21. Explain the requirements for night-lights in bedrooms and bathrooms.
22. Name several types of lamps used in modern bedroom and bathroom installations.
23. What are the general requirements for attic lighting?
24. Why may specialized wiring be installed in an attic?
25. Give a brief list of requirements for good basement lighting.
26. Where would three-way switches be installed in a basement lighting system?

27. How is a garage lighting system planned? What is the most important requirement?
28. Name a special wiring installation that may be used in a garage.
29. Explain the operation of an arc-fault interrupter.
30. Why would an electrician avoid using oversize wire with a given outlet or switch?
31. How does a glow-type starter differ from a manual starting switch?

Chapter 13

Electric Heating

Electricians are called upon to make many types of electric heating installations. Space heaters are the most commonly used. A space heater is defined as a heater without a reflector or other device for directing the heat output; the heat is produced by electric current flow through resistive elements and is distributed into the surrounding space chiefly by convection and conduction. Transfer of heat by *conduction* means the flow of heat through a solid substance such as iron. Transfer of heat by *convection* means the carrying of heat by air rising from a heated surface. Both conduction and convection are different from *radiation* of heat because *radiation* takes place in the absence of matter, as in the passage of heat through the vacuum inside the bulb of an incandescent lamp.

Electricians work with two types of space heaters: (a) the steel-sheath type, which can be operated up to 750°F; and (b) the porcelain-enameled type, which can be operated up to 1200°F. Space heaters with alloy-steel sheaths are similar to the porcelain-enameled types in that they can also be operated up to 1200°F. Both of the basic types of heaters are used in appliances such as ovens. In room heating, both convection and radiation provide distribution of heat into the surrounding space. As a rough rule of thumb, from 1 to 2 watts per cubic foot is required to heat the air in a room when the temperature is near the freezing point outside.

General Installation Considerations

Electric heaters may be built into walls or ceilings or may be mounted along the baseboard. Switching facilities are usually automatic and include a thermostat. Typical units are shown in Figure 13-1. A wall heater may be provided with one or more fans to provide forced-air circulation. This feature gives more rapid distribution of warm air when the heater is first turned on. A typical wall heater of this type fits in a $7^{1}/_{4} \times 14$-inch wall opening and consumes 1500 watts. Larger wall-heater units consume up to 4000 watts. In many designs, fans can be operated with the heater element turned off so that air circulation is provided in summer.

Baseboard heaters are manufactured in lengths from 28 to 107 inches and consume from 300 to 2000 watts. In some installations, an electrician may use portable baseboard heaters. They are provided with carrying handles and are plugged into convenience outlets. A portable baseboard heater is often desired for use in sun porches and other seasonally occupied rooms. This type of heater is

338 Chapter 13

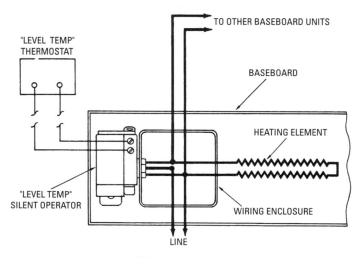

(A) Baseboard unit.

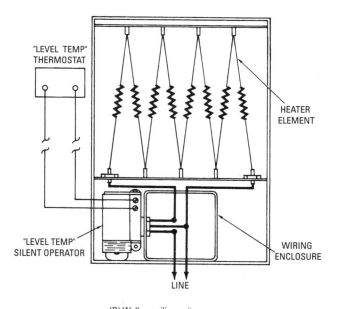

(B) Wall or ceiling unit.

Figure 13-1 Various types of electric heating units.

Electric Heating 339

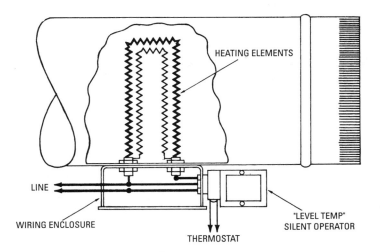

(C) Electric duct unit.

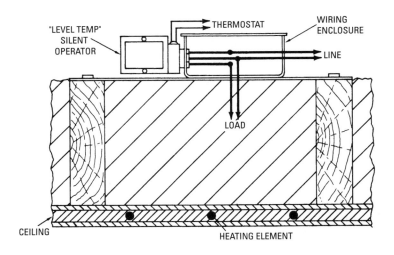

(D) Ceiling cable unit.

Figure 13-1 (*continued*)

340 Chapter 13

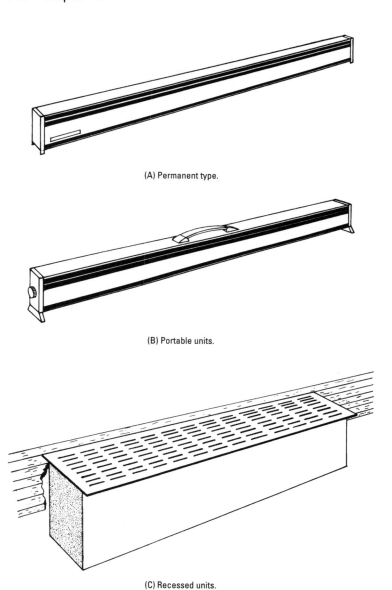

(A) Permanent type.

(B) Portable units.

(C) Recessed units.

Figure 13-2 Baseboard electric heating units.

manufactured in lengths from 27 to 71 inches and consumes from 300 to 1500 watts. Baseboard heaters are usually between 9 and 10 inches high and between 3 and 4 inches deep.

Figure 13-2 shows the exterior appearance of permanent baseboard, portable baseboard, and recessed heaters. A recessed electric heater is basically a drop-in floor unit that may be used where baseboard or wall heaters are unsuitable. Lengths of recessed heaters range from approximately 35 to 107 inches and consume from 300 to 2000 watts. Baseboard and recessed heaters generally operate by convection and do not have forced-air fans.

Circuiting

Basic circuiting for electric heaters is shown in Figure 13-1. A bimetallic thermostat is used to turn the heater on automatically when the room temperature falls below a preset level. Similarly, the thermostat turns the heater off automatically when the room temperature rises above the preset level. A thermostat is not suitable for switching heavy currents, such as those that must be switched in operation of an electric heater. Therefore, the thermostat is used in a relay circuit, as shown in Figure 13-3. Thus, a small current in the thermostat branch can switch a heavy current in the heater branch.

The thermostat cannot switch the heater circuit on or off directly because the thermostat contacts move together or move apart slowly. Arcing at the contacts would quickly burn out the thermostat. Therefore, low voltage and small current are used in the thermostat branch. A current of 0.2 ampere is typical. This thermostat current is provided by a small voltage step-down transformer. In turn, the thermostat current operates a power relay, commonly called an *operator*.

An operator also contains a thermostatic device called a *thermal-type relay*. It has a bimetal heater that is warped when heated by a resistive element connected in series with the thermostat. In turn, warping of the bimetal blade suddenly trips a snap switch and closes the power circuit to the heater. The snap switch permits the power circuit to be opened or closed quickly, thereby minimizing arcing at the contacts. Thus, a small current in the thermostat branch effectively controls a large current in the heater element (load) branch.

Figure 13-4 shows the circuiting details for three heater loads using one thermostat and three operators. This is a *sequence* system in which the loads are switched on in succession at 45-second intervals and are switched off successively at 45-second intervals. Sequence operation is desirable when several loads are to be switched because the thermostat carries only 0.2 ampere in a sequence system.

342 Chapter 13

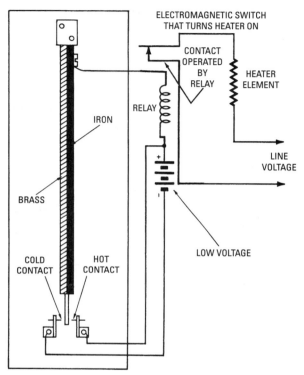

Figure 13-3 Compound metal thermostat and circuit.

If sequencing were not used, the thermostat would have to carry 1 ampere for controlling five loads; this would tend to shorten the life of the thermostat.

The sequence of operation in Figure 13-4 is as follows. When the contacts of the low-voltage thermostat close, 0.2 ampere flows through the heating element in the operator (thermal-type relay). In approximately 45 seconds, the bimetal blade warps sufficiently to suddenly close the snap switch for load No. 1. At this time, the step-down transformer in the second operator is energized, and its thermal-type relay is tripped about 45 seconds later. As soon as the snap switch for load No. 2 is closed, the step-down transformer in the third operator is energized, and the snap switch for load No. 3 closes about 45 seconds later. When the room comes up to the preset temperature, the low-voltage thermostat opens its circuit, and load

Electric Heating 343

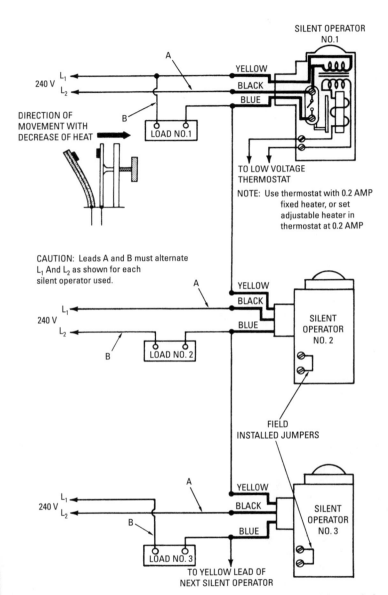

Figure 13-4 A sequence system for one low-voltage relay and three thermal power relays. *(Courtesy White-Rodgers Co.)*

No. 1 switches off after 45 seconds, followed by switch-off of load No. 2 and load No. 3 at 45-second intervals.

In some installations, an electrician may need to install a mechanical switch to supplement control by a thermostat. A *limit switch* is used for this purpose, as shown in Figure 13-5. This type of switch is different from a manually operated switch in that the switch is automatically operated by a moving object such as a cabinet door, which deenergizes the heater when the door is closed. Since the limit switch is connected in series with the heating elements, the heater cannot be energized by the thermostat as long as the door is closed.

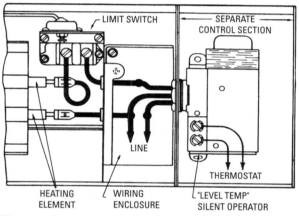

Figure 13-5 Limit switch included in a baseboard heater. *(Courtesy White-Rodgers Co.)*

AC-Operated Relays

In the foregoing examples, the relay contacts were closed or opened by means of mechanical forces. Thermal-type relays use the mechanical force produced by warping of a bimetallic blade. Limit switches use mechanical force from a moving object. In addition to these devices, electricians may use relays that operate from electromagnetic force in an AC circuit. These are somewhat similar to relays used in DC circuits, except that special construction is employed in an AC relay to avoid armature vibration (chattering). In other words, a 60-Hz AC current rises to a peak and falls to zero 120 times a second. This vibration in electromagnetic force will cause a simple iron armature to chatter.

Chatter is objectionable, not only because it makes a loud buzzing sound, but also because of aggravated contact arcing. Therefore, *shading coils* are used in AC relays, as shown in Figure 13-6. Another problem in AC relays is heating due to eddy current losses in the magnetic circuit. Accordingly, the core of an AC relay is generally laminated. In Figure 13-6, the shading coil consists of a heavy copper loop. Current is induced in the loop by transformer action. Therefore, the electromagnetic forces on the armature are produced both by the relay-coil current and by the shading-coil current.

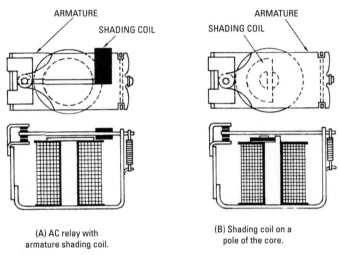

(A) AC relay with armature shading coil.

(B) Shading coil on a pole of the core.

Figure 13-6 Shading coil.

We know that the current lags in an inductive circuit. In turn, although a single-phase current flows through the relay coil in Figure 13-6, a current with lagging phase flows in the shading coil. Thus, the combined magnetic force produced by the relay coil and the shading coil rises to a peak and falls to zero 240 times a second. Since the armature cannot vibrate to any appreciable extent at a rate of 240 times a second, the chatter is minimized, and the AC relay operates in much the same manner as a DC relay. Vibration cannot be eliminated entirely, but it is reduced to a very small amount by the shading coil, which makes AC operation practical.

Hot-Water Electric Heat

Another type of electric space heating uses hot water without a plumbing installation. Figure 13-7 shows the essential features of this heating method. An electrical element A inside the copper tubing heats a permanently sealed-in water and antifreeze solution. Operation of the electrical element is thermostatically controlled. Since heat tends to rise, the hot water circulates upward and through the finned heat distribution area B where heat is transferred from the fin surfaces to the surrounding air. This cools the water, which causes it to circulate down and back over the electrical element. Expansion space is provided by chamber C.

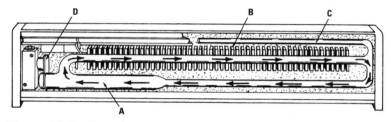

Figure 13-7 Basic function of a hot-water electric heater. *(Courtesy International Oil Burner Co.)*

In case the water becomes overheated for any reason, the enclosure cannot explode due to limit switch D. As the temperature of the copper tube increases, the tube expands and increases in length. When a certain length is exceeded, the limit switch is tripped, which opens the circuit of the electrical element. Antifreeze is used in the water to avoid damage due to freezing in case the unit is exposed to low winter temperatures.

Radiant Heater

A radiant heater is defined as an electric heating unit that has an exposed incandescent heating element. The electrical element glows red in operation, and a reflector is provided that directs heat radiation out through the grille much as a mirror reflects light. Heat reflectors are generally made from bright metal. It is basically the reflector that distinguishes a radiant heater from a space heater. Figure 13-8 shows a small radiant heater, such as might be installed in a bathroom. This is a ceiling-type heater that contains a fan to supplement heat radiation by convection. Units of this type usually consume 1250 watts.

Electric Heating 347

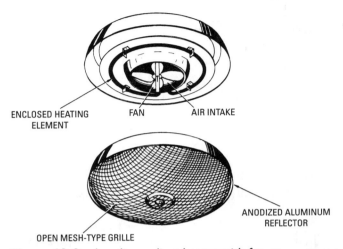

Figure 13-8 A ceiling radiant heater with fan. *(Courtesy Nutone Inc.)*

Radiant heaters may also be installed in a wall. If an exhaust fan is provided, the heater can also serve as a ventilator in summer. Radiant heaters mounted in a wall are suitable for either kitchen or bathroom. They operate from either 120-volt or 240-volt lines, and a typical heater consumes 1600 watts. A radiant wall heater is usually provided with a manual switch but may also be installed with an automatic thermostat control if desired. Thermostats operate as previously explained for space heaters.

Electric Water Heaters

Electric water heaters use *immersion heaters* that are designed to operate in water. Figure 13-9 shows some typical immersion heater elements. A snap-action thermostat is generally used with an immersion heater, as shown in Figure 13-10. The snap action provides sudden opening or closing of the contacts, thereby minimizing arcing at the contacts. Installation of the pipes and the heater is the plumber's job; the electrician is concerned only with installation of the electrical system.

Since heat tends to rise, a tall heater may be manufactured with one immersion heater at the bottom and another at the top. In such a case, the upper heater is often operated by a double-throw snap-action thermostat, as shown in Figure 13-11. If the lower heater unit consumes 1000 watts, the upper heater unit typically consumes

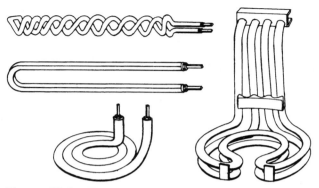

Figure 13-9 Electric heating elements for water heaters.

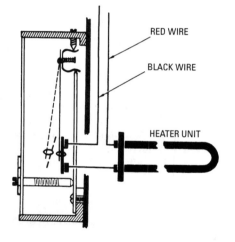

Figure 13-10 A snap-action thermostat for an immersion heater.

1500 watts. Each of the heater units is separately connected to the line through fuses or circuit breakers in a connection box. Note in Figure 13-11 that a double-throw thermostat energizes the upper heater first. After the water in the top of the tank comes up to the preset temperature, the left-hand contacts are closed so that the lower heater unit can be energized. In turn, the water at the bottom of the tank comes up to the preset temperature.

The use of two heater units provides a better supply of hot water under conditions of varying demand. With a small demand, only

Electric Heating 349

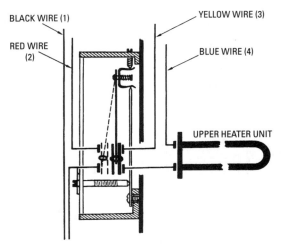

Figure 13-11 A double-throw snap-action thermostat.

the lower heater unit may be switched into operation. With a large demand, both heater units will be switched into operation; after the demand ceases, the upper heating unit will switch off first. Thermostats may be set to cut out at about 150°F in a typical installation. Of course, under conditions of very heavy demand, the temperature of the hot water may drop.

Before making an installation, the electrician should check local codes and practices. For example, a separate watt-hour meter may be required for an electric water heater. In some locations, operation may be forbidden during certain hours, and the electrician must install an electric time switch to open the heater circuit during these hours. Only the reserve supply of hot water in the tank is available while the heater circuit is open. Hence, a comparatively large tank may be used, or a small supplementary tank that does not require an electric time switch may be installed.

Time switches may be connected on the supply side of the disconnecting means, according to the *National Electrical Code*. Taps from service conductors to supply time switches may be installed as separate conductors, in cables approved for the purpose, or enclosed in rigid conduit; electrical metallic tubing may also be used. The service-entrance conductors must not be run within the hollow spaces of frame buildings unless fuses or circuit breakers are installed at the outer end of the conductors.

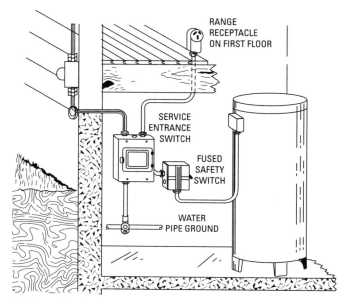

Figure 13-12 A safety switch is installed in addition to the service-entrance switch.

Means must also be provided for disconnecting the conductors from the service-entrance conductors (see Figure 13-12). This provision must be located at a readily accessible point nearest the entrance of the conductors, either inside or outside the building wall. One may use either an approved manually operated switch or a circuit breaker of the air-break or oil-immersed type with a handle that is marked and identified. A typical circuit breaker with an external handle is shown in Figure 13-13. An external handle is required. Where electrical remote control is desired, a push-button control can be used in addition to the manual handle.

Note that a common enclosure, or a group of separate enclosures, must not contain more than six switches or six circuit breakers. Two or three single-pole switches or breakers, capable of individual operation, may be installed on multiwire circuits, with one pole for each ungrounded conductor, as a unitary multipole disconnect, *provided they are equipped with "handle ties"* or equivalent approved arrangement. If the circuit breaker does not open the grounded conductor, another switch must be provided in the service cabinet for disconnecting the grounded conductor from the interior wiring.

Electric Heating 351

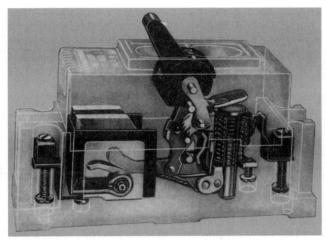

Figure 13-13 A single-pole circuit breaker.

A unitary multipole disconnect switch has holes drilled through the individual switch handles through which a metal rod is passed. The ends of the rod are secured by a rectangular metal stamping that surrounds the handles. Thus, all of the individual switch handles are operated simultaneously and cannot be operated separately. This simultaneous disconnect installation is required for all two-phase or three-phase circuits but is not required for single-phase circuits.

As seen in Figure 13-13, circuit breakers are basically electromagnetic relays that open automatically when the current demand is excessive. This is the most common type of breaker, and it must be reset manually after it has been tripped due to an overload. A breaker will immediately trip again if the cause of the overload has not been corrected. When new equipment is used in an installation, the most likely cause of overload is an error in wiring connections. In case of difficulty, it is advisable to have the connections checked by an assistant to get new ideas in following the wiring diagram.

Automatic fuseless circuit breakers are available in ratings up to 200 amperes and in 10 different box sizes. A two-circuit service panel is used with water heaters. For general-purpose installations, 12-, 14-, or 24-circuit service panels may be used. Circuit breakers are separate from the panels and are plugged in; never install a circuit breaker with a rating higher than the allowable current-carrying capacity of the wires used in its circuit. There are six breaker ratings in common use: 15- or 20-ampere single-pole breakers are used for general lighting installations, and 20-, 30-, 40-, or 50-ampere double-pole breakers are used for heavy loads.

Ground Circuit

The *National Electrical Code* requires that metal frames of water heaters operating on circuits above 150 volts to ground shall have a ground circuit as shown in Figure 13-12. If grounding happens to be impractical in a particular installation, special permission must be obtained, and the metal frames must then be permanently and effectively insulated from the ground. It is recommended that frames be grounded, even on circuits operating at less than 150 volts above ground. This is always good practice because 120 volts can, in some situations, give a person a fatal shock.

The white (neutral) wire of all AC installations must be grounded. A No. 6 or No. 4 copper ground wire is the size normally used. If No. 8 wire is used, it must be armored provided there is no danger of mechanical damage. Note that in a rural installation the ground wire does not run through the entrance switch but is tapped off the neutral overhead wire, brought down the side of the house or yard pole, and connected to a ground rod or to an underground water pipe system (see Figure 13-14) if allowed by the local water company.

If a ground rod must be used, either a copper rod or a galvanized iron or steel pipe is suitable. A copper ground rod must be at least $1/2$ inch in diameter and a galvanized pipe at least $3/4$ inch in diameter according to most codes. The rod or pipe must be at least 8 feet long and located at least 2 feet from any building, and the top must be driven at least 1 foot below the surface. Thus, the top of the rod and the connection to the ground clamp are buried below the surface. Most codes recommend two such grounding rods, spaced at least 6 feet apart.

Grounding is extremely important, and both the *NEC* and local codes are paying more attention to this. Always consult the latest versions of all codes and be sure to abide strictly by their requirements.

There is an increasing trend toward the use of *polarized* devices with any electric heater or similar appliance. A polarized device has two current-carrying contacts plus one grounding contact. Polarized devices guard against dangers from current leakage due to faulty insulation or exposed wiring and help prevent accidental shock. Typical polarized receptacles are shown in Figure 13-15.

Electric Heating 353

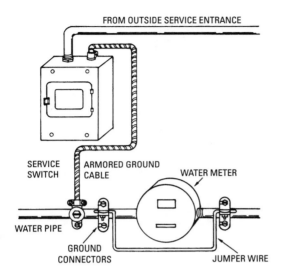

(A) Usual method of grounding city and town systems.

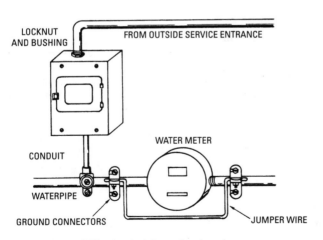

(B) Usual method of grounding city and town system using conduit.

Figure 13-14 A ground rod is used if a water pipe ground is unavailable. Two such rods, spaced 6 feet apart or more, are recommended.

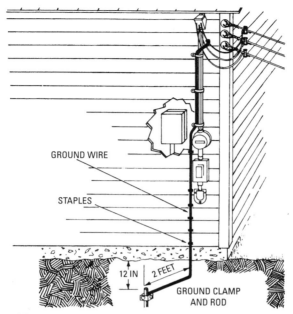

(C) Approved REA method of grounding wire system with ground rod.

Figure 13-14 (*continued*)

If an existing installation has standard receptacles, polarized plugs are used with an adapter, as shown in Figure 13-16. If the wires that connect the outlet to the service are run in conduit or armored cable, the lug on the lead wire of the adapter is connected to one of the screws holding the receptacle to the box. If the wires to the receptacle are run in nonmetallic cable with a bare ground wire, the lug on the lead wire of the adapter is connected to this bare wire. However, if the wires to the receptacle are in nonmetallic cable without a bare ground wire, the lug on the lead wire of the adapter must be specially connected to an approved ground such as a water pipe.

Heating Cables

A special type of electric heater is manufactured in cable form, as shown in Figure 13-17. Heater cable is generally available in lengths from 20 to 60 feet. It has a flexible lead-covered construction. This type of electric heating is used to prevent frozen water pipes, to keep plants from being damaged by cold, and in gutters, troughs, and

Electric Heating 355

Cord sets are available in different lengths for connection to high-speed dryers, ranges, etc. Permits easy disconnection of equipment when redecorating, cleaning, and servicing. Connect with 3-wire cable... For equipment using 240 volts —50 amperes or less.

(A) Surface-type receptacle.

Cord sets are available to complete the connection to standard dryers, etc., to permit easy disconnecting of dryer for cleaning and servicing. Connect with 3-wire cable... For equipment using 240 volts—30 amperes or less.

(B) Surface-type receptacle with L-shaped ground.

Also available in duplex, but some areas will not permit a duplex receptacle of this type. For standard switch box and receptacle plate... For equipment using 240 volts—15 amperes or less, such as small air conditioners, etc.

(C) Single receptacle with crow-foot blade.

Fits any standard switch box and uses a standard single receptacle plate. Connect with 3-wire cable... For use with equipment using 240 volts—20 amperes or less, such as larger air conditioners, power tools, garden equipment, etc.

(D) Single receptacle with tandem blades.

Figure 13-15 Typical polarized receptacles.

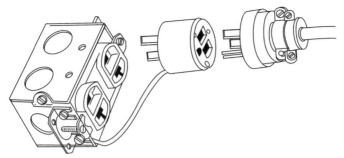

Figure 13-16 Adapter used to provide a ground circuit to a standard receptacle.

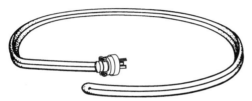

Figure 13-17 A short length of heater cable.

animal drinkers. Heater cable may be used under porches, garage driveways, and sidewalks to melt ice and snow. Its lead-covered construction makes the cable weatherproof.

Summary

Three types of heat transfer are conduction, convection, and radiation. Conduction is the flow of heat through a solid substance such as iron. Transfer of heat by convection refers to the carrying of heat by air rising from a heated surface. Radiation takes place in the absence of matter, as in the passage of heat through the vacuum inside the bulb of an incandescent lamp.

Electric heating may be built into walls or ceilings or may be mounted along the baseboard. A recessed electric heater is basically a drop-in floor unit that may be used where baseboard or wall heaters are unsuitable. Baseboard and recessed heaters generally operate by convection and do not have forced-air fans.

A bimetallic thermostat is used to turn the heaters on automatically when the room temperature falls below a preset level. Thermostats are not suitable for switching heavy currents, such as those that must be switched in the operation of an electric heater. Thermostats are used in a relay circuit where small currents can operate branch circuits to switch heavy current in the heater units.

Another type of electric space heating uses hot water without plumbing installations. An electrical element inside the copper tubing heats a permanently sealed solution. When heated, the solution tends to expand, sending the water through the finned-head distribution area, where the heat is transferred from the fin surface to the surrounding air. This action also cools the water, which causes it to circulate down and back over the electrical heating elements.

Radiant heating is defined as an electric heating unit that has an exposed incandescent heating element. The electrical element glows red in operation, and a reflector is provided that directs heat radiation out through a grille much as a mirror reflects light.

Radiant heaters may also be installed in a wall. If an exhaust fan is provided, the heater can also serve as a ventilator in the summer. Radiant wall heaters are usually provided with a manual switch but may be installed with an automatic thermostat control if desired.

Electric water heaters use immersion-type heater elements designed to operate in water. Because hot water tends to rise, some manufacturers use only one element at the top and one element at the bottom. With a small demand, only the lower heater unit may be used, whereas with a large demand, both heater units may operate together.

Test Questions

1. Define a space heater.
2. How does heating by convection differ from heating by conduction? By radiation?
3. What is the difference between a heater thermostat and a heater operator?
4. Explain the operation of a thermal-type relay.
5. Why is a sequence system preferred for operation of several heater loads?
6. Discuss the principle of a limit switch.
7. How does a shading coil function in an AC relay?
8. Why are laminated cores used in AC relays?
9. Describe the operating cycle of a hot-water electric heater.
10. Explain the features of a radiant heater.
11. How does a snap-action thermostat operate?

12. When is a double-throw snap-action thermostat used in an electric water heater?
13. Define an immersion heater.
14. What is an electric time switch? Why might this device be required in an electric water heater installation?
15. How can an electrician determine whether a separate watt-hour meter is required with an electric water heater?
16. Where should a fused safety switch for an electric water heater be installed?
17. Why must metal frames of electric water heaters always be grounded?
18. How does an electrician provide a ground connection in rural areas?
19. Define a polarized electrical device.
20. Why are polarized plugs and receptacles being used extensively for all appliances?
21. How is a polarized adapter connected to a standard receptacle?
22. Discuss several types of polarized receptacles.
23. What is the function of heating cable?
24. Where does an electrician usually install heating cable?
25. Why is heating cable enclosed in a lead sheath?

Chapter 14

Alarms and Intercoms

Alarms and intercoms are common, useful, and important technologies. These are also technologies that change continually. Formerly they featured vacuum tubes and relays; now they are completely solid-state. Intercoms were previously hard-wired and are now being replaced with wireless systems. Many alarm systems are using wireless technology as well.

Alarm Devices

Electricians often install various types of alarm devices. For example, most business establishments have a burglar alarm system. Infrared light units (Figure 14-1) are in wide use. These are similar to photoelectric bell-control units (Figure 14-2) except that an infrared light source is used, which is practically invisible. The receiver unit controls an electric bell, which is usually mounted on the outside of the building so that it will attract the attention of police and passersby. Installation must be made in accordance with applicable electrical codes. Instructions for complete electrical connections are generally included with the equipment.

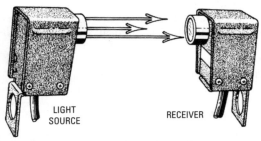

Figure 14-1 An infrared alarm device.

Other common alarm systems employ a wiring system with contacts that automatically open or close when a door or window is opened. These are classified into the closed-circuit and open-circuit systems, as shown in Figure 14-3. The alarm bell may be in the building with the installation, or it may be located at a remote point such as a guardian station or police station. The more elaborate systems include annunciators to indicate the office, shop, or room that has been entered. A simple annunciator wiring system with relays

360 Chapter 14

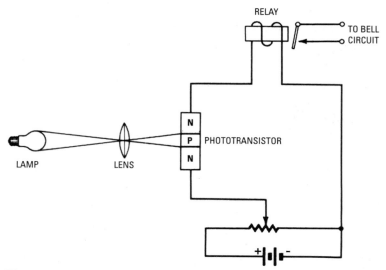

Figure 14-2 Phototransistor bell-control circuit.

to close lighting circuits, test switches, master switch, and bell is shown in Figure 14-4.

Many offices and shops are equipped with fire and/or smoke alarm systems. These installations include a heat sensor such as a thermostat or a number of thermostats located at suitable points and connected to bells, annunciators, or both. Relays are used when the thermostat must switch a substantial amount of power. Smoke alarms are similar to the photoelectric systems used with doorbell installations. When the light beam is reduced in intensity by smoke, bells and/or annunciators are energized. The alarm indicators may be either locally or remotely located.

Photoelectric systems are also used to automatically switch night-lights on and off in business institutions and factories. They are occasionally installed in residence wiring systems. Night-lights are switched on after sunset and switched off after sunrise. Wiring instructions are included with the equipment, and the electrician must make sure that the installation complies with the electrical code for that particular area.

All of these systems, regardless of how simple or complex, can be divided into three basic parts:

Sensors. These are the devices that sense or respond to certain conditions in and around the protected area. (There are many types of sensors for many different applications.)

Alarms and Intercoms 361

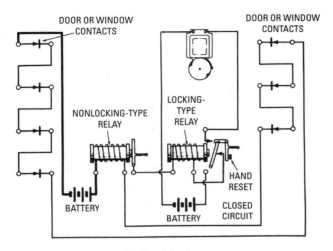

(A) Closed circuit.

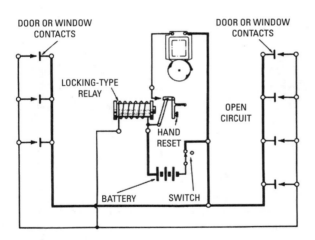

(B) Open circuit.

Figure 14-3 A burglar alarm system.

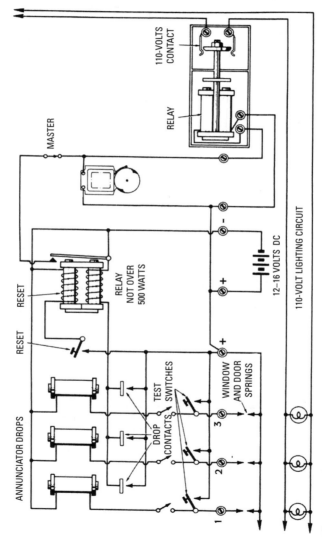

Figure 14-4 Burglar alarm circuit having annunciator with relay.

Controls. The brains of the system are these controls that respond to the input from the sensors according to the desires of whoever set up the system.

Signaling devices. These devices give out some type of signal (typically a siren or buzzer) when an alarm condition is reached.

The primary control for a security system is a central control panel. This panel contains the electronic processing equipment that makes the system operate. While some systems still incorporate electromechanical relays and controls to carry out their operations, they are rapidly changing over to totally solid-state systems. Of course, in any security system, there are parts that must be electromechanical, but those parts are kept to a bare minimum in newer designs. The new designs provide for a much more thorough system, with many more conveniences than the older ones.

One of the major advantages of the new systems is that they are far more intelligent than their predecessors. These controllers supervise the security system; many of them will check twice to make sure that an alarm signal is not a false alarm. As with so many of our new innovations, this capability is based on microprocessor technology. In effect, the controller is a miniature computer. Not only does it receive and react to various inputs, but it also analyzes the data and checks it twice to make sure that it is correct. Some of the most advanced systems can even detect and ignore faulty sensors. Not coincidentally, some of these systems call their control panel a supervisor panel. Most of these identify problems and perform self-checks.

Another difference with the new systems is that they need to be programmed after they are installed, much the same as a computer is programmed. This programming sets up the system for the owner's specific needs. For example, all of the perimeter detectors except for door switches could be programmed so that they are operating at all times, whether the facility is occupied or not. The interior detectors, however, would be activated only when the facility is not operating. Fire detectors would be on at all times.

One other important setting for these systems is a time delay that begins when the system is armed and expires when the person arming the system must be out of the building. Usually, a time delay of 45 seconds to $1\frac{1}{2}$ minutes is allowed for the person arming the system to get out the door. Remember that these specific times and arrangements are used only for examples; these systems can be programmed in any way desired.

Once the system is installed and programmed, it is usually controlled by a keypad. The keypad puts digital electronic commands into the system and has several colored LED lights to monitor the various stages of the system's operations. In order to control the system through the keypad, a code number must first be entered. This is, of course, done for security reasons so that a person who doesn't know the secret number cannot give any instructions to the system. There can also be more than one keypad used. One keypad is almost always installed just inside the front entrance so that it can be quickly accessed when entering or leaving. Many systems allow for up to 10 separate arming and disarming codes.

A standby battery is also an important part of a security system's control. This battery provides power to the system in the event of a power outage, which is a time when a lot of break-ins typically occur. These batteries are usually installed inside of the controller and provide power for up to 72 hours when the normal power goes out. The batteries recharge themselves and are rated for 4 to 5 years of service.

Other control devices include the following:

Keypads. The most common method of accessing a controller is through a keypad. This device allows you to see system status, arm and disarm the system, make user code changes, or reconfigure programming to meet a specific need or function of the alarm. Keypads are either LED or LCD types. LED types have only lights that tell the user what is going on. LCD-style keypads have a display that informs the user or technician what is going on with the alarm system: status, alarm memory, program data values, and so on.

Power Supplies. The power supply provides the necessary voltage to operate not only the controller but also the keypads, protective devices, and signaling devices. Most alarm panels are connected to line power through a transformer that takes 120 V AC and converts it to 12–16 V AC. The power supply further converts the 12–16 V AC to 12 V DC. The power supply provides constant auxiliary power to all system components and provides a charging circuit for the system backup battery (or batteries). In the event of AC power failure, the backup battery instantly takes over the duty of providing power to the entire alarm system. The standby service time is determined by the total current draw of all components against the capacity of the backup battery. If the system were to go into alarm during a power failure, service time would be greatly reduced due to the extra drain on the battery (or batteries) from bells, sirens, or other warning devices.

Monitored alarm systems should always be programmed to report power failure and low battery conditions. When power is restored, the power supply will automatically recharge the battery. The shelf life of a rechargeable battery is 3 to 5 years. Excessive power failure or deep discharge due to power failure may require more frequent battery replacement. Routine testing under full load of your battery backup system is essential to proper operation of your alarm in an emergency.

Some of the newest and most popular security systems are wireless systems. Conventional security systems send signals from the sensors to the controller via copper wires. The wireless systems use radio communication instead of wires. The benefit gained by going wireless is that the system is less labor intensive to install. In particular, it is the cost of installing cables that is expansive, especially in existing homes. By sending signals via radio waves, this cost is completely avoided. Although a good deal of money is saved by avoiding the costly installation of wires, the wireless devices cost quite a bit more than the conventional devices. The reason is that each device must have two parts, a sensor and a transmitter. Additionally, they must have some type of built-in power source. Because of this, the wireless devices are not only more expensive but also physically larger than the standard components.

Wireless window switches are designed to be wired directly into a transmitter, which then sends a signal to the receiver when necessary. Most systems have the receiver built right into the control unit, although some use a separate receiver. Sometimes separate receivers are required for outdoor or special transmitters. A typical transmitter has a range of about 200 feet. If the transmitter is farther than this from the main receiver, a separate receiver must be installed within the transmitter's range.

Intercoms

An intercommunication (intercom) system provides speaker communication between two or more locations. Figure 14-5 shows an older style of master station base for a modern intercom set. A single amplifier using transistors is generally employed in an intercom set to amplify the voice signal so that it can be heard on a speaker at the receiving end. Thus, an intercom system is basically a simplified and private telephone system.

System Planning

The station at which the amplifier is located is called the *master*, and each remote station is called a *substation*. Figure 14-6 shows

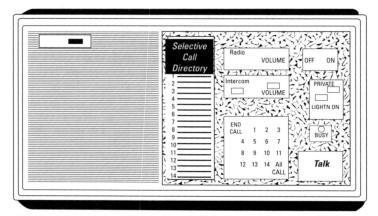

Figure 14-5 Typical master station base for intercom.

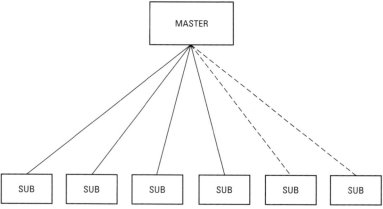

Figure 14-6 System employing one master and any number of subs.

the plan of a master station connected to a number of substations, while Figure 14-7 shows the plan of a complete master installation for all stations. In other arrangements, some stations may be master installations and some may be substations. In *nonprivate* installations, the master station's push-to-talk switch functions for both the master and the substation. Thus, only the master location can originate a call and listen in on a selected substation.

A nonprivate installation is commonly used for a front-door station. When the doorbell rings, the master station can be switched to

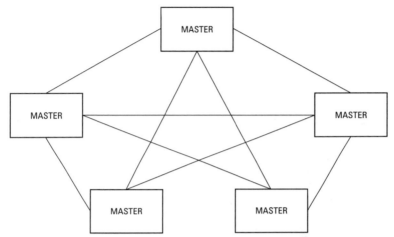

Figure 14-7 Intercom system using master units at each sub.

the front porch to allow one to carry on a conversation with a visitor before opening the door. Modern home intercom systems may be expanded and diversified to operate as music distribution systems, security installations, answering services, and numerous other functions. With the development of ever smaller and more complex silicon chips, the boundaries of such systems seem endless. One rather modest installation is illustrated in Figure 14-8. These elaborate systems employ better-quality speakers for improved music fidelity and usually a switching function at each substation to permit selection of a particular music channel. Deluxe systems may even include a dual-speaker installation in each room for stereo reproduction of music.

You must first familiarize yourself with the equipment that you plan to install and then discuss its details with the office manager or homeowner. After determining the number and location of stations, a floor plan is drawn up, such as the one shown in Figure 14-9. This is a plan for a sales office and stockroom; however, the same type of plan is drawn up for a residence, apartment building, or rural installation. Keep in mind that the master station will need a source of AC power (unless a battery-operated transistor system is used).

The *National Electrical Code* requires that interconnecting cable be run not closer than 2 inches to any AC line, even if the line is enclosed in conduit. This not only is a safety precaution but also helps to minimize AC hum pickup. Interconnecting cables should

368 Chapter 14

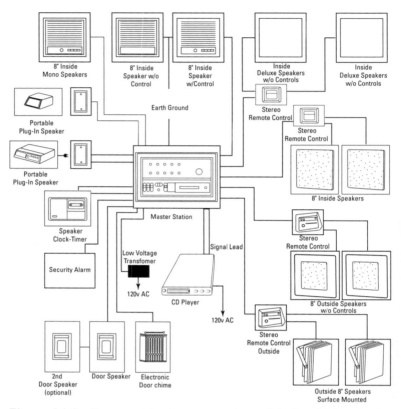

Figure 14-8 System wiring (representation only) of a middle-price intercom system with typical add-on features. These complex systems include complete wiring diagrams with instructions for installing the system.

be run away from telephone wires to avoid pickup of telephone signals. Interference due to stray field pickup can be minimized by using intercom sets that work into low-impedance lines.

There are various other factors to be considered in the layout of an intercom system. For example, the electrician must observe the construction of the building or house in which the installation is to be made. A brick house may have two courses of brick and only $3/4$-inch furring strips between the inner layer of brick and the plaster wall, which gives little space through which to draw cable. Only the inside walls, constructed of $2'' \times 4''$ studs, have adequate clearance

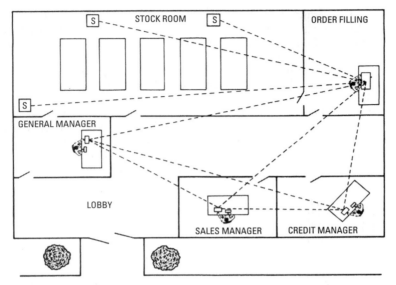

Figure 14-9 Floor plan showing position of stations and communication between each station.

for cables. A brick veneer house has one course of brick, with 2″ × 4″ studs in the outer as well as the inner walls. Frame houses have 2″ × 4″ studs in all the walls.

Connecting an intercom system from room to room requires running the cables up the space between studs to the attic, across the attic, and then down between the studs in another wall. Another method is to run the cable down into the basement or crawl space, across and up into the wall between the studs. The bottoms of the studs are usually nailed to 2″ × 4″ sole plates, which in turn are nailed to the flooring. Therefore, the electrician must drill through both the sole plate and the flooring to get into the space between studs. Similar construction is used at the tops of the studs so that about the same amount of drilling is necessary to go up to the attic and over.

The most difficult part of the installation is running the cable from floor to floor in a two-story house. This kind of installation requires removing the baseboard on the second floor and drilling down to the space between the studs in the room below. Outlet holes to the stations should be drilled through the baseboard. A suitable snake is usually required to draw the cables. After the snake has been run through the space for the intercom wires, the wires are twisted through the hook and secured with tape. Then, the snake is pulled out, drawing the wires with it.

Installation is much easier in homes or buildings under construction. If junction boxes are used, leave about 3 inches of intercom cable hanging out for final connection. On the other hand, if the cable is to be run directly to the stations, it is necessary to allow sufficient length according to the floor plan. Roll up the excess cable close to where it comes out of the baseboard and enclose it in a plastic bag, as shown in Figure 14-10. This precaution keeps the cable clean during subsequent plastering and painting.

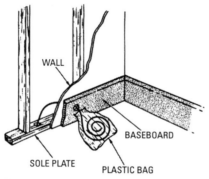

Figure 14-10 A plastic bag protects cable.

Planning Details

It is necessary to determine whether the complete intercom installation is to be in-wall or on-wall, or whether only the master station is to be built in with the substations on-wall. For built-in installations in new homes, the electrician must discuss the plan with the plasterer so that he or she will know where the wall outlets are to be located. Otherwise, the electrician must return to make the cutouts after the plastering is completed and before painting is started. Most in-wall intercom stations hang from the plaster or wallboard, but some of the heavier units will require wood framing for support. Some intercom systems use metal boxes that can be put in by the plasterer.

If a front-door station is to be part of an intercom system installed in a brick house, it must be put in when the bricks are laid. Therefore, the architect should include the box in his masonry plans. When a single in-wall master amplifier is centrally located and is not a part of the master station, the electrician must provide an AC outlet where the amplifier is to be located. This should be a behind-the-wall outlet for a permanent type of connection. The same consideration also applies to in-wall master stations with a built-in amplifier.

Since the signal voltages from intercom amplifiers are low, it is usually adequate to install open wiring. However, on long runs that pass through strong stray fields, such as in heavy manufacturing areas, a conduit installation may be required to minimize noise interference. The basic intercom cable used between a master station and a substation has three conductors. In many systems, one wire is a common line, with the second wire used for incoming signals and the third wire used for outgoing signals. In any case, the electrician should check the installation instructions provided with the intercom system to determine the number of conductors that will be required.

For installation convenience, most intercom systems are designed to use multiconductor cable that can serve a number of stations. Some intercom systems are designed for use with *balanced* lines and use *twisted pairs*. Note that neither side of a balanced line is grounded. Figure 14-11 shows examples of balanced and unbalanced line connections. Other intercom systems use shielded wire for the incoming lead; however, most systems do not use shielded wire. Unless an electrician has had previous experience with a particular system, it is advisable to use the cable specified by the manufacturer even if it costs more than ordinary cable. This precaution will ensure against unsatisfactory installations.

A good intercom system has reserve power so that an additional speaker can usually be operated in parallel with the original speaker if desired. As explained previously, maximum signal power transfer is obtained when the speaker load matches the source impedance. The impedance of the intercom line is seldom a matter for concern because the lines are comparatively short with respect to voice signal wavelengths. In an ideal installation using long lines, in which maximum fidelity and maximum power transfer are desirable, both the source and load impedances should be the same as (i.e., should match) the line impedance. Coaxial cable is supplied in a wide range of impedances, from about 50-ohms to 200-ohms. Twisted cable is also available in a wide range of impedances.

Wireless Intercom Systems

A wireless intercom requires no cable runs between stations. The 120-volt wiring system is used to conduct the voice signals from one

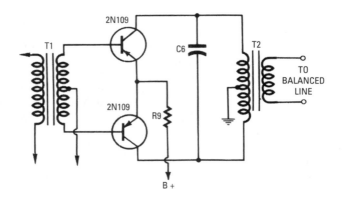

(A) Balanced.

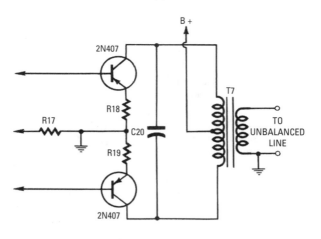

(B) Unbalanced.

Figure 14-11 Illustrating balanced and unbalanced outputs.

station to another. This is done by means of a comparatively high-frequency *carrier* current that can be easily separated from the low-frequency 120-volt current. A typical all-master wireless intercom system can use up to 12 master units (see Figure 14-12). Each master can originate calls to any other master. It can also receive calls on any channel (carrier frequency). By switching several masters to the same channel, a conference system is provided.

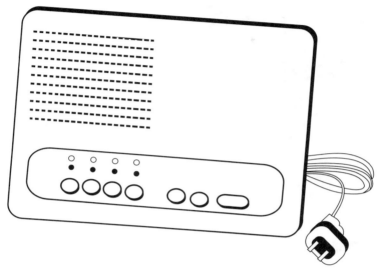

Figure 14-12 A wireless intercom master unit.

A 12-station system can accommodate six separate conversations simultaneously. Auxiliary stations (see Figure 14-13) are used in locations where calls need not be originated but where replies must be made. An auxiliary station operates on only one channel and hence provides communication with a single master location. Although a wireless intercom system is very easy to install (each unit is merely plugged into a convenience outlet), the system has certain disadvantages. The chief disadvantage is a tendency to reproduce line noises and hum. Interference can be minimized by careful adjustment of the signal filters in the units. However, complete elimination can seldom be obtained.

Wireless intercom systems occasionally pick up broadcast signals of operator code signals from nearby radio stations, with resulting low-level background interference. The interference level often changes when switches in the 120-volt line are opened or closed and when various numbers of intercom units are switched into operation. Another disadvantage of a wireless intercom system is that communication is limited to the 120-volt circuit supplied by the local line transformer. In other words, high-frequency carrier currents are stopped by a line transformer.

374 Chapter 14

Figure 14-13 A wireless intercom auxiliary unit.

Privacy cannot be obtained with ordinary wireless intercom systems. If an apartment building is powered by a single line transformer, all occupants can hear any conversation with a suitable intercom unit. A wireless intercom system can be compared to a party-line telephone system in which a subscriber can listen in on conversations between other subscribers. Although elaborate scramblers can be used, these are quite expensive and are seldom installed.

Music Distribution Systems

A music distribution system is nearly always an in-wall installation that requires additional plaster cutouts, not only for the control amplifier/tuner/record player but also for the speakers in each room. It is often desired to have patio or other outdoor speakers installed with a music distribution system. In such cases, suitable cables must be run to the outdoor locations.

The electrician must plan an outdoor installation in the same way that he or she would plan an outdoor lighting installation. Exposed receptacles and units must be of the weatherproof type. All applicable electrical codes should be checked before a wiring plan is finalized. If a homeowner wishes to install a music distribution system that uses a radio or high-fidelity source that is not designed for cable operation, the electrical requirements should be carefully drawn up by a highly experienced electrician or a specialist. Otherwise, the completed installation may leave much to be desired.

Industrial Installations

Intercom installations in offices or plants are often easier than ones in homes. The manager will often permit the electrician to tack the cables on the outside of walls. This type of intercom wiring will usually meet electrical codes. However, the electrician must run the wires through the walls to get from one room or office to another. Either a star drill and hammer or an electric drill with a carbide-tipped bit can be used. This part of an intercom installation follows the same methods used in a lighting installation, where cables are run up the space between the wall studs to the attic, across the attic, and then down between the studs in another wall.

When an installation is made in a plant, it is good practice for the manager and electrician to go over the entire area carefully during working hours. The noise levels can be noted, and in turn, an experienced electrician can give good advice concerning the number of stations that should be provided and the best location for each. This precaution is good insurance against misunderstandings that can result from installations that do not provide the intended utility and convenience for communication.

School Installations

School intercom installations usually require a system with an all-call function so that the principal can make a general announcement to all classes at the same time. This added function requires a few changes from ordinary cable runs, but installation is otherwise the same as in homes or small offices. Note that schools may have their own electrical codes, which go beyond the requirements of codes applying to homes and offices. For example, intercom cables may have to be installed in conduit to meet a school code. Therefore, it is necessary to check with the school board before finalizing a school wiring plan.

The *National Electrical Code* states that intercom conductors may be run in the same shaft with light and power conductors if the intercom conductors are separated at least 2 inches or where the conductors of either system are encased in noncombustible tubing. Conductors bunched together in a vertical shaft must have a fire-resistant covering capable of preventing fire from being carried from floor to floor, except where the conductors are encased in noncombustible

376 Chapter 14

tubing or when they are located in a fireproof shaft that has fire stops at each floor.

Bells and Chimes

Electric bells, chimes, or buzzers operate from comparatively low voltage, such as 6 volts. Either AC or DC may be used, although AC operation is most common. Figure 14-14 shows a basic bell circuit using a step-down transformer. Note that the primary of the transformer is permanently connected to a 117-volt AC supply line. The secondary circuit is closed by a pushbutton to ring the bell. The primary has substantial inductance so that little idling current is drawn by the transformer. A bell-ringing transformer is designed to have considerable magnetic leakage so that it will not burn out in case of a short circuit across the secondary. For example, if circuit points 2 and 3 become short-circuited, there would be no danger of fire damage.

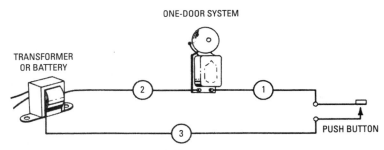

Figure 14-14 An electric bell circuit.

A door chime is wired in the same way as a bell. In the case of a two-chime unit, such as the one shown in Figure 14-15, a three-wire installation is used. The transformer is the same as for a single bell or chime. However, the secondary terminals are connected to two branch circuits controlled by pushbuttons at individual doors. One branch circuit consists of conductors 2, 3, and 5; the other branch circuit consists of conductors 1, 3, and 4. Each chime has a distinctive tone to indicate which pushbutton has been operated.

A tone signal system is shown in Figure 14-16. This is the equivalent of a three-tone chime system, but it employs a simpler wiring installation. The unijunction transistor operates in an oscillator circuit that supplies a tone signal to the speaker (or speakers). Note that a different value of resistance is connected in series with each

Alarms and Intercoms 377

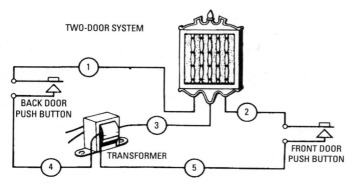

Figure 14-15 A two-door chime circuit.

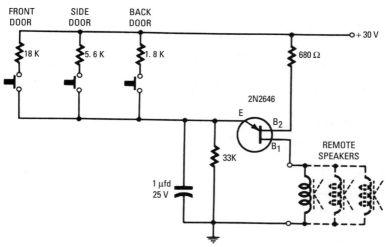

Figure 14-16 A three-tone signal system.

pushbutton. This results in a distinctive tone, depending on which button is operated. This system requires a DC supply, usually obtained from a transformer followed by a selenium rectifier and filter.

Bells, buzzers, chimes, and signal systems must be installed in accordance with applicable electrical codes. Therefore, the electrician should check with local codes and with any special codes such as those specified by schools and other public institutions. Hospital installations are comparatively complex and require careful planning.

Special types of bells are used in outdoor installations, such as schoolyards, and are usually connected to electric timers.

Photoelectric Control of Bell Circuits

Electric bells installed in small business offices are often controlled by a photoelectric circuit, such as the one shown in Figure 14-2. When a person passes through the doorway, a light beam is interrupted, which causes the relay contacts to close. In turn, the bell circuit is closed and the bell (or chime) rings. The phototransistor is a device that changes light energy into electrical energy and produces amplified current flow in the relay circuit. When the light beam is interrupted, practically no current flows in the relay circuit. The potentiometer is a sensitivity control that is adjusted to obtain proper operation at the prevailing level of ambient light.

Phototransistors operate at comparatively low voltages. Although a battery is shown in Figure 14-2 for simplicity, a commercial unit usually has a transformer, selenium rectifier, and filter to provide the DC supply. Installation of a photoelectric bell-control system must be made in such a way that excessive ambient light cannot strike the lens in front of the phototransistor. This often requires a hood or tube to be mounted around the lens so that most of the light that enters is provided by the lamp.

Summary

Various alarm devices are used in homes and businesses. One of the most common sensors is the infrared light beam, though magnetic switches and a variety of seismic and glass-break sensors are very widely used. Infrared units are similar to the photoelectric cell except that an infrared light source is invisible.

Other common alarm systems employ a wiring system with contacts that automatically open or close when a door or window is opened. Many offices, shops, and homes are equipped with fire or smoke alarm systems. When the light beam is reduced in intensity by smoke, bells are energized, setting off the alarm system.

An intercom provides speaker communication between two or more persons at different locations. An intercom system is basically a simplified private telephone system.

The location that houses the amplifier is called the master unit, and each remote location is called a substation. Each intercom system is designed to carry any number of substations, depending on individual requirements.

A nonprivate installation is commonly used for a front-door station. When the doorbell rings, the master station can be switched to the front door to allow one to carry on a conversation with the visitor before opening the door. A home installation can also be operated as a music distribution system to various rooms. In this case, a radio tuner and/or CD player would be added. These elaborate systems use better-quality speakers for improved music fidelity. Deluxe systems even include a dual-speaker installation in each room for stereo reproduction of music.

There are various factors to be considered when laying out an intercom system. The installer must observe the construction of the building in which the installation is to be made. Connecting the intercom from room to room requires running the cable between wall studs, across the attic, and under floors. Installation is much easier in homes or buildings under construction.

It is necessary to determine whether the system is to be an in-wall or on-wall installation. If the front door is to be included in the system, a substation must be put in when the bricks are laid or wood siding is installed. A good intercom system has reserve power so that additional speakers can usually be operated in parallel with the original speakers.

A wireless intercom requires no cable runs between stations. The 120-volt wiring system is used to conduct the signal from one station to another. The chief disadvantage is a tendency to reproduce line noises and hum. Interference can be minimized by careful adjustment of the signal filters in the unit. Wireless intercom systems can also pick up stray broadcast signals from nearby radio stations.

Test Questions

1. Explain the function of an intercommunication system.
2. What is the difference between a master station and a substation?
3. How does a music distribution system operate?
4. Discuss the difference between a private and a nonprivate intercom installation.
5. Explain how an AC power supply operates.
6. What type of cable is commonly used in an intercom installation?
7. Why should the electrician check with the plasterer before an installation is made in a new home?
8. Describe the operation of a wireless intercom system.
9. What is the chief advantage of a wireless intercom system?

10. Why does a line transformer block intercom signals?
11. State some precautions to be observed in the installation of an industrial intercom system.
12. What special requirements may an electrician encounter in a school installation?
13. Explain the operation of the three-tone signal system.
14. How does a phototransistor bell-control system operate?
15. Why is a sensitivity control required in a photoelectric bell-control circuit?
16. Where would an infrared alarm device ordinarily be installed?
17. Explain the difference between a closed-circuit and an open-circuit alarm system.
18. What is an annunciator?
19. How does a fire alarm system operate?
20. How does a smoke alarm system operate?
21. Why are relays used in many alarm systems?
22. Explain how night-lights are automatically switched on and off.

Chapter 15
Generating Stations and Substations

A *generating station* is a plant where electric energy is produced from some other form of energy by means of suitable apparatus. An *automatic station* is a station (usually unattended) that, under predetermined conditions, goes into operation by an automatic sequence and that maintains the required type of service; it goes out of operation by automatic sequence under other predetermined conditions and provides protection against operating emergencies. An automatic station may be either an automatic generating station or an automatic substation.

An electric power *substation* is an assembly of equipment used for purposes other than generation or utilization, through which electricity passes for the purpose of switching or modifying its characteristics. Service equipment, distribution transformer installations, and other minor distribution equipment are not classified as substations. A substation is of such size that it includes one or more buses and circuit breakers and usually serves as a receiving point for more than one supply circuit or sectionalizes the transmission circuits that pass through by means of circuit breakers.

Generating Stations

The plan of a generating station deals with the installation of generating units and all the mechanical and auxiliary electrical equipment required to produce mechanical energy, change the mechanical energy into electrical energy, and deliver the electrical energy to the associated system. There are four classifications of generating stations:

1. Gas power
2. Oil power
3. Water power (hydroelectric)
4. Steam generating

However, the electrical features are similar in all four classes of stations. Figure 15-1 illustrates a simplified section of a small hydroelectric generating station. Figure 15-2 shows the plan of a generator and directly connected turbine installation. The turbines, alternators, exciters, and controlling switchboard are housed in one large

382 Chapter 15

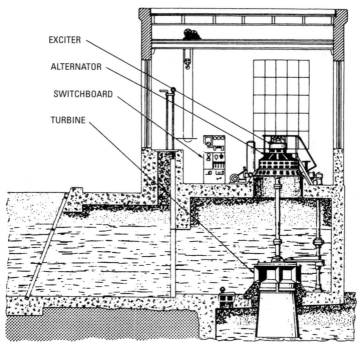

Figure 15-1 A small hydroelectric generating station.

room of a hydroelectric generating station. Prime movers are arranged in a single row to simplify the penstock and tailrace design. The *turbine* is a rotary engine actuated by water under high pressure. An *exciter* is an auxiliary generator that supplies energy for the field excitation of another electric machine. A *prime mover* is an initial source of motive power that is applied to other machines. Figure 15-3 shows a cutaway view of a water reaction turbine.

The exciter voltage is 250 volts in large plants and 125 volts for smaller installations. An exciter is generally a part of the generator assembly, as seen in Figure 15-1. However, exciters driven by separate prime movers are also used. Generators that supply voltages up to 22,000 volts are used. Power outputs range up to 125,000 kilovolt-amperes (kVA). Armature speeds range from 10 to 3600 rpm. Provisions are made to avoid damage in case of accidental short circuits and to prevent overheating.

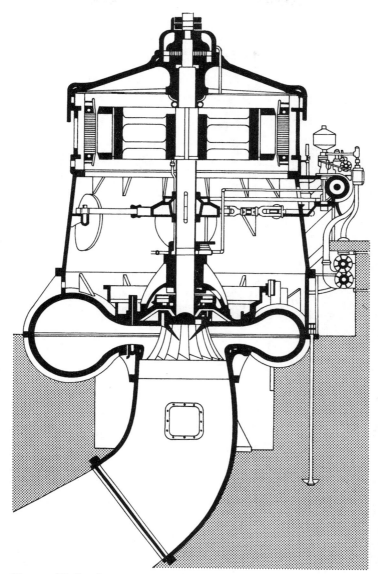

Figure 15-2 Cutaway view of a typical generator.

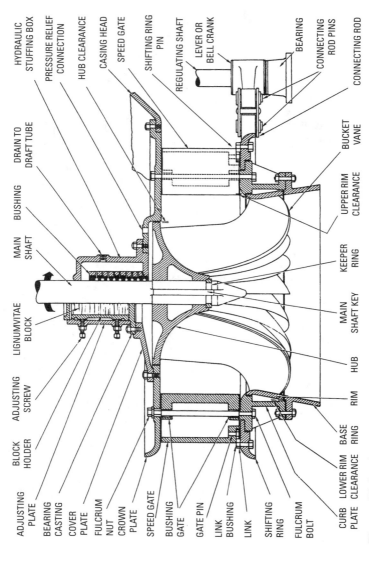

Figure 15-3 Cutaway view of a water reaction turbine.

Single-Bus System

The main electrical connections in a generating station depend on the size and type of generator, the feeder arrangement, the function of the generating station in the system, and so on. Figure 15-4 shows the simplest connection arrangement, called the *single-bus system*. This system is used only in small generating stations where the possibility of service interruptions is permissible. It provides adequate switching facilities and protection of apparatus in case of failure, but it lacks flexibility. If any alternator circuit fails, the corresponding machine and circuit breaker must be withdrawn with its feeder circuits. The feeders are taken off between the alternators, and in case of insulation failure of a bus bar support, a complete shutdown is necessary until the defect has been corrected.

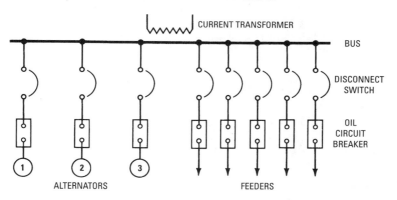

NOTE: Bus has a bar construction.
Disconnect switches are air switches.
Feeders connect generating stations or a generating station and a substation.

Figure 15-4 Single-bus, single-circuit-breaker system.

Circuit Breakers

By use of sectionalizing disconnecting switches in the bus bars of a single-bus system, a complete shutdown of the station can be partly guarded against. A circuit breaker is a device for interrupting a circuit between separable contacts under normal or abnormal conditions. Circuit breakers are generally required to operate only infrequently, although some types are suitable for frequent operation. *Normal* operation means the interruption of currents not in excess of the rated continuous current of the circuit breaker. *Abnormal* operation means the interruption of currents in excess of rated values, such as those caused by accidental short circuits.

An oil circuit breaker has its contacts immersed in an oil bath. On the other hand, an air circuit breaker has its contacts exposed to surrounding air. A small air circuit breaker is shown in Figure 15-5. The plan of a small oil circuit breaker is shown in Figure 15-6A. Any circuit breaker must be designed to open the circuit rapidly and to quickly kill the arc between its contacts. Elaborate means are provided to extinguish arcs in high-power circuit breakers. For example, a high-velocity blast of gas may be used to blow out an arc. Figure 15-6B shows the appearance of a vacuum circuit breaker. The vacuum circuit breaker is a comparatively recent development.

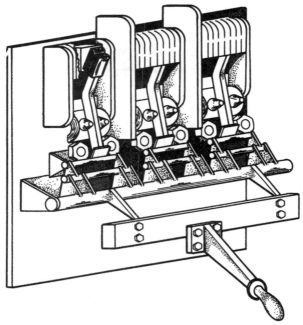

Figure 15-5 Typical air circuit breaker.

Instrument Transformers

An *instrument transformer* is a special type of step-down transformer used for connection to a voltmeter or ammeter. A current transformer has its primary winding connected in series with the circuit in which current is to be measured. Its secondary winding is connected to an ammeter. On the other hand, a potential or voltage transformer has its primary connected across the circuit in which

Generating Stations and Substations 387

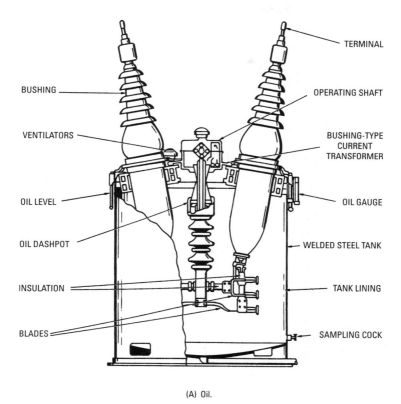

(A) Oil.

Figure 15-6 Small oil and vacuum circuit breakers.
(Courtesy Westinghouse Electric Corp.)

voltage is to be measured. Its secondary is connected to a voltmeter. Figure 15-7 shows the connections for current and voltage transformers. Instrument transformers are used for safety reasons, and the secondary circuit is always grounded. It is very important never to open the secondary circuit of a live current transformer because of the high induced voltage.

Double-Bus System

To obtain greater flexibility and to reduce the possibility of service interruptions, two buses with additional disconnect switches or manually operated oil circuit breakers may be used in small generating stations. One bus is called the main and the other is called the spare bus, as shown in Figure 15-8. Ordinarily, the spare circuit

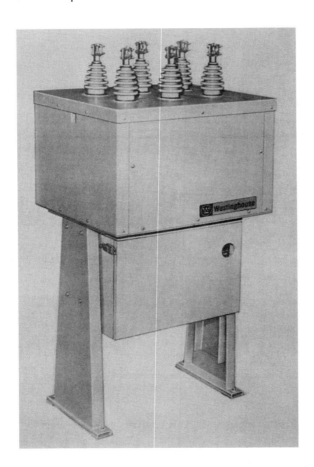

(B) Vacuum.

Figure 15-6 (*continued*)

breaker and spare bus are not used. However, if one of the regular circuit breakers develops a fault, it can be switched out of the circuit for repair or replacement, and the spare circuit breaker can be temporarily switched into the circuit. However, if a fault occurs in the bus insulation, complete station shutdown is necessary until the fault is corrected.

Another connection arrangement that employs two buses is shown in Figure 15-9. This is called the double-bus, single-breaker

Generating Stations and Substations **389**

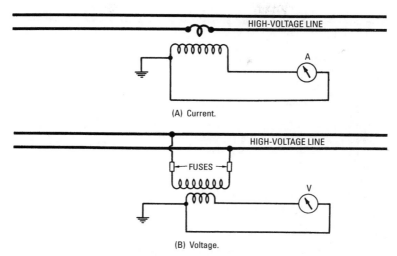

Figure 15-7 Current and voltage transformer connections.
(Courtesy Westinghouse Electric Corp.)

system. It practically eliminates the possibility of a prolonged shutdown in case of a bus failure. It also permits continuance of service when maintenance or repair work is required on either bus. However, a service outage is unavoidable in case of a fault on the corresponding circuit breaker. Its principal advantage over the

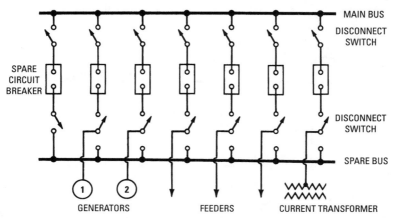

Figure 15-8 Spare-bus, spare-circuit-breaker system.

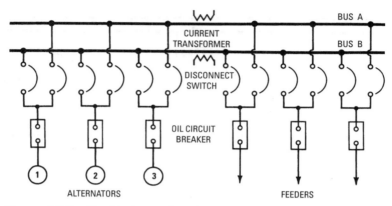

Figure 15-9 Double-bus and single-circuit-breaker system.

single-bus, single-circuit-breaker system is that when a feeder trips out, it can first be tested on the other bus before it is returned to normal service on the first main bus. This system is used in small generating stations.

Quite often, a tie-bus breaker is provided, as shown in Figure 15-10. It facilitates line testing or quick transferring of power from one bus to the other. With the tie-bus breaker closed, thereby energizing both sets of bus bars, the transfer of a circuit carrying power from one to the other can be accomplished without danger of interruption to service by means of the disconnecting switches. Compare the tie-bus system in Figure 15-10 with the spare-bus,

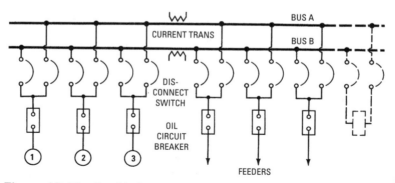

Figure 15-10 Double-bus system with the tie bus in dotted line.

spare-circuit-breaker system shown in Figure 15-8. The tie-bus system is also confined chiefly to small generating stations.

Maximum flexibility is provided by the double-bus, double-breaker system, shown in Figure 15-11. It has all the advantages of the double-bus, single-breaker system, with added assurance against shutdown of any particular circuit due to circuit breaker faults. It is a comparatively expensive arrangement, used chiefly in large generating stations where continuity of service is of prime importance and the high cost is justified. Electricians in generating stations also work with other arrangements, which differ slightly from the double-bus, double-breaker system and the single-bus system.

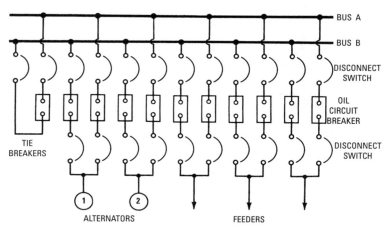

Figure 15-11 Double-bus, double-circuit-breaker system.

The variation in connection systems has the purpose of providing the degree of flexibility that is warranted by local conditions, while minimizing the cost of the system. For example, Figure 15-12 shows a modification of the double-bus, single-breaker system called a ring bus. The two buses are tied together by means of bus-tie circuit breakers and disconnecting switches. In the H system, two feeder circuits are served from a pair of selector switches to either of two buses, as shown in the examples of Figure 15-13. This arrangement requires two breakers per feeder and three breakers per generator.

For systems that distribute all or part of their power through step-up transformers, we will find more complex bus-bar connection arrangements. For example, Figure 15-14 shows a service arrangement

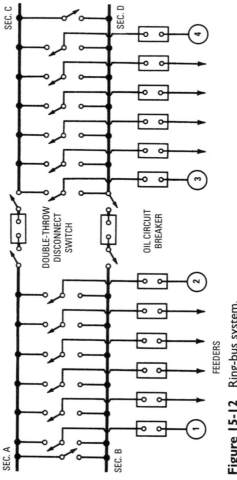

Figure 15-12 Ring-bus system.

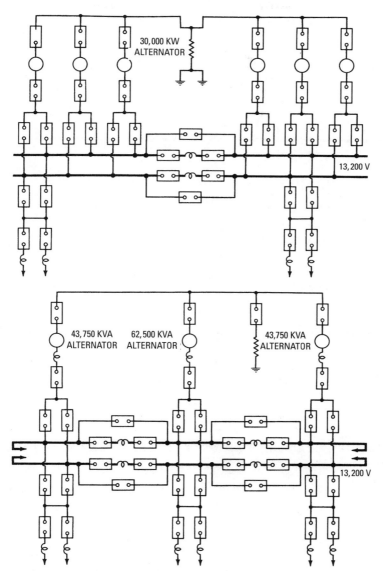

Figure 15-13 A typical H system.

394 Chapter 15

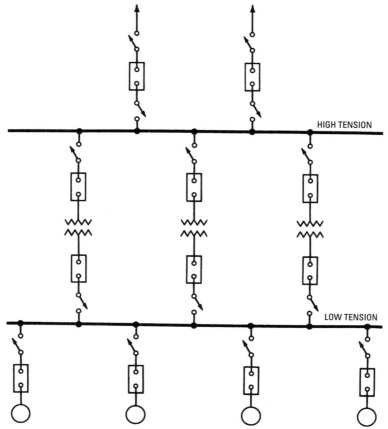

Figure 15-14 System using single low-tension and single high-tension bus.

from a single low-tension bus to a single high-tension bus. High-tension (high-voltage) buses are constructed from copper, aluminum, or galvanized-steel tubing, bars, or other shapes and are supported by insulators. Large buses are surrounded by fireproof barriers with considerable clearance space. Smaller buses may be mounted on a pipe or structural steel framework located in an open space.

A step-up transformer arrangement using double buses and double breakers on both high- and low-tension circuits is shown in

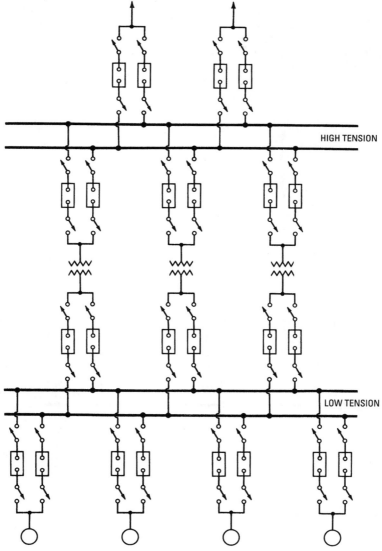

Figure 15-15 Bus system using double buses and double circuit breakers.

Figure 15-15. If the station is located at some distance from the load center, particularly in the case of a hydroelectric station, a commonly used bus arrangement is employed as shown in Figure 15-16. Each alternator and step-up transformer is treated as a unit, and all power is transmitted over two or more lines. The load center (or load area) is not a generating station but a substation fed from one or more generating stations.

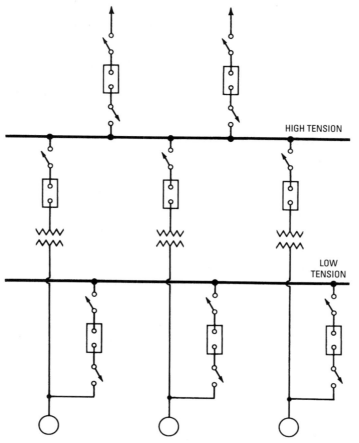

Figure 15-16 Common arrangement when station is at a distance from the lead center.

Paralleling Generators

Auxiliary power for the station is obtained from a low-tension bus that may be connected to any one of the alternators. Normally, alternators are not paralleled on the low-tension bus. This arrangement is economical but lacks flexibility. Let us briefly consider some of the factors that are involved in parallel operation of generators. The generator voltages must be equal, and their instantaneous polarities must be the same. Generators connected in parallel must operate in phase with each other. Electricians describe this requirement as *phasing out* and *synchronizing*.

Three-phase generators are commonly used, with individual phases identified as A, B, and C. *Synchronizing* means bringing the generators into step so that their positive and negative alternation peaks occur at exactly the same time. If the alternators were connected in parallel when the phase difference is more than a few degrees, heavy current surges would flow in the conductors between the alternators. Hence, phase-indicating instruments are used to determine when a pair of generators are synchronized and may be switched into a system for parallel operation.

A phase-angle meter is illustrated in Figure 15-17A. The dial is calibrated in electrical degrees and corresponding power factor. Circuit no. 1, which is generally employed as the current circuit, has rated ranges of 1, 3, 10, and 30 amperes, supplemented by voltage ranges of 60 and 120 volts. Circuit no. 2, which is generally utilized as the voltage circuit, has switch-selected ranges of 15, 30, 60, 120, 240, and 480 volts, supplemented by a 3-ampere current range. In turn, the instrument can be used to measure the phase angle between a current and a voltage, between two voltages, or between two currents.

With circuit No. 2 energized alone, the pointer slowly rotates on the 360° scale. Energization of circuit No. 1 stops the pointer rotation and indicates the phase angle between the sources, as well as whether circuit No. 1 is leading or lagging. An instrument of this type has the following applications:

- Measurement of phase angles.
- Connecting and checking directional and differential relays and their associated circuits.
- Connecting and checking current transformers or voltage transformers.
- Connecting and checking polyphase transformers, rectifier transformers, and so on.

398 Chapter 15

(A) Meter.

Figure 15-17 Phase-angle meter.

- Checking and connecting watt-hour meters, wattmeters, power-factor meters, synchroscopes, and their related circuits. (A synchroscope is an instrument that indicates when one generator is brought exactly into phase with another generator for parallel operation.)
- Checking phase rotation (clockwise or counterclockwise) in polyphase systems.
- Phasing buses for correct connection.
- Use as a portable power-factor meter or synchroscope. Terminal connection arrangements are depicted in Figure 15-17B.

Distribution
Each alternator in Figure 15-16 can be used only as a unit on the corresponding transformer, and failure of either of the conductors between them will result in a shutdown of both. Furthermore, a failure of the high-tension bus will result in a complete shutdown of

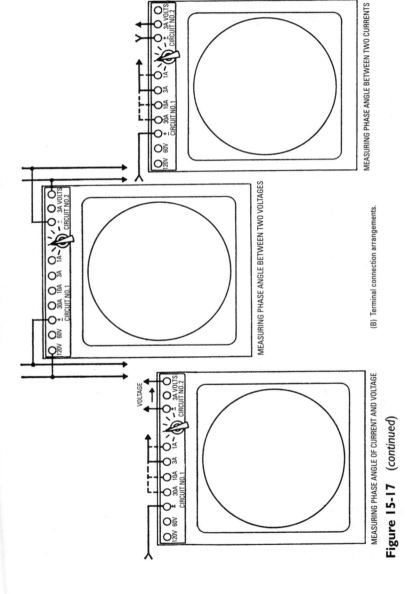

(B) Terminal connection arrangements.

Figure 15-17 (continued)

the plant until repairs are completed. To obtain somewhat greater flexibility, the method of high-tension connection shown in Figure 15-18 is often used. It treats the transformer bank as part of the transmission line rather than as a unit with the alternator. A *transmission line* is defined as a line used for electric power transfer.

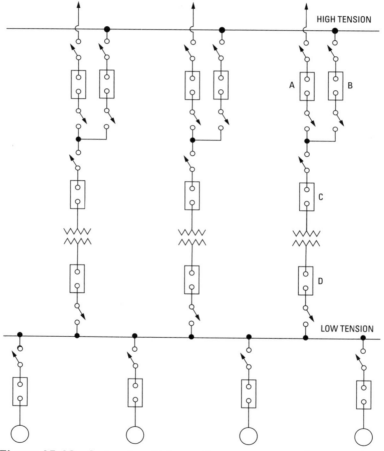

Figure 15-18 System in which transformers are treated as part of the transmission line.

In Figure 15-18, the transformer capacity (rating) is chosen with respect to the capacity of the transmission line. This arrangement shows three high-tension breakers per group. Often, to reduce cost, the two oil breakers *A* and *B* are replaced by three-pole air-break disconnect switches that may be operated electrically or manually. When this substitution is made, breaker *C* operates as both a line breaker and a transformer breaker. The low-tension breaker *D* is often used to trip out the circuit, as the transformer would be considered as part of the line. The advantage of this arrangement is that, when operating on the low-tension side, the magnitude of voltage surges resulting from high-tension switching is minimized.

Two of the disadvantages of the arrangement shown in Figure 15-18 are that it does not work well in a network system and it is not economical when generating stations supply widely separated loads. A network primary distribution system is an arrangement in which the primaries of the distribution transformers are connected to a common network supplied from the generating station. A primary distribution system is simply an arrangement for supplying the primaries of distribution transformers from a generating station. The term *distribution* includes all parts of an electric utility system.

The arrangement shown in Figure 15-18 has feasibility in a station where power is to be transmitted over a number of lines to a single substation. In this example, the line and transformer banks are identical. If the line becomes lost due to a fault, the corresponding transformer cannot be used. Banking of distribution transformers means the tying together of secondary mains of adjacent transformers that are supplied by the same primary feeder. One circuit supplies all the transformers that have banked secondaries.

Figure 15-19 shows an arrangement used in large steam-generating plants where all power is fed into a high-tension network that is distributed over a considerable area. This station is called the *base load* plant. A base load is defined as the minimum load over a given period of time. With this arrangement, each alternator and step-up transformer is treated as a unit, with no switching devices between them. The high-tension bus is a straight double bus with a double breaker arrangement, which provides maximum flexibility. The power for the station auxiliary is obtained from a high-tension step-down transformer bank.

Generating-station auxiliaries are the accessory units of equipment necessary for plant operation, such as pumps, stokers, and fans. Essential auxiliaries are those that do not sustain service interruptions of more than 1 minute, such as boiler feed pumps, forced-draft fans, and fuel feeders. Nonessential auxiliaries, such as air pumps, clinker grinders, and coal crushers, may sustain service interruptions of up to 3 minutes or more. Generating-station auxiliary

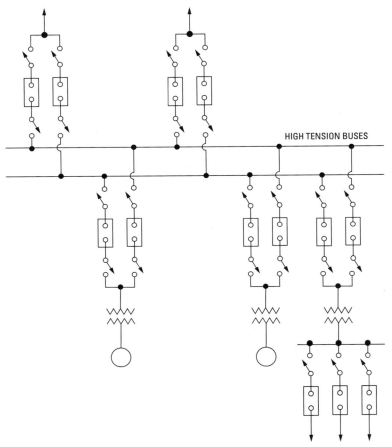

Figure 15-19 Bus system for large steam stations where power is fed into a high-tension network.

power is defined as the power that is required for operation of the auxiliaries.

Figure 15-20 shows another station load plant in which all power is delivered to a high-tension bus. In this example, the transformer banks are of exceptionally large size, and the alternators are in two or three units, the steam end consisting of one high-pressure and one or two low-pressure turbines. A maximum degree of flexibility

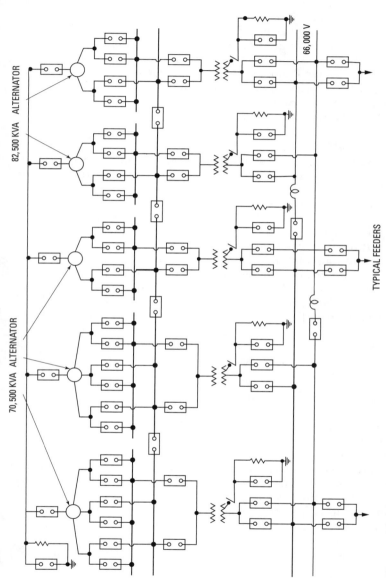

Figure 15-20 Bus system where all power is delivered to a high-tension bus.

is provided, inasmuch as double buses are provided in both the high- and low-tension sections.

Figures 15-21 through 15-24 show some of the more important and commonly used arrangements of main circuits in generating stations. Primary distribution mains are the conductors that feed from the center of distribution to direct primary loads or to transformers that feed secondary circuits. A distribution center is a point where equipment is located consisting generally of automatic overload protective devices connected to buses.

Substations

A substation consists of equipment used for purposes other than generation or utilization of electricity, through which electric energy is passed in bulk for the purpose of switching or modification of electrical characteristics. A substation includes one or more buses and various circuit breakers and is the receiving point for one or more supply circuits. It may also sectionalize the transmission circuits that pass through it by means of circuit breakers.

Substations may be manually operated, semiautomatic, automatic, portable, or supervisory controlled. Figure 15-25 shows the layout of a small substation. Substation transformers produce considerable heat, which must be dissipated by suitable means. Small transformers radiate heat from their shells; medium sizes have corrugated shells to increase the surface area. Small distribution substations may have no reserve transformer capacity because transformer failures are rare and replacement can be made rapidly when necessary. A spare transformer is provided in a large substation.

Automatic Substations

In order to eliminate the uncertainty and expense of manual operation, unattended or automatic substations are often used. An automatic substation goes into operation under predetermined conditions by an automatic sequence. It automatically maintains the required characteristics of service. It goes out of operation by automatic sequence under other predetermined conditions and provides protection against usual operating emergencies. An automatic substation is usually started by a load demand on that part of the system within its particular district. This is accomplished by a voltage relay.

An automatic substation is stopped by the operation of an undervoltage relay when the load diminishes to an uneconomical point. Starting and stopping may also be accomplished by a remote-control system or a time switch. The sequence of the various operations is

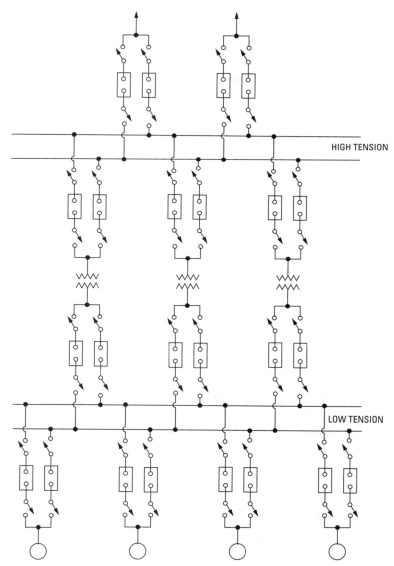

Figure 15-21 Single sectionalized bus system.

406 Chapter 15

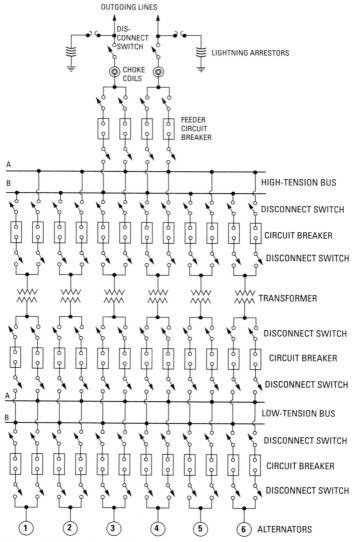

Figure 15-22 Double-bus, double-circuit-breaker system.

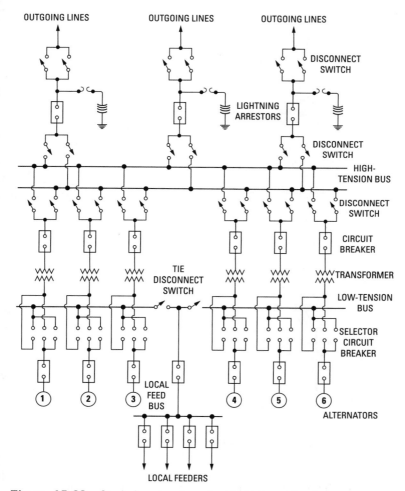

Figure 15-23 Single low-tension, double high-tension bus.

determined by a motor-driven master switch. It makes and breaks circuits and actuates contactors and relays that act directly on the machine circuits. This ensures correct sequence of operation and also eliminates a large number of interlocks. An *interlock* is a device actuated by operation of another device, with which it is directly associated, to govern succeeding operations. Interlocks may be either electrical or mechanical.

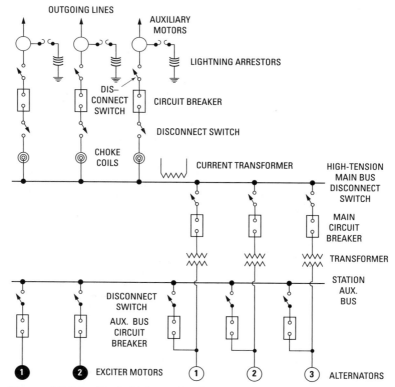

Figure 15-24 Single high-tension bus.

Semiautomatic Substations

A *semiautomatic substation* is started manually and runs until shut down, according to some schedule, by one of a number of different methods. These methods include a time switch, momentary interruption of the AC supply, or manual shutdown by an attendant. Figure 15-26 shows the layout of a semiautomatic substation. Because a semiautomatic substation is attended only during the starting and possibly during the shutting-down period, it must be equipped with all protective devices included with fully automatic equipment. These devices prevent open-phase running, excessive temperature of machine or transformer windings, overheated bearings, operation with open shunt-field winding, and armature overspeed. Complete automatic operation of the DC equipment is essential.

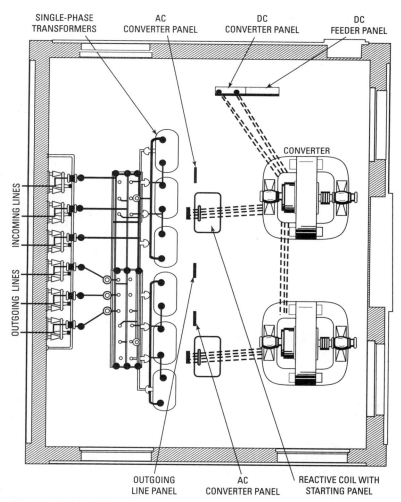

Figure 15-25 Plans for a small substation.

To meet emergency conditions, it is essential that means be provided for quickly opening all feeders. In many installations, it is also desirable to shut down and lock out certain automatic substations during light-load periods. There is, accordingly, a requirement for supervising unattended automatic substations from a central point or dispatcher's station. Automatic supervisory equipment provides

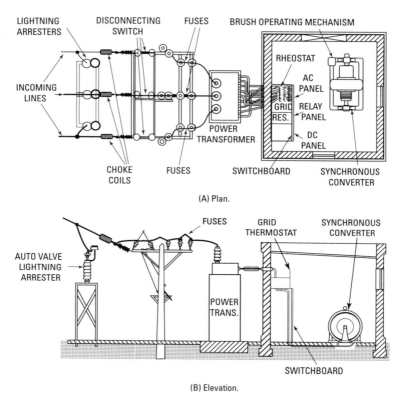

Figure 15-26 Plan and elevation of a typical outdoor semiautomatic substation.

the dispatcher with a means of selectively controlling devices in the substations and automatically gives the dispatcher a visual indication of the substation apparatus by means of indicating lamps located in cabinets at his or her office.

The dispatcher's office equipment consists of control keys, indicating lamps, and necessary supervisory devices for receiving control impulses that indicate the positions of the various supervised units. The substation equipment consists of the supervisory devices that transmit the control impulses to the auxiliary control relays and send back indication impulses to the dispatcher's office. Typical supervisory equipment is illustrated in Figure 15-27.

Generating Stations and Substations 411

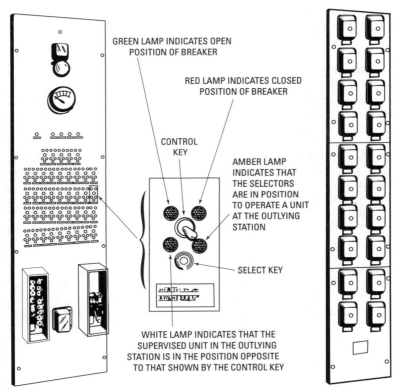

Figure 15-27 Dispatcher's control panel.

At the dispatcher's office, each supervised unit has a key and lamp combination consisting of a standard two-position turnkey for control, a red light for indicating the closed position, and a white light for indicating an automatic operation of the corresponding breaker unit. Each combination has a two-position push-and-pull selecting key to stop the selectors at a point corresponding to the unit it is desired to control and an associated amber lamp for indicating when the selectors are connected to that particular unit.

The dispatcher controls the supervised unit in the substation over the control circuit, and the indications from the supervised units are returned to the dispatcher over the indication circuit. The equipment at the substation is made to operate in synchronism with the equipment in the dispatcher's office by means of current impulses sent over the synchronizing circuit.

Summary

A generating station is a plant where electric energy is produced. A substation is an assembly of equipment through which electricity passes for the purpose of switching or modifying its characteristics. A substation is usually a receiving point for more than one supply circuit.

A generating station deals with the installation of generating units and mechanical and auxiliary electrical equipment required to utilize mechanical energy, changing the mechanical energy into electrical energy. This energy is then delivered to the associated systems.

There are four classifications of generating stations: gas powered, oil powered, water powered, and steam generating. The electrical features are similar in all four classes of stations. The turbines, alternators, exciters, and controlling switchboards are all housed in one large room of the hydroelectric generating station.

The main electrical connections in a generating station depend on the size and type of generator, the feeder arrangement, and the function of the generating station in the system. The simplest connection arrangement is called the single-bus system, which is used in small stations where interrupted service is permissible.

To obtain greater flexibility and reduce the possibility of service interruptions, two buses with additional disconnect switches or manually operated oil circuit breakers may be used in small generating stations. One bus is called the main and the other is called the spare bus. Another connection arrangement that employs two buses is called the double-bus, single-breaker system. It practically eliminates the possibility of a prolonged shutdown in case of a bus failure.

Substations may be manually operated, semiautomatic, automatic, portable, or supervisory controlled. In order to eliminate the uncertainty and expense of manual operation, unattended or automatic substations are often used. An automatic substation goes into operation under predetermined conditions by an automatic sequence.

Semiautomatic substations are started manually and run until shut down by one of a number of different methods. These methods include a time switch, momentary interruption of the AC supply, or shutdown by an attendant's action. Because a semiautomatic station is attended only during the start-up and shutdown periods, it must be equipped with all protective devices included in fully automatic substations.

Test Questions

1. Define a generating station.
2. What is meant by an automatic station?
3. Explain the purpose of a prime mover.
4. How does an exciter operate?
5. Describe a single-bus system for a generating station.
6. When does a circuit breaker open its circuit?
7. Discuss the operation of an instrument transformer.
8. What are the chief features of a spare-bus, spare-circuit-breaker system?
9. Describe a double-bus, single-circuit-breaker system.
10. Define a tie-bus breaker.
11. Explain the features of a double-bus, double-breaker system.
12. Discuss a ring-bus system.
13. What is meant by an H system?
14. How are transformers banked?
15. Define a base load.
16. Describe a single sectionalized bus system.
17. What is an advantage of the single low-tension, double high-tension, single-circuit-breaker arrangement?
18. Explain the arrangement of a single high-tension bus system.
19. Define an electric power substation.
20. What are the chief features of an automatic substation?
21. Explain what is meant by a semiautomatic substation.
22. State several functions of protective devices used in semiautomatic substations.
23. Describe the purposes served by automatic supervisory equipment.
24. How does a dispatcher determine the position of supervised units?

Appendix

Design Data for R. L. M. Dome Reflectors
(Direct-Lighting Luminaires)

The values in this table have been calculated using the following data: Reflector efficiency (white bowl lamp) 65%; Allowance for depreciation 30% to 40%.

These data apply to R. L. M. Dome reflectors equipped with white bowl Mazda lamps or slightly etched glass cover plate or glass ring when used with High-Intensity Mercury Vapor Lamp. For clear or inside-frosted Mazda lamps, increase the average foot-candle values by 10%.

Min. Mount. Ht. (Feet)	Approx. Luminaire Spacing (Feet)	Floor Area/ Outlet (Sq. Ft.)	Room Factor	Average Foot-candles									
				Mazda Lamps								Mercury Lamps	
				100-W	150-W	200-W	300-W	500-W	750-W	1000-W	1500-W	250-W	400-W
8	8 × 8	60–70	A	8.5–12	15–19	21–27	35–44	56–75				42–56	
			B	6.5–8.5	11–15	16–21	26–35	43–56				32–42	
			C	5.0–6.5	8.5–11	12–16	19–26	33–43				25–32	
8½	9 × 9	70–90	A	7.5–10	12–17	17–23	27–38	47–65				35–48	
			B	5.5–7.5	8.5–12	13–17	20–27	35–47				26–35	
			C	4.0–5.5	6.5–8.5	9.5–13	15–20	26–35				19–26	
9½	10 × 10	90–110	A		10–13	13–18	22–30	35–50	52–73			26–37	57–80
			B		7.5–10	10–13	16–22	27–35	40–52			20–26	44–57
			C		5.5–7.5	7.5–10	12–16	21–27	31–40			16–20	34–44
10½	11 × 11	110–130	A	5.0–6.5	8–11	11–15	18–24	30–41	44–60			22–31	48–66
			B	4–5	6–8	8.5–11	13–18	23–30	34–44			17–22	37–48
			C	3–4	4.5–6.0	6.5–8.5	10–13	18–23	26–34			13–17	29–37
11	12 × 12	130–150	A	4.5–5.5	7–9	10–13	16–20	26–35	38–50	55–72		19–26	42–55
			B	3.5–4.5	5.5–7.0	7.5–10	12–16	20–26	29–38	42–55		15–19	32–42
			C	2.5–3.5	4.0–5.5	5.5–7.5	9–12	15–20	22–29	32–42		11–15	24–32
11½	13 × 13	150–180	A		6–8	8–11	13–18	22–30	33–44	44–62		17–23	36–48
			B		4.5–6.0	6–8	10–13	17–22	25–33	34–44		13–17	27–36
			C		3.5–4.5	4.5–6.0	7.5–10	13–17	19–25	26–34		9.5–13	21–27
12	14 × 14	180–210	A		5.5–6.5	7.5–9.0	11–15	18–25	27–36	39–52		14–19	30–40
			B		4.0–5.5	5.5–7.5	8.5–11	14–18	21–27	30–39		11–14	23–30
			C		3–4	4.0–5.5	6.5–8.5	11–14	16–21	23–30		8–11	20–23
12½	15 × 15	210–240	A		4.5–5.5	6–8	9–13	17–22	23–31	34–44	53–70	13–16	25–34
			B		3.5–4.5	4.5–6.0	7–9	13–17	18–23	26–34	40–53	10–13	20–25
			C		2.5–3.5	3.5–4.5	5.5–7.0	9.5–13	14–18	20–26	31–40	7–10	16–20
13½	16 × 16	240–270	A		3.5–5.0	5–7	8.5–11	15–19	21–27	30–39	47–61	11–14	23–30
			B		2.5–3.5	4–5	6.5–8.5	11–14	16–21	23–30	36–47	8–11	18–23
			C		2.0–2.5	3–4	5.0–6.5	8.5–11	12–16	18–23	28–36	6.5–8.0	13–18
14	17 × 17	270–300	A			5–6	8–10	13–18	18–25	27–35	42–54	10–13	20–26
			B			4–5	6–8	9.5–13	14–18	21–27	32–42	7–10	16–20
			C			3–4	4.5–6.0	7.5–9.5	11–14	16–21	25–32	5.5–7.0	12–16
15	18 × 18	300–340	A			4.5–5.5	6.5–9.0	11–15	17–22	23–31	37–49	8–11	19–24
			B			3.5–4.5	5.0–6.5	9–11	13–17	18–23	28–37	7–8	15–19
			C			2.5–3.5	4–5	7–9	10–13	14–18	22–28	5–7	11–15
15½	19 × 19	340–380	A			3.5–5.0	6–8	10–13	15–19	22–27	33–43	7.5–10	17–21
			B			2.5–3.5	4.5–6.0	8–10	12–15	17–22	26–33	6.0–7.5	13–17
			C			2.0–2.5	3.5–4.5	6–8	9–12	13–17	20–26	4.5–6.0	10–13
16	20 × 20	380–420	A				5.5–7.0	9–12	13–17	18–25	30–39	7–9	15–19
			B				4.0–5.5	7–9	10–13	14–18	23–30	5–7	11–15
			C				3–4	5.5–7.0	8–10	11–14	18–23	4–5	9–11
17	21 × 21	420–460	A				5.0–6.5	8.5–11	13–16	17–22	26–35	6.5–8.0	15–18
			B				4–5	6.5–8.5	10–13	13–17	20–26	5.0–6.5	11–15
			C				3–4	5.0–6.5	7.5–10	10–13	16–20	4–5	8–11
18	22 × 22	460–500	A				4.5–6.0	8–10	11–14	16–20	24–32	6.0–7.5	12–16
			B				3.5–4.5	6–8	8.5–11	12–16	19–24	4.5–6.0	9.5–12
			C				2.5–3.5	4.5–6.0	6.5–8.5	9.5–12	15–19	3.5–4.5	7.0–9.5
19	23 × 23	500–550	A				4.0–5.5	6.5–9.0	10–13	14–19	23–29	5–7	11–15
			B				3–4	5.0–6.5	8–10	11–14	18–23	4–5	9–11
			C				2.5–3.0	4–5	6–8	8.5–11	14–18	3–4	6.5–9.0

Design Data for Semidirect Lighting Luminaires

The values in this table have been calculated using the following data: Luminaire efficiency 75% to 90%; Allowance for depreciation 30%.				Prismatic Glass			Opal Glass			
Min. Mount Ht. (Feet)	Approx. Luminaire Spacing (Feet)	Floor Area/ Outlet (Sq. Ft.)	Room Factor	Average Foot-candles						
				100-W	150-W	200-W	300-W	500-W	750-W	1000-W	1500-W
8	8 × 8	60–70	C	7.5–11	12–18	19–26	29–42	50–72			
			D	5.0–7.5	9–12	13–19	20–29	34–50			
			E	3.5–5.0	6–9	8.5–13	14–20	23–34			
8½	9 × 9	70–90	C	6.5–9.5	11–16	15–22	24–36	40–62	56–90		
			D	4.5–6.5	7–11	9.5–15	16–24	27–40	38–56		
			E	3.0–4.5	4.5–7.0	6.5–9.5	11–16	18–27	26–38		
9½	10 × 10	90–110	C	5.0–7.5	7.5–12	12–17	19–28	32–48	46–70	65–100	
			D	3.5–5.0	5.0–7.5	8–12	13–19	22–32	31–46	44–65	
			E	2.5–3.5	3.5–5.0	5.5–8.0	8.5–13	15–22	21–31	30–44	
10½	11 × 11	110–130	C	4.5–6.0	6.5–10	9.5–14	16–23	26–39	38–57	54–81	
			D	3.0–4.5	4.5–6.5	6.5–9.5	11–16	18–26	26–38	37–54	
			E	2–3	3.0–4.5	4.5–6.5	7.5–11	12–18	18–26	25–37	
11	12 × 12	130–150	C		6.0–8.5	9–12	14–20	23–33	35–48	47–68	75–108
			D		4–6	6–9	9.5–14	16–23	24–35	32–47	51–72
			E		3–4	4–6	6.5–9.5	11–16	16–24	22–32	35–51
11½	13 × 13	150–180	C		5.0–7.5	6.5–10	12–17	19–29	28–42	38–59	63–93
			D		3.5–5.0	4.5–6.5	8–12	13–19	19–28	26–38	43–63
			E		2.5–3.5	3.0–4.5	5.5–8.0	9–13	13–19	18–26	29–43
12	14 × 14	180–210	C			6.0–8.5	9.5–14	16–24	23–35	35–49	54–78
			D			4.5–6.0	6.5–9.5	11–16	16–23	24–35	37–54
			E			3.0–4.5	4.5–6.5	7.5–11	11–16	16–24	25–37
12½	15 × 15	210–240	C			5.0–7.5	9–12	14–20	19–30	29–42	47–67
			D			3.5–5.0	6–9	10–14	13–19	20–29	32–47
			E			2.5–3.5	4–6	7–10	9.5–13	14–20	22–32
13½	16 × 16	240–270	C				7.5–12	18–26	26–37	41–58	
			D				5.0–7.5	8.5–12	12–18	18–26	28–41
			E				3.5–5.0	8.0–8.5	8.5–12	12–18	19–28
14	17 × 17	270–300	C				6.5–9.5	11–16	16–23	23–33	37–52
			D				4.5–6.5	7.5–11	11–16	16–23	25–37
			E				3.0–4.5	5.5–7.5	8–11	11–16	17–25
15	18 × 18	300–340	C				6.0–8.5	10–14	15–21	21–30	32–47
			D				4.5–6.0	7–10	10–15	15–21	22–32
			E				3.0–4.5	5–7	7–10	10–15	15–28
15½	19 × 19	340–380	C				5.0–7.5	9–13	12–18	19–26	29–41
			D				3.5–5.0	6.5–9.0	9–12	13–19	20–29
			E				2.5–3.5	4.5–6.5	6–9	8.5–13	14–20
16	20 × 20	380–420	C					7.5–11	11–16	16–23	26–37
			D					5.5–7.5	8–11	11–16	18–26
			E					4.0–5.5	5.5–8.0	8–11	12–18
17	21 × 21	420–460	C					7–10	11–15	15–21	23–33
			D					5–7	7.5–11	10–15	16–23
			E					3.5–5.0	5.0–7.5	7–10	11–16
18	22 × 22	460–500	C					6.5–9.5	9.5–14	13–19	21–30
			D					4.5–6.5	6.5–9.5	9.5–13	15–21
			E					3.0–4.5	4.5–6.5	6.5–9.5	10–15
19	23 × 23	500–550	C					6.0–8.5	9–12	12–18	19–28
			D					4.5–6.0	6–9	9–12	13–19
			E					3.0–4.5	4–8	6–9	6.5–13

Glossary

Access Fitting. A fitting that permits access to conductors in concealed or enclosed wiring, elsewhere than at an outlet.

Active Electrical Network. A network that contains one or more sources of electrical energy.

Admittance. The reciprocal of impedance.

Air-Blast Transformer. A transformer cooled by forced circulation of air through its core and coils.

Air Circuit Breaker. A circuit breaker in which the interruption occurs in air.

Air Switch. A switch in which the interruption of the circuit occurs in air.

Alive. Electrically connected to a source of emf, or electrically charged with a potential different from that of the earth. Also, practical synonym for *current-carrying* or *hot*.

Alternating Current. A periodic current, the average value of which over a period is zero.

Alternator. Synchronous generator; a synchronous alternating-current machine that changes mechanical power into electrical power.

Ambient Temperature. The temperature of a surrounding cooling medium, such as gas or liquid, that comes into contact with the heated parts of an apparatus.

Ammeter. An instrument for measuring electric current.

Ampere. A charge flow of one coulomb per second.

Annunciator. An electromagnetically operated signaling apparatus that indicates whether a current is flowing or has flowed in one or more circuits.

Apparent Power. In a single-phase, two-wire circuit, the product of the effective current in one conductor multiplied by the effective voltage between the two points of entry.

Appliance. Current-consuming equipment, fixed or portable, such as heating or motor-operated equipment.

Arc-Fault Circuit Interrupter (AFCI). An electrical device that detects the unique electronic characteristics of electrical arcs. If an arc is sensed, the device further deenergizes the circuit to which it is connected.

Arcing Contacts. Contacts on which an arc is drawn after the main contacts of a switch or circuit breaker have parted.

Arcing Time of Fuse. The time elapsing from the severance of the fuse link to the final interruption of the circuit under specified conditions.

Arc-Over of Insulator. A discharge of power current in the form of an arc, following a surface discharge over an insulator.

Armor Clamp. A fitting for gripping the armor of a cable at the point where the armor terminates, or where the cable enters a junction box or other apparatus.

Armored Cable. A cable provided with a wrapping of metal, usually steel wires, primarily for the purpose of mechanical protection.

Arrester, Lightning. A device that reduces the voltage of a surge applied to its terminals and restores itself to its original operating condition.

Autotransformer. A transformer in which part of the winding is common to both the primary and secondary circuits.

Back-Connected Switch. A switch in which the current-carrying conductors are connected to studs in back of the mounting base.

Bank. An assemblage of fixed contacts in a rigid unit over which wipers or brushes may move and make connection with the contacts.

Bank, Duct. An arrangement of conduit that provides one or more continuous ducts between two points.

Benchboard. A switchboard with a horizontal section for control switches, indicating lamps, and instrument switches; may also have a vertical instrument section.

Bidirectional Current. A current that has both positive and negative values.

Bond, Cable. An electrical connection across a joint in the armor or lead sheath of a cable, between the armor or sheath to ground, or between the armor or sheath of adjacent cables.

Box, Conduit. A metal box adapted for connection to conduit for installation of wiring, making connections, or mounting devices.

Box, Junction. An enclosed distribution panel for connection or branching of one or more electric circuits without making permanent splices.

Box, Junction (Interior Wiring). A metal box with blank cover for joining runs of conduit, electrical metallic tubing, wireway, or raceway and for providing space for connection and branching of enclosed conductors.

Box, Pull. A metal box with a blank cover which is used in a run of conduit or other raceway to facilitate pulling in the conductors; it may also be installed at the end of one or more conduit runs for distribution of the conductors.

Branch Circuit. That portion of a wiring system extending beyond the final automatic overload protective device.

Branch Circuit, Appliance. A circuit supplying energy either to permanently wired appliances or to attachment-plug receptacles such as appliance or convenience outlets and having no permanently connected lighting fixtures.

Branch Circuit Distribution Center. A distribution circuit at which branch circuits are supplied.

Branch Circuit, Lighting. A circuit supplying energy to lighting outlets only.

Branch Conductor. A conductor that branches off at an angle from a continuous run of conductor.

Break. The break of a circuit-opening device is the minimum distance between the stationary and movable contacts when the device is in its open position.

Breakdown. Also termed *puncture*, denoting a disruptive discharge through insulation.

Breaker, Line. A device that combines the functions of a contactor and a circuit breaker.

Buried Cable. A cable installed under the surface of the soil in such a manner that it cannot be removed without digging up the soil. (Type UF is commonly used for home wiring.)

Bus. A conductor or group of conductors that serves as a common connection for three or more circuits in a switchgear assembly.

Bushing. Also termed *insulating bushing*; a lining for a hole for insulation and/or protection from abrasion of one or more conductors passing through it.

Cabinet. An enclosure for either surface or flush mounting, provided with a frame, mat, or trim.

Cable. The package of wires, insulating material, sheathing, and whatever else is necessary for the type being installed. It is usually purchased in large spools.

Cable Fault. A partial or total local failure in the insulation or continuity of the conductor.

Cable Joint. Also termed a *splice*; a connection between two or more individual lengths of cables, with their conductors individually connected, and with protecting sheaths over the joint.

Cable, Service. Service conductors arranged in the form of a cable (see **Service**).

Cable Sheath. The protective covering, such as lead or plastic, applied over a cable.

Charge, Electric. An inequality of positive and negative electricity in or on a body. The charge stored in a capacitor (condenser) corresponds to a deficiency of free electrons on the positive plate and to an excess of free electrons on the negative plate.

Choke Coil. A low-resistance coil with sufficient inductance to substantially impede ac or transient currents.

Circuit, Electric. A conducting path through which electric charges may flow. A dc circuit is a closed path for charge flow; an ac circuit is not necessarily closed and may conduct in part by means of an electric field (displacement current).

Circuit, Earth (Ground) Return. An electric circuit in which the ground serves to complete a path for charge flow.

Circuit, Magnetic. A closed path for establishment of magnetic flux (magnetic field) that has the direction of the magnetic induction at every point.

Cleat. An assembly of a pair of insulating material members with grooves for holding one or more conductors at a definite distance from the mounting surface.

Clip, Fuse. Contacts on a fuse support for connecting a fuse holder into a circuit.

Closed-Circuit Voltage. The terminal voltage of a source of electricity under a specified current demand.

Closed Electric Circuit. A continuous path or paths providing for charge flow. In an ac closed circuit, charge flow may be changed into displacement current through a capacitor (condenser).

Coercive Force. The magnetizing force at which the magnetic induction is zero at a point on the hysteresis loop of a magnetic substance.

Coil. A conductor arrangement (basically a helix or spiral) that concentrates the magnetic field produced by electric charge flow.

Composite Conductor. A conductor consisting of two or more strands of different metals, operated in parallel.

Concealed. To be made inaccessible by the structure or finish of a building; also, wires run in a concealed raceway.

Condenser. Also termed *capacitor*; a device that stores electric charge by means of an electric field.

Conductance. A measure of permissiveness to charge flow; the reciprocal of resistance.

Conductor. A substance that has free electrons or other charge carriers that permit charge flow when an emf is applied across the substance.

Conduit. A structure containing one or more ducts; commonly formed from iron pipe or electrical metallic tubing (EMT).

Conduit Fittings. Accessories used to complete a conduit system, such as boxes, bushings, and access fittings.

Conduit, Flexible Metal. A flexible raceway of circular form for enclosing wires or cables; usually made of steel wound helically and with interlocking edges and a weather-resistant coating. Sometimes called *Greenfield*.

Conduit, Rigid Steel. A raceway made of mild steel pipe with a weather-resistant coating.

Conduit Run. A duct bank; an arrangement of conduit with a continuous duct between two points in an electrical installation.

Contactor. An electric power switch, not operated manually, designed for frequent operation.

Contacts. Conducting parts that employ a junction that is opened or closed to interrupt or complete a circuit.

Control Relay. A relay used to initiate or permit a predetermined operation in a control circuit.

Coulomb. An electric charge of 6.28×10^{18} electrons. One coulomb is transferred when a current of one ampere continues past a point for one second.

Counter EMF. CEMF; the effective emf within a system which opposes current in a specified direction.

Current. The rate of charge flow. A current of one ampere is equal to a flow rate of one coulomb per second.

Cycle. The complete series of values that occurs during one period of a periodic quantity. The unit of frequency, the hertz, is equal to one cycle per second.

Dead. Functionally conducting parts of an electrical system that have no potential difference or charge (voltage of zero with respect to ground).

Degree, Electrical. An angle equal to 1/360 of the angle between consecutive field poles of like polarity in an electrical machine.

Diagram, Connection. A drawing showing the connections and interrelations of devices employed in an electrical circuit.

Dielectric. A medium or substance in which a potential difference establishes an electric field that is subsequently recoverable as electric energy.

Direct Current. A unidirectional current with a constant value. Constant value is defined in practice as a value that has negligible variation.

Direct EMF. Also termed *direct voltage*; an emf that does not change in polarity and has a constant value (one of negligible variation).

Discharge. An energy conversion involving electrical energy. Examples include discharge of a storage battery, discharge of a capacitor, and lightning discharge of a thundercloud.

Displacement Current. The apparent flow of charge through a dielectric such as in a capacitor; represented by buildup and/or decay of an electric field.

Disruptive Discharge. A rapid and large current increase through an insulator due to insulation failure.

Distribution Center. A point of installation for automatic overload protective devices connected to buses where an electrical supply is subdivided into feeders and/or branch circuits.

Divider, Voltage. A tapped resistor or series arrangement of resistors, sometimes with movable contacts, providing a desired IR drop. (A voltage divider is not continuously and manually variable as in a potentiometer).

Drop, Voltage. An IR voltage between two specified points in an electric circuit.

Duct. A single enclosed runway for conductors or cables.

Effective Value. The effective value of a sine-wave ac current or voltage is equal to 0.707 of peak. Also called the root-mean-square (rms) value, it produces the same I^2R power as an equal dc value.

Efficiency. The ratio of output power to input power, usually expressed as a percentage.

Electrical Units. In the practical system, electrical units comprise the volt, the ampere, the ohm, the watt, the watt-hour, the coulomb, the mho, the henry, the farad, and the joule.

Electricity. A physical entity associated with the atomic structure of matter that occurs in polar forms (positive and negative) and that are separable by expenditure of energy.

Electrode. A conducting substance through which electric current enters or leaves in devices that provide electrical control or energy conversion.

Electrolyte. A substance that provides electrical conduction when dissolved (usually in water.)

Electrolytic Conductor. Flow of electric charges to and from electrodes in an electrolytic solution.

Electromagnetic Induction. A process of generation of emf by movement of magnetic flux that cuts an electrical conductor.

Electromotive Force (EMF). An energy-charge relation that results in electric pressure, which produces or tends to produce charge flow (see **Voltage**).

Electron. The subatomic unit of negative electricity; it is a charge of 1.6×10^{-19} coulomb.

Electronics. The science dealing with charge flow in vacuum, gases, and crystal lattices.

Electroplating. The electrical deposition of metallic ions as neutral atoms on an electrode immersed in an electrolyte.

Electrostatics. A branch of electrical science dealing with the laws of electricity at rest.

Energy. The amount of physical work a system is capable of doing. Electrical energy is measured in watt-seconds (the product of power and time).

Entrance, Duct. An opening of a duct at a distributor box or other accessible location.

Equipment, Service. A circuit breaker or switches and fuses with their accessories, installed near the point of entry of service conductors to a building.

Exciter. An auxiliary generator for supplying electrical energy to the field of another electrical machine.

Farad. A unit of capacitance defined by the production of one volt across the capacitor terminals when a charge of one coulomb is stored.

Fault Current. An abnormal current flowing between conductors or from a conductor to ground due to an insulation defect, arc-over, or incorrect connection.

Feeder. A conductor or a group of conductors for connection of generating stations, substations, generating stations and substations, or a substation and a feeding point.

Ferromagnetic Substance. A substance that has a permeability considerably greater than that of air; a ferromagnetic substance has a permeability that changes with the value of applied magnetizing force.

Filament. A wire or ribbon of conducting (resistive) material that develops light and heat energy due to electric charge flow; light radiation is also accompanied by electron emission.

Fixture Stud. A fitting for mounting a lighting fixture in an outlet box and which is secured to the box.

Flashover. A disruptive electrical discharge around or over (but not through) an insulator.

Fluorescence. An electrical discharge process involving radiant energy transferred by phosphors into radiant energy that provides increased luminosity.

Flux. Electrical field energy distributed in space, in a magnetic substance, or in a dielectric. Flux is commonly represented diagrammatically by means of flux lines denoting magnetic or electric forces.

Force. An elementary physical cause capable of modifying the motion of a mass.

Frequency. The number of periods occurring in unit time of a periodic process such as in the flow of electric charge.

Frequency Meter. An instrument that measures the frequency of an alternating current.

Fuse. A protective device with a fusible element that opens the circuit by melting when subjected to excessive current.

Fuse Cutout. An assembly consisting of a fuse support and holder, which may also include a fuse link.

Fuse Element. Also termed *fuse link*; the current-carrying part of a fuse that opens the circuit when subjected to excessive current.

Fuse Holder. A supporting device for a fuse that provides terminal connections.

Galvanometer. An instrument for indicating or measuring comparatively small electric currents. A galvanometer usually has zero-center indication.

Gap. Spark gap; a high-voltage device with electrodes between which a disruptive discharge of electricity may pass, usually through air. A sphere gap has spherical electrodes; a needle gap has sharply pointed electrodes; a rod gap has rods with flat ends.

Ground. Also termed *earth*; a conductor connected between a circuit and the soil. A chassis ground is not necessary at ground potential but is taken as a zero-volt reference point. An accidental ground occurs due to cable insulation faults, an insulator defect, and so on.

Ground-Fault Interrupter (GFI). A device installed in circuits where current leakage can be especially dangerous, such as outdoor or bathroom circuits. It shuts off current flow within 0.025 second at the onset of a leak as small as 5 milliamperes.

Grounding Electrode. A conductor buried in the earth for connection to a circuit. The buried conductor is usually a cold-water pipe, to which connection is made with a ground clamp.

Ground Lug. A lug for convenient connection of a grounding conductor to a grounding electrode or device to be grounded.

Ground Outlet. An outlet provided with a polarized receptacle with a grounded contact for connection of a grounding conductor.

Ground Switch. A switch for connection or disconnection of a grounding conductor.

Guy. A wire or other mechanical member having one end secured and the other end fastened to a pole or structural part maintained under tension.

Hanger. Also termed *cable rack*; a device usually secured to a wall to provide support for cables.

Heat Coil. A protective device for opening and/or grounding a circuit by switching action when a fusible element melts due to excessive current.

Heater. In the strict sense, a heating element for raising the temperature of an indirectly heated cathode in a vacuum or gas tube. Also applied to appliances such as space heaters and radiant heaters.

Henry. The unit of inductance; it permits current increase at the rate of 1 ampere per second when 1 volt is applied across the inductor terminals.

Hickey. A fitting for mounting a lighting fixture in an outlet box. Also, a device used with a pipe handle for bending conduit.

Horn Gap. A form of switch provided with arcing horns for automatically increasing the length of the arc and thereby extinguishing the arc.

Hydrometer. An instrument for indicating the state of charge in a storage battery.

Hysteresis. The magnetic property of a substance which results from residual magnetism.

Hysteresis Loop. A graph that shows the relation between magnetizing force and flux density for a cyclically magnetized substance.

Hysteresis Loss. The heat loss in a magnetic substance due to application of a cyclic magnetizing force to a magnetic substance.

Impedance. Opposition to ac current by a combination of resistance and reactance; impedance is measured in ohms.

Impedances, Conjugate. A pair of impedances that have the same resistance values, and that have equal and opposite reactance values.

Impulse. An electric surge of unidirectional polarity.

Indoor Transformer. A transformer that must be protected from the weather.

Induced Current. A current that results in a closed conductor due to cutting of lines of magnetic force.

Inductance. An electrical property of a resistanceless conductor, which may have a coil form and which exhibits inductive reactance to an ac current. All practical inductors also have at least a slight amount of resistance.

Inductor. A device such as a coil with or without a magnetic core which develops inductance, as distinguished from the inductance of a straight wire.

Instantaneous Power. The product of an instantaneous voltage by the associated instantaneous current.

Instrument. An electrical device for measurement of a quantity under observation or for presenting a characteristic of the quantity.

Interconnection, System. A connection of two or more power systems.

Interconnection Tie. A feeder that interconnects a pair of electric supply systems.

Interlock. An electrical device whose operation depends on another device for controlling subsequent operations.

Internal Resistance. The effective resistance connected in series with a source of emf due to resistance of the electrolyte, winding resistance, and so on.

Ion. A charged atom, or a radical. For example, a hydrogen atom that has lost an electron becomes a hydrogen ion; sulphuric acid produces H^+ and SO_{-4} ions in water solution.

IR Drop. A potential difference produced by charge flow through a resistance.

Isolating Switch. An auxiliary switch for isolating an electric circuit from its source of power; it is operated only after the circuit has been opened by other means.

Joule. A unit of electrical energy, also called a watt-second. One joule is the transfer of one watt for one second.

Joule's Law. The rate at which electrical energy is changed into heat energy is proportional to the square of the current.

Jumper. A short length of conductor for making a connection between terminals, around a break in a circuit, or around an electrical instrument.

Junction. A point in a parallel or series-parallel circuit where current branches off into two or more paths.

Junction Box. An enclosed distribution panel for the connection or branching of one or more electrical circuits without using permanent splices. In the case of interior wiring, a junction box consists of a metal box with a blank cover; it is inserted in a run of conduit, raceway, or tubing.

Kirchhoff's Law. The voltage law states that the algebraic sum of the drops around a closed circuit is equal to zero. The current law states that the algebraic sum of the currents at a junction is equal to zero.

Knockout. A scored portion in the wall of a box or cabinet which can be removed easily by striking with a hammer; a circular hole is provided thereby for accommodation of conduit or cable.

kVA. Kilovolt-amperes; the product of volts and amperes divided by 1000.

Lag. Denotes that a given sine wave passes through its peak at a later time than a reference sine wave.

Lamp holder. Also termed *socket* or *lamp receptacle*; a device for mechanical support of and electrical connection to a lamp.

Lay. The lay of a helical element of a cable is equal to the axial length of a turn.

Lead. Denotes that a given sine wave passes through its peak at an earlier time than a reference sine wave.

Leakage, Surface. Passage of current over the boundary surfaces of an insulator as distinguished from passage of current through its bulk.

Leg of a Circuit. One of the conductors in a supply circuit between which the maximum supply voltage is maintained.

Lenz's Law. States that an induced current in a conductor is in a direction such that the applied mechanical force is opposed.

Limit Switch. A device that automatically cuts the power off at or near the limit of travel of a mechanical member.

Load. The load on an ac machine or apparatus is equal to the product of the rms voltage across its terminals and the rms current demand.

Locking Relay. A relay that operates to make some other device inoperative under certain conditions.

Loom. See **Tubing, Flexible.**

Luminosity. Relative quantity of light.

Magnet. A magnet is a body that is the source of a magnetic field.

Magnetic Field. A magnetic field is the space containing distributed energy in the vicinity of a magnet and in which magnetic forces are apparent.

Magnetizing Force. Number of ampere-turns in a transformer primary per unit length of core.

Magnetomotive Force. Number of ampere-turns in a transformer primary.

Mass. Quantity of matter; the physical property that determines the acceleration of a body as the result of an applied force.

Matter. Matter is a physical entity that exhibits mass.

Meter. A unit of length equal to 39.37 inches; also, an electrical instrument for measurement of voltage, current, power, energy, phase angle, synchronism, resistance, reactance, impedance, inductance, capacitance, and so on.

Mho. The unit of conductance defined as the reciprocal of the ohm.

Mounting, Circuit Breaker. Supporting structure for a circuit breaker.

Multiple Feeder. Two or more feeders connected in parallel.

Multiple Joint. A joint for connecting a branch conductor or cable to a main conductor or cable to provide a branch circuit.

Multiplier, Instrument. A series resistor connected to a meter mechanism for the purpose of providing a higher voltage-indicating range.

Mutual Inductance. An inductance common to the primary and secondary of a transformer, resulting from primary magnetic flux that cuts the secondary winding.

Negative. A value less than zero; an electric polarity sign indicating an excess of electrons at one point with respect to another point; a current sign indicating charge flow away from a junction.

Network. A system of interconnected paths for charge flow.

Network, Active. A network that contains one or more sources of electrical energy.

Network, Passive. A network that does not contain a source of electrical energy.

No-Load Current. The current demand of a transformer primary when no current demand is made on the secondary.

Normally Closed. Denotes the automatic closure of contacts in a relay when deenergized (not applicable to a latching relay).

Normally Open. Denotes the automatic opening of contacts in a relay when deenergized (not applicable to a latching relay).

Ohm. The unit of resistance; a resistance of one ohm sustains a current of one ampere when one volt is applied across the resistance.

Ohmmeter. An instrument for measuring resistance values.

Ohm's Law. States that current is directly proportional to applied voltage and inversely proportional to resistance, reactance, or impedance.

Open-Circuit Voltage. The terminal voltage of a source under conditions of no current demand. The open-circuit voltage has a value equal to the emf of the source.

Open-Wire Circuit. A circuit constructed from conductors that are separately supported on insulators.

Oscilloscope. An instrument for displaying the waveforms of ac voltages.

Outdoor Transformer. A transformer with weatherproof construction.

Outlet. A point in a wiring system from which current is taken for supply of fixtures, lamps, heaters, and so on.

Outlet, Lighting. An outlet used for direct connection of a lamp holder, lighting fixture, or a cord that supplies a lamp holder.

Outlet, Receptacle. An outlet used with one or more receptacles that are not of the screw-shell type.

Overload Protection. Interruption or reduction of current under conditions of excessive demand, provided by a protective device.

Ozone. A compound consisting of three atoms of oxygen, produced by the action of electric sparks or specialized electrical devices.

Peak Current. The maximum value (crest value) of an alternating current.

Peak Voltage. The maximum value (crest value) of an alternating voltage.

Peak-to-Peak Value. The value of an ac waveform from its positive peak to its negative peak. In the case of a sine wave, the peak-to-peak value is double the peak value.

Pendant. A fitting suspended from overhead by a flexible cord that may also provide electrical connection to the fitting.

Pendant, Rise-and-Fall. A pendant that can be adjusted in height by means of a cord adjuster.

Period. The time required for an ac waveform to complete one cycle.

Permanent Magnet. A magnetized substance that has substantial retentivity.

Permeability. The ratio of magnetic flux density to magnetizing force.

Phase. The time of occurrence of the peak value of an ac waveform with respect to the time of occurrence of the peak value of a reference waveform. Phase is usually stated as the fractional part of a period.

Phase Angle. An angular expression of phase difference; it is commonly expressed in degrees and is equal to the phase multiplied by 360°.

Plug. A device inserted into a receptacle for connection of a cord to the conductor terminations in the receptacle.

Polarity. An electrical characteristic of emf that determines the direction in which current tends to flow.

Polarization (Battery). Polarization is caused by development of gas at the battery electrodes during current demand and has the effect of increasing the internal resistance of the battery.

Pole. The pole of a magnet is an area at which its flux lines tend to converge or diverge.

Positive. A value greater than zero; an electric polarity sign denoting a deficiency of electrons at one point with respect to another point; a current sign indicating charge flow toward a junction.

Potential Difference. A potential difference of one volt is produced when one unit of work is done in separating unit charges through a unit distance.

Potentiometer. A resistor with a continuously variable contact arm; electrical connections are made to both ends of the resistor and to the arm.

Power. The rate of doing work, or the rate of converting energy. When one volt is applied to a load and the current demand is one ampere, the rate of energy conversion (power) is one watt.

Power, Real. Real power is developed by circuit resistance, or effective resistance.

Primary Battery. A battery that cannot be recharged after its chemical energy has been depleted.

Primary Winding. The input winding of a transformer.

Proton. The subatomic unit of positive charge; a proton has a charge that is equal and opposite to that of an electron.

Pull Box. A metal box with a blank cover for insertion into a conduit run, raceway, or metallic tubing, which facilitates the drawing of conductors.

Pulsating Current. A direct current that does not have a steady value.

Puncture. A disruptive electrical discharge through insulation.

Quick-Break. A switch or circuit breaker that has a high contact-opening speed.

Quick-Make. A switch or circuit breaker that has a high contact-closing speed.

Raceway. A channel for holding wires or cables; constructed from metal, wood, or plastics, rigid metal conduit, electrical metal tubing, cast-in-place, underfloor, surface metal, surface wooden types, wireways, busways, and auxiliary gutters.

Rack, Cable. A device secured to the wall to provide support for a cable raceway.

Rating. The rating of a device, apparatus, or machine states the limit or limits of its operating characteristics. Ratings are commonly stated in volt, amperes, watts, ohms, degrees, horsepower, and so on.

Reactance. Reactance is an opposition to ac current based on the reaction of energy storage, either as a magnetic field or as an electric field. No real power is dissipated by a reactance. Reactance is measured in ohms.

Reactor. An inductor or a capacitor. Reactors serve as current-limiting devices such as in motor starters, for phase-shifting applications as in capacitor start motors, and for power-factor correction in factories or shops.

Receptacle. Also termed *convenience outlet*; a contacting device installed at an outlet for connection externally by means of a plug and flexible cord.

Rectifier. A device that has a high resistance in one direction and a low resistance in the other direction.

Regulation. Denotes the extent to which the terminal voltage of a battery, generator, or other source decreases under current demand. Commonly expressed as the ratio of the difference of the no-load voltage and the load voltage to the no-load voltage under rated current demand; usually expressed as a percentage.

Relay. A device operated by a change in voltage or current in a circuit, which actuates other devices in the same circuit or in another circuit.

Reluctance. An opposition to the establishment of magnetic flux lines when a magnetizing force is applied; usually measured in rels.

Remanence. The flux density that remains in a magnetic substance after an applied magnetomotive force has been removed.

Resistance. A physical property that opposes current and dissipates real power in the form of heat. Resistance is measured in ohms.

Resistor. A positive component; may be wire-wound, carbon-composition, thyrite, or other design.

Rheostat. A variable resistive device consisting of a resistance element and a continuously adjustable contact arm.

Rosette. A porcelain or other enclosure with terminals for connecting a flexible cord and pendant to the permanent wiring.

Safety Outlet. Also termed *ground outlet*; an outlet with a polarized receptacle for equipment grounding.

Secondary Battery. A battery that can be recharged after its chemical energy is depleted.

Sequence Switch. A remotely controlled power-operated switching device.

Series Circuit. A circuit that provides a complete path for current and has its components connected end-to-end.

Service. The conductors and equipment for supplying electrical energy from the main or feeder or from the transformer to the area served.

Serving, of Cable. A wrapping over the core of a cable before it is leaded or over the lead if it is armored.

Shaded Pole. A single heavy conducting loop placed around one half of a magnetic pole that develops an ac field, in order to induce an out-of-phase magnetic field.

Sheath, Cable. A protective covering (usually lead) applied to a cable.

Shell Core. A core for a transformer or reactor consisting of three legs, with the winding located on the center leg.

Short Circuit. A fault path for current in a circuit that conducts excessive current; if the fault path has appreciable resistance, it is termed a leakage path.

Shunt. Denotes a parallel connection.

Sine Wave. Variation in accordance with simple harmonic motion.

Sinusoidal. Having the form of a sine wave.

Sleeve, Splicing. Also termed *connector*; a metal sleeve (usually copper) slipped over and secured to the butted ends of conductors to make a joint that provides good electrical connection.

Sleeve Wire. A circuit conductor connected to the sleeve of a plug or jack.

Sliding Contact. An adjustable contact arranged to slide mechanically over a resistive element, over turns of a reactor, over series of taps, or around the turns of a helix.

Snake. A steel wire or flat ribbon with a hook at one end, used to draw wires through conduit, et cetera.

Socket. A device for mechanical support of a device (such as a lamp) and for connection to the electrical supply.

Solderless Connector. Any device that connects wires together without solder; wire nuts are the most common type.

Solenoid. A conducting helix with a comparatively small pitch; also applied to coaxial conducting helices.

Spark Coil. Also termed *ignition coil*; a step-up transformer designed to operate from a dc source via an interrupter that alternately makes and breaks the primary circuit.

Sparkover. A disruptive electrical discharge between the electrodes of a gap; generally used with reference to measurement of high-voltage values with a gap having specified types and shape of electrodes.

Splice. Also termed *straight-through joint*; a series connection of a pair of conductors or cables.

Standard Cell. A highly precise source of dc voltage, also called a Weston cell; standard cells are used to check voltmeter calibration and for highly precise measurement of dc voltage values.

Station, Automatic. A generating station or substation that is usually unattended and performs its intended functions by an automatic sequence.

Surge. A transient variation in current and/or voltage at a given point in a circuit.

Switch. A device for making, breaking, or rearranging the connections of an electric circuit.

Symbol. A graphical representation of a circuit component; also, a letter or letters used to represent a component, electrical property, or circuit characteristic.

Glossary 435

Tap. In a wiring installation, a T joint (Tee joint), Y joint, or multiple joint. Taps are made to resistors, inductors, transformers, and so on.

Terminal. The terminating end(s) of an electrical device, source, or circuit, usually supplied with electrical connectors such as terminal screws, binding posts, tip jacks, snap connectors, or soldering lugs.

Three-Phase System. An ac system in which three sources energize three conductors, each of which provide a voltage that is 120° out of phase with the voltage in the adjacent conductor.

Tie Feeder. A feeder connected at both ends to sources of electrical energy. In an automatic station, a load may be connected between the two sources.

Time Delay. A specified period of time from the actuation of a control device to its operation of another device or circuit.

Tip, Plug. The contacting member at the end of a plug.

Torque. Mechanical twisting force.

Transfer Box. Also termed *pull box*; a box without a distribution panel containing branched or otherwise interconnected circuits.

Transformer. A device that operates by electromagnetic induction with a tapped winding, or two or more separate windings, usually on an iron core, for the purpose of stepping voltage or current up or down, for maximum power transfer, for isolation of the primary circuit from the secondary circuit, and in special designs for automatic regulation of voltage or current.

Transient. A nonrepetitive or arbitrarily timed electrical surge.

Transmission (AC). Transfer of electrical energy from a source to a load or to one or more stations for subsequent distribution.

Troughing. An open earthenware channel, wood, or plastic in which cables are installed under a protective cover.

Tubing, Electrical Metal(lic) (EMT). A thin-walled steel raceway of circular form with a corrosion-resistant coating for protection of wire or cables.

Tubing, Flexible. Also termed *loom*; a mechanical protection for electrical conductors; a flame-resistant and moisture-repellent circular tube of fibrous material.

Twin Cable. A cable consisting of two insulating and stranded conductors arranged in parallel runs and having a common insulating covering.

Underground Cable. A cable designed for installation below the surface of the ground or for installation in an underground duct.

Undergrounded System. Also termed *insulated supply system*; an electrical system that floats above ground, or one that has only a very high impedance conducting path to ground.

Unidirectional Current. A direct current or a pulsating direct current.

Units. Established values of physical properties used in measurement and calculation; for example, the volt unit, the ampere unit, the ampere-turn unit, the ohms unit.

Value. The magnitude of a physical property expressed in terms of a reference unit, such as 117 volts, 60 Hz, 50 ohms, 3 henrys.

VAR. Denotes volt-amperes reactive; the unit of imaginary power (reactive power).

Variable Component. A component that has a continuously controllable value, such as a rheostat or movable-core inductor.

Vector. A graphical symbol for an alternating voltage or current, the length of which denotes the amplitude of the voltage or current, and the angle of which denotes the phase with respect to a reference phase.

Ventilated. A ventilated component is provided with means of air circulation for removal of heat, fumes, vapors, and so on.

Vibrator. An electromechanical device that changes direct current into pulsating direct current (direct current with an ac component).

Volt. The unit of emf; one volt produces a current of one ampere in a resistance of one ohm.

Voltage. In a circuit, the greatest effective potential difference between a specified pair of circuit conductors.

Voltmeter. An instrument for measurement of voltage values.

Watt. The unit of electrical power, equal to the product of one volt and one ampere in dc values, or in rms ac values.

Watt-hour. A unit of electrical energy, equal to one watt operating for one hour.

Wattmeter. An instrument for measurement of electrical power.

Wave. An electrical undulation, basically of sinusoidal form.

Weatherproof. A conductor or device designed so that water, wind, or usual vapors will not impair its operation.

Wind Bracing. A system of bracing for securing the position of conductors or their supports to avoid the possibility of contact due to deflection by wind forces.

Wiper. An electrical contact arm.

Wire Nut. The most commonly used type of solderless connector.

Work. The product of force and the distance through which the force acts; work is numerically equal to energy.

Working Voltage. Also termed *closed-circuit voltage*; the terminal voltage of a source of electricity under a specified current demand; also, the rated voltage of an electrical component such as a capacitor.

X Ray. An electromagnetic radiation with extremely short wavelength, capable of penetrating solid substances; used in industrial plants to check the perfection of device and component fabrication (detection of flaws).

Y Joint. A branch joint used to connect a conductor to a main conductor or cable for providing a branched current path.

Y Section. Also termed *T section*; an arrangement of three resistors, reactors, or impedances that are connected together at one end of each, with their other ends connected to individual circuits.

Zero-Adjuster. A machine screw provided under the window of a meter for bringing the pointer exactly to the zero mark on the scale.

Zero-Voltage Level. A horizontal line drawn through a waveform to indicate where the positive excursion falls to zero value, followed by the negative excursion. In a sine wave, the zero-voltage level is located halfway between the positive peak and the negative peak.

Index

A

abnormal operation, 386
AC circuits
  capacitive reactance, 163–166
  impedance in, 161–162
  inductive, 153–163
  Ohm's law in, 133–136
  parallel, capacitance and resistance in, 177–179
  parallel, inductance, capacitance, and resistance in, 179–182
  parallel, inductance and resistance in, 175–177
  power in an impedance, 162–163
  resistance in, 159–161
  series, capacitive reactance and resistance in, 166–171
  series, inductance, capacitance, and resistance in, 172–175
  transformers in, 142–148, 151
AC generator, 129
AC-operated relays, 344–345
AC resistance, 150
AC voltage. *See* alternating current
active electrical network, 417
active network, 429
access fitting, 417
accidental ground, 425
adapter
  conduit, 297
  for grounding, 356
adjustable core coils, 153
admittance, 417
aiding magnetic fields, 9–10
air, conduction of electricity by, 45–50
air circuit breaker, 386, 387, 417
air conditioning, planning circuitry for, 311
air core, 16, 154
air switch, 417
air-blast transformer, 417
alarm devices, 359–365
  annunciator systems, 360, 362
  closed-circuit systems, 360, 361
  controls, 363
  infrared light beam systems, 359
  keypad, 364
  open-circuit systems, 360
  photoelectric bell control units, 359, 360, 378
  power supplies, 364
  sensors, 360
  signaling devices, 363
  wireless systems, 365
alarms
  annunciator wiring systems, 360, 362
  automatic night-lights, 360
  backup battery, 364
  bells and chimes, 376–378
  burglar alarm system, sample schematic for, 361
  central control panel, 363
  closed-circuit systems, 360, 361
  controls, 363
  devices, 359–365, 378
  false, automatic handling of, 363
  fire, 360
  infrared light beam devices, 359
  installation, 359
  keypad control of, 363–364
  new systems, advantages of, 363
  open-circuit systems, 360
  photoelectric bell control units, 359, 360, 378
  power failure, 365
  power supply, 364
  sensors, 360
  signaling devices, 363
  smoke alarms, 360
  time delay, 363
  wireless systems, 365
alive, defined, 417
alloys, 5, 33
Alnico alloy, 5, 9, 33
alternating current (AC), 129–151
  circuits. *See* AC circuits
  core loss, 148
  core lamination, 148–150

439

440 Index

alternating current (AC) (*continued*)
 defined, 129, 417
 and direct current compared, 132
 frequency of, 129, 151
 instantaneous and effective voltages, 129–132
 Ohm's law and, 133–136
 resistance, DC vs. AC, 150
 resistive AC currents, power laws in, 136–137
 sine curve during cycle, illustration of, 130
 three-phase system, 434
 transformer action in AC circuits, 142–148
 voltages, combining 137–142
 voltmeter, 129, 131–132
alternator, 129, 417
aluminum wire, 43–44
ambient temperature, defined, 417
American Standards Association, 39
American Wire Gauge, 39–43
ammeter
 connecting, 61–62, 86
 defined, 417
 troubleshooting with, 33
 use of, 17, 19, 94
ampere, 17, 423
 defined, 417
ampere-turns, 19–20
angle
 measurement of, 214
 phase, 139, 398, 399, 431
 sines and tangents of, 215–216
angle connectors, 297
angstroms, 185
annunciator, defined, 417
annunciator wiring alarm systems, 360, 362
apparent power, 159, 162, 417
appliance(s)
 average power consumed by, 252
 defined, 417
 fixed, 311
 portable, 311

appliance circuits
 branch, 419
 high-wattage, individual, 246
 National Electric Code for, 246
 planning for, 311
 regular, 246
applied voltage, 121, 182
aqueous solution, 50
arc lamps, 188
arc-fault circuit interrupter (AFCI), 331–332
 defined, 417
arcing contacts, 418
arcing time of fuse, 418
arc-over of insulator, 418
arcs, 47–48
argon gas, 205
armature, 125
armor clamp, 418
armored cable, 235
 advantages, 279, 297
 bending, 284, 287
 and conduit compared, 279, 297
 connecting to junction box, 281–282, 285–286
 cutting, 279–280
 defined, 418
 fishing in old houses, 282
 installing 280–284
 and nonmetallic sheathed cable compared, 279
 preinstallation examination, 280
 protective bushing, 280, 283, 284
 removing the metal casing, 281–282
 restrictions on, 297
 for service entrances, 279–280
 uses of, 279
arrester, lightning, 119, 418
artificial light, 185
atom, 11
attic, lighting for, 324
attractive force, 4
automatic station, 381, 434

automatic substation, 381,
    407–410, 413
  interlocks and, 409
  operation of, 408
  plans for, 409
  starting and stopping, 408–409
automobile ignition system,
    diagram of, 117
autotransformer, 418

B

back-connected switch, 418
back emf, 121
back-wire outlet receptacles, 271
balanced lines, 371
ballast, 197, 331
bank
  defined, 418
  duct, 418
bar magnets, 5, 9
baseboard heater, 337–338, 339, 356
base load, defined, 402–403
base load plant, 402
basement, lighting for, 324–326, 335
basic house wiring. *See* house wiring
bathroom, lighting for, 323–324, 335
battery
  backup, alarms, 364
  defined, 25
  polarization, 431
  primary, 66, 431
  resistance of, 63–65, 86
  secondary, 66, 433
  storage, 66, 68
  testers, 63–64, 65
bayonet candelabra lamp bases, 191
bayonet lamp bases, 191
bedroom
  circuit breakers required for, 332
  lighting for, 320–323, 335
bells and chimes, 376–378
  in hospitals, 377
  installation of, 377
  outdoor installations, 378
  photoelectric control of, 360, 378
  tone signal system, 376–377
  wiring of, 376
benchboard, 418
B-H curve, 23–25
bidirectional current, 418
bimetallic thermostat, 341, 356
bipost lamp bases, 191
black insulated wire, 286, 307
blower installation, kitchens, 319
bond, cable, 418
box(es)
  ceiling, 274
  conduit, 284, 418
  connecting cable to, 262–263
  exact location of, 258
  junction, 281–282, 285–286, 418–419, 427
  maximum number of conductors per, 255
  mounting. *See* box mounting
  pull, 284, 419, 431
  transfer, 435, *See also* pull box
box mounting
  metal plates, 261
  old work, 256–258
  running boards, 261–262
  running sheathed cable, 259–260
  special devices for, 258–259
  step-by-step instructions for, 259
branch circuit, 69, 70–71
  appliance, 419
  defined, 69, 86, 419
  lighting, 419
branch circuit distribution center, 419
branch conductor, 419
break, defined, 419
breakdown, 419
breaker, line, 419
brightness, units of, 225
brushes, 125
building feeder, 229
built-in lighting, 311, 313

bulbs, lamp
  blackening of, 188–189, 190, 210
  light-center length, 193
  light diffusion, 193
  three-way, 316
  types of, 193–194
buried cable, 419
burglar alarm system, sample schematic for, 361
bus, defined, 419
bushing(s)
  defined, 419
  insulated, 419
  for protecting cable, 280, 283, 284
  purpose of, 237, 240

C
cabinet, 419
cable(s)
  armored, 235, 279–284
  bond, 418
  buried, 419
  bushings for protecting, 280, 283, 284
  coaxial, 371
  connecting to boxes, 262–263
  defined, 419
  fault, 420
  fishing, in old work, 263–268
  heating, 354, 356
  for intercom installation, 369–370, 371
  joint, 420
  lay, 428
  nonmetallic sheathed, 55, 235, 259–260
  plastic-insulated sheathed, 55
  raceway, 432
cable rack, 432
  service, 420
  serving of, 433
  sheath, 420, 433
  twin, 436
  twisted pairs, 371
  type AC, 235, 265, 279, 280
  type ACT, 279
  type ASE, 279
  type ACV, 279
  type BX, 235
  type NM, 235, 259–260, 262, 265
  type UF, 262, 419
  type USE, 279
  underground, 279–280, 436
candelabra lamp bases, 191
candle per square inch, 225
candlepower, 30, 213
candlepower distribution curve, 213
capacitance, 164, 182
capacitive reactance, 182
  in AC circuits, 163–166
  and resistance in series, 166–171
capacitors, 116, 164, 182
  Ohm's law for, 166
  Static, 169, 170
  types of, 169
capacity, 68, 164
carbon arcs, 48
ceiling box, 274
ceiling cable heating unit, 339
cell(s)
  dry, 12, 25, 64, 65, 66
  parallel connection of, 84–86
  series-parallel connection of, 92–93
  standard, 434
central control panel, alarm systems, 363
charge, electric, 116, 420
charge flow, 18–19
charge separator, 12
choke coil, 119–122, 420
circuit(s), 12
  AC, 133–136, 153–182
  appliance, 246
  arc-fault interrupter, 331–332
  branch, 69, 70—71, 86, 419
  with capacitance and resistance connected in parallel, 177–179

Index

with capacitive reactance and resistance connected in series, 166–171
closed, 25
closed electric, 420
contacts, 421
earth (ground) return, 420
electric, 420
equivalent, 60
general-purpose, residential, 246, 250
ground, 45, 47, 352–354, 420
ground-fault interrupter, 331, 425
improving, 251
with inductance and resistance connected in parallel, 175–177
with inductance, capacitance, and resistance connected in parallel, 179–182
with inductance, capacitance, and resistance connected in series, 172–175
leg of, 428
magnetic, 20–25, 420
open, 27–28
open-wire, 430
outdoor, 425
parallel, 69–73, 86–87
parallel-resonant, 179
power-factor correction, 180–181
probable consumption on, calculating, 251
reduction, 94–98
resistive and inductive compared, 154
series, 59, 60–62, 69, 166–175, 433
series-opposition voltages, 66–69
series-parallel, 89–106
series-resonant, 175, 183
short, 433
circuit breaker(s)
air, 386, 387, 417
for bedroom, 332
defined, 385
electric water heaters, 350–351
generating station, 385–386
mounting, 429
quick-break, 432
quick-make, 432
tripping, reason for, 331
vacuum, 386, 389
circuit reductance
by conductances, 98
by pairs, 97
by product-and-sum formula, 97
by reciprocals, 97–98
circuit reduction, 94–98
circuiting, 244–246. *See also* home circuiting
for air conditioning, 311
appliances, planning for, 311
for electric heaters, 341–344
example of typical circuitry for farm home with outbuildings, 333–334
functional planning, 310, 332
lamp control from electrolier switch, 303, 305–307
possible circuit arrangements, 301–303
preplanning, importance of, 301, 332
safety considerations, 310, 332
symbols used to denote, 247–250
wiring plan for two-story home, 312
circular mil, 36, 56
converting to square mil, 37
circular mil-foot, 37–38
class A insulation, 53
class B insulation, 53
class C insulation, 53
class O insulation, 53
cleat, 420
clip, fuse, 420
closed circuit, 25
closed electric circuit, 420
closed-circuit alarm systems, 360, 361
closed-circuit voltage, 420

closed-core transformer, 151
closets, lighting for, 323
coefficient of utilization, 218–219
coercive force, 420
coil(s)
  adjustable core, 153
  choke, 119–122, 420
  current buildup in, 124
  defined, 421
  flip, 108, 126
  heat, 425
  ignition, 113, 124. *See also* spark coils
  inductance, 153
  iron core, advantage of, 119, 120, 154
  spark, 113, 116, 117, 434
  winding resistance of, 150
collector grid, 189
color-coded switch wiring, 307, 309–310
colored lighting, 317, 332
compass, 1
composite conductor, 421
concealed, defined, 421
condenser(s), 164, 421
conductance
  defined, 421
  measuring, 73
conductance values, 73–78
conduction
  earth, 45
  of electricity by air, 45–50
  of electricity by liquids, 50–52
  mercury-vapor arc, 50
  sparks and arcs, 46–48
  transfer of heat by, 337
conductor(s)
  air as, 45–50
  aluminum wire, 43–44
  American Wire Gauge, 39–43
  branch, 419
  circular mil-foot, 37–38
  classes of, 35–36
  composite, 421
  defined, 35, 55, 421
  earth (ground) conduction, 45
  electrolytic, 423
  gases as, 49–50
  jumper, 427
  line drop, 44
  liquids as, 50–52
  maximum per box, 255
  resistance in, 35, 55
  stranded wires, 43
  temperature coefficient of resistance, 44–45
  wires used as, 36–37, 56
conduit
  adapter, 297
  and armored cable compared, 279, 297
  bending, 290, 291, 297
  box, 284, 418
  connectors, 296–297, 298
  continuous run, 292
  cutting to length, 291, 299
  defined, 421
  duct bank, 418
  elbows, 284, 288, 290–291, 297
  electrical metallic tubing (EMT), 235, 284, 291–292, 297, 435
  fishing, 293
  fittings, 284–293, 294, 295, 297, 421
  flexible metal, 293–296, 298, 299, 421
  grounding, 291
  hanger support, 292
  reaming, 291, 297
  restrictions on, 299
  rigid steel, 297, 421
  run, 421
  service entrance, 236–237, 274
  size, selection of, 287–288, 289, 297
  threaded couplings, 291
  threading, 291
  threadless fittings, 291–292
  weatherproof, 240, 241
  wiring, 292–293
conjugate impedance, 426
connectors
  angle, 297
  conduit, 296–297, 298

duplex, 297
solderless, 268–271, 434
squeeze, 296–297
connection diagram, defined, 422
conservation of energy, law of, 31
constant-current rating, dry cell, 64
contactor, 421
contacts, 421
continuity tester, 32, 33
control relay, 421
controls, alarm systems, 363
convection, transfer of heat by, 337
convenience outlets
in the basement, 324
in the bathroom, 324
in the bedroom, 323
in the garage, 326
in the kitchen, 319
in the living room, 314
load, 229
in offices, 231
planning for, 311
copper loss, transformers, 150
copper sulfate, 50
copper wire, 36, 40–42, 56
cord, 43
core, 16
air, 16, 154
flux density, 23
lamination, 148–150
loss, 148
reluctance of, 20
shell, 433
core-type transformer, 145, 151
coulomb, 163, 421, 423
counter emf (cemf), 121, 421
cove lighting, 311, 314
crest voltage, 129, 151
current
bidirectional, 418
buildup of in coils, 124
defined, 34, 421
displacement, 422
eddy, 148, 151, 345
fault, 424
induced, 426
inductive, 182
instantaneous, 137
no-load, 429
Ohm's law and, 17
peak, 430
pulsating, 431
starting surge of, 189–190
unidirectional, 436
current-carrying, defined, 417
current limitation, 118
current transformer, 386, 390
cycle, defined, 422

D

DC resistance, 150
DC voltage
and AC voltage compared, 132
combining, 137
dead, defined, 422
degree, electrical, 422
diagram, connection, defined, 422
dielectric, 422, 424
diffusion, 194
dining room, lighting for, 316–318, 332, 335
direct current (DC)
defined, 422
resistance, 150
voltage, 132, 137
direct emf, 422
direct lighting, 224–225, 232
direct-lighting luminaires, design data for, 415
direct voltage, 422
discharge, 118, 422
disruptive, 422
flashover, 424
sparkover, 434
displacement current, 422
disruptive discharge, 422
distribution, generating stations, 400–403
defined, 401
distribution center, defined, 403, 422
divider, voltage, 422
dome reflectors, design data for, 415–416

double-bus system, generating station, 387–392
  with double-circuit breaker, 392, 396
  H system, 391, 394
  ring-bus system, 391, 393
  with single-circuit breaker, 391
  with single low- and high-tension bus, 392, 395
  with tie bus, 391
double-strip flexible metal conduit, 294
drop, voltage, 30, 44, 60, 63, 227, 422
dry cell, 66
  constant-current rating of, 64
  electromotive force of, 12
  intermittent duty rating of, 64
  internal resistance of, 65
  series-connected, 25
duct
  bank, 418
  defined, 422
  entrance, 423
duplex connectors, 297
duplex receptacles, 231
dyne, 4

E

earth conduction, 45
eddy current, 148, 151, 345
effective power, 137
effective value, 422–423
effective voltage, 129, 151
efficiency
  calculating, 65–66, 86, 99
  defined, 65, 423
  of a power distribution system, 175
eight watts per square foot, illumination provided by, 232
elbows, conduit, 284, 288, 290–291, 297
electric bell, operation of, 113
electric charge, 116, 420
electric circuits, 25–30, 59–87
  circuit voltages in opposition, 66–69
  closed, 25
  conductance values, 73–78
  defined, 59, 86, 420
  efficiency and load power, 65–66
  open, 27–28
  parallel, 69–73, 86–87
  parallel connection of cells, 84–86
  picture diagrams, 59
  principles of parallel, 69–72
  resistance of a battery, 63–65
  schematic diagrams, 59–60
  series circuit, 25, 59, 60–62, 86
  voltage measurements with respect to ground, 62–63
  voltage polarities, 60–62
electric current
  as electron flow, 11, 17, 33
  gases as conductors of, 49–50
  magnetic field produced by, 13–14
  as rate of charge flow, 18–19
electric duct heating unit, 339
electric fence, operation of, 118–119
electric heating
  AC-operated relays, 344–345
  circuiting, 341–344
  conduction, 337, 356
  convection, 337, 356
  electric water heaters, 347–352, 357
  general installation considerations, 337–341, 356
  ground circuit, 352–354
  heating cables, 354–356
  heat transfer, methods of, 337, 356
  hot-water electric heat, 346, 357
  radiant heater, 346–347, 357
  radiation, 337, 356
  shading coils, 345
  space heaters, 337
  thermal-type relays, 341, 344
  thermostats, 341–344, 356–357
electric lighting, 185–210
  aging characteristics, 190–191

Index **447**

electroluminescent panels, 209
flash tubes, 206–208
high-pressure sodium (HPS)
  lamps, 203
illumination, measurement of,
  186–188
infrared lamps, 208
lamp bases, 191–193
lamp bulbs, 193–194
light sources, 185–186
low-pressure sodium (LPS)
  lamps, 203–204
metal halide lamps, 202–203
miniature lamps, 192
ozone-producing lamps,
  205–206
sodium lamps, 197, 203–204
starting surge of current,
  189–190
sun lamps, 204–205
troubleshooting, 32–33
tungsten filaments, 188–189
vapor-discharge lamps, 195–202
electric range, wiring for, 320
electric water heaters, 347–352
  circuit breakers for, 350–352
  demand, and operation of,
    348–349, 357
  grounding, 352–354
  ground rod, 352, 353
  immersion heaters in, 347–348,
    357
  installation of, 349–352
  polarized devices, use of, 352,
    354, 355
  time switches, 349
electrical codes. See National
  Electrical Code
electrical degree, 422
electrical energy, 31–32
  discharge, 118, 422
  measurement of, 423
electrical metallic tubing (EMT),
  235, 297
  and conduit compared, 284,
    291–292
electrical power, 30–32
  basic formulas of, 31–32

  law of conservation of energy, 31
  measurement of, 30–31
electrical pressure, 11
electrical symbols, 59–60,
  247–250, 386, 434
electrical troubleshooting, 32–33
electrical units
  ampere, 14, 417, 423
  coulomb, 163, 421, 423
  defined, 423
  farad, 163, 423
  henry, 122, 153, 154, 423, 426
  joule, 423, 427
  mho, 73, 423, 429
  ohm, 17, 122, 153, 161, 423,
    429
  volt, 12, 17, 423, 436
  watt, 31, 99, 162, 165, 423, 436
  watt-hour, 32, 423
electricity
  applications of, xiii
  conduction of by air, 45–50
  conduction of by liquids, 50–52
  defined, 423
  electric circuits, 19–20, 25–30
  electromagnetism, 11–17
  importance of, xiii
  left-hand rule of, 14, 16, 122
  magnetism and, 1–34
  total quantity of, 110
  volts, amperes, and ohms, 17–19
electrode, 423
  grounding, 425
electroluminescent (EL) panels,
  209
electrolyte, 63, 423
electrolytic conductor, 423
electromagnet
  ampere-turns, 19–20
  defined, 14
  flywheel effect of, 112
  solenoid type, 16
  strength of, 16–17, 20
electromagnetic induction
  choke coils, 119–122
  defined, 423
  flip coil, 108, 126
  galvanometer, 108, 114, 126

electromagnetic induction (*continued*)
  generator principle, 124–126
  induced secondary voltage, reversal of, 122
  iron core, 119, 126
  laws of, 109–110
  principle of, 107–108
  self-induction, 110–113
  switching surges, 122–124
  transformers, 113–119, 126
electromagnetism, 11–17, 33
  defined, 11
electromotive force (emf) 33, 63. *See also* voltage
  back, 121
  counter, 121, 421
  defined, 11, 423
  direct, 422
  and magnetomotive force compared, 20
  measurement of, 11
  produced by the source of electricity, 30
  and voltage compared, 30
electron, 423
electron flow, 11, 17, 34
electronics, defined, 423
electroplating, 50–51, 423
electrostatics, 423
ells (conduit elbows), 284, 297
  measuring, 288, 290–291
emf. *See* electromotive force (emf)
energy, defined, 423
entrance, duct, 423
entrance panel, making connections to, 241–244
equipment, service, 423
equivalent circuit, 60
exciter, 382, 412, 424
extension cords, 318

F

false alarms, automatic handling of, 363
farad(s), 163, 423, 424
fault current, 424

feeder
  building, 229
  defined, 232, 424
  multiple, 429
  positive, 100, 105
  service, 229, 232
  tie, 434
ferromagnetic substance, 424
filament(s)
  defined, 424
  helical, 188, 210
  resistance, 190
  tungsten, 188–189, 208, 210
fire alarms, 360
fish tape, 265, 268, 293
fish wire, 265–266
fishing cable in old work, 263–268, 276
  bypassing one floor of a two-story home, 266–268
  step-by-step instructions for, 267
  two-person technique, 265–266
fishing conduit 293
fitting(s)
  access, 417
  conduit, 284–293, 294, 295, 297, 421
  threadless, 291–292
fixed appliances, 311
fixture stud, 424
fixtures, mounting, 274, 275
flash tubes, 206–208
flashover, 424
flexible metal conduit
  advantages/disadvantages of, 293–294
  connecting to outlet box, 298
  defined, 293, 421
  double-strip, 294
  fittings, 294, 295
  requirements for, 294
  restrictions on, 294
  single-strip, 294
  uses of, 294
  wiring in, 293–296
flexible tubing, 435
flip coil, 108, 126

floodlights, outdoor, 311
fluorescence, 424
fluorescent lamps, 199–200, 210, 327–331
  ballast, 331
  high-voltage, 195
  instant-starting, 330
  in kitchens, 319
  starters, 199–200, 330
  stroboscopic effect in, 200, 201
  types of, 327, 328–329
  uses of, 195
flux, 5, 33, 121, 205, 424
flux density, 4, 23
fluxmeter, 108
flywheel effect, 112
foot-candle, 186, 210, 217–218
foot-lambert, 225
force, 4
  defined, 424
four watts per square foot, illumination provided by, 232
frequency
  of an AC voltage, 129, 150–151
  defined, 424
frequency meter, 424
furnace rooms, illumination of, 324–325
fuse
  arcing time of, 418
  clip, 420
  cutout, 424
  defined, 28, 424
  element, 425
  holder, 425
  link, 425
  operation of, 28–30, 331

G

galvanometer, 108, 114, 126, 425
gap, 121, 425
garages
  door openers, wiring for, 326
  lighting for, 326
garden lighting, 314
gases, as electrical conductors, 49–50
gas-filled lamps, 188, 198, 194, 199
gauss, defined, 3–4
gaussmeter, 108
general-purpose residential circuits, 246, 250
general-service lamps, 311
generating station, 381–403
  automatic, 381, 434
  auxiliaries, 403
  auxiliary power, 403
  base load plant, 402
  circuit breakers, 385–386
  classifications of, 381, 412
  defined, 381, 411
  distribution, 400–403
  double-bus system, 387–392, 413
  electrical connections in, 385, 412
  exciter, 382, 412, 424
  generator, plan of, 381, 383
  hydroelectric, 382
  instrument transformers, 386
  main bus, 413
  main circuits, arrangement of, 405–408
  parallelling generators, 392–400
  prime mover, 382
  purpose of, 412
  single-bus system, 385, 413
  spare bus, 413
  turbines, 382, 384, 412
generator(s), 125
  AC, 129
  connected in parallel, 392, 395, 397–400
  principle, 124–126
  typical, cutaway view of, 381, 383
geographic poles, 3
getters, 188
glare, 224, 311, 332
gold cyanide, 52
Greenfield conduit. *See* flexible metal conduit

ground
  accidental, 425
  defined, 425
  voltage measurements with respect to, 62–63
ground circuit
  defined, 420
  electric water heaters, 352–354
  use of, 45, 47
ground fault circuit interrupters (GFCIs), 331, 425
ground lug, 425
ground outlet, 425, 433
ground rod, 352, 353
ground switch, 425
grounded wire, 286
grounding, 352
  adapter for, 356
  conduit, 291
  electric water heaters, 352–354
grounding electrode, 425
guy, 425

H

H system, 391, 394
halls, lighting for, 312–314
handle ties, 350
hanger
  defined, 425
  types of, 292
heat, defined, 185
heat coil, 425
heater, defined, 426
heating. *See also* Electric heating
  conduction, 337, 356
  convection, 227, 256
  installation, general considerations, 337–341, 356
  planning circuitry for, 311
  radiation, 337, 356
  space heaters, 337
  thermostats, 341–344, 357
heating cables, 354, 356
heating units, types of, 337–341
heat lamps, 208, 210

heat transfer
  by conduction, 337
  by convection, 337
  by radiation, 337
helical filaments, 188, 210
henry, 122, 153, 154, 423, 426
hertz, 151
hickey, 290, 291, 297, 426
high-pressure sodium (HPS) lamps, 203
high-voltage, cold-cathode vapor-discharge lamps, 195, 210
high-voltage fluorescent lamp, 195
home circuiting
  arc-fault circuit interrupters (AFCIs), 331–332
  attic, 324
  basement, 324–325
  bathroom, 323–324
  bedroom, 320–323
  built-in lighting, 311, 314
  cove lighting, 311, 314
  dining room, 316–318
  entrance lighting, 311
  floodlights, outdoor, 311
  fluorescent lights, 327–331
  functional planning, 310, 332
  garage, 326
  garden lighting, 314
  ground circuit fault interrupters, 331
  halls, 312–314
  kitchen, 318–3320
  lamp control, preplanning, 301, 332
  lamp control from electrolier switch, 303, 305–307
  lighting, general requirements for, 311
  living room, 314–316
  location of lamps from more than one location with three- and four-way switches, 307
  night lights, 311
  outdoor lighting, 311–312

pilot lights, 311
planning of equipment, 311–312
possible circuit arrangements, 301–303
safety considerations, 310, 332
stairway lamp control lighting, 307
switch wiring, color-coding, 307, 309–310
wiring plan for two-story home, 315
horn gap, 426
horsepower, 99–100, 105
horseshoe magnets, 5, 33
hot, defined, 417
hot-cathode, low-voltage neon lamp, 195
hot-filament, low-voltage vapor-discharge lamps, 195, 210
hot-water electric heat, 346
house wiring, 235–276
  appliance circuits, 246, 274
  boxes, mounting, 255–262, 276
  cable, connecting to boxes, 262–263
  ceiling box, 274
  circuitry, 244–246, 274
  efficiency in, 320
  entrance panel, making connections to, 241–244, 274
  fishing cable in old work, 263–268, 276
  fixture mounting, 274, 275
  general-purpose circuits, 246, 250
  improving circuits, 251
  National Electrical Code, standards for, 235, 274
  new work in old houses, 250–251
  outlet receptacles, installing, 271–273, 276
  outlets, placement of, 229, 251–254, 274
  roughing-in, 254–255, 274, 276

  service connections, 236–240, 274
  service entrance, 236–237, 238, 239, 274
  single-pole switches, 273–274
  solderless connectors, 270–271
  splicing, 268–270, 276, 285
  standard electrical symbols, 59–60, 247–250, 386, 434
  switches, placing, 254, 276
  three-wire line installation through service entrance, 320, 321
  underground service, 240–241, 242
hydrometer, 426
hysteresis, 426
hysteresis loop, 420, 426
hysteresis loss, 426

I

ignition coils, 113, 124. *See also* spark coils
illumination
  average level of, 217
  eight watts per square foot, 232
  four watts per square foot, 232
  intensity, 186
  lamp wattage, 226–227
  level, classifications of, 226
  lumen method of calculating, 217–223
  measurement of, 186–188
  one watt per square foot, 231
  point-by-point method of calculating, 213–217
  six watts per square foot, 232
  two watts per square foot, 231–232
  uniform, obtaining, 220–223, 332
  on vertical to horizontal plane, ratio of, 216
  watts per square foot, 213, 231–232
image-retaining panel, 209
immersion heaters, 347–348, 357

impedance
  in AC circuits, 161–162
  defined, 161, 192, 426
  conjugate, 426
  measurement of, 161
  power in, 162–163
impedance ratio, 146
impregnated insulation, 53
impulse, 426
incandescent lamps, 186
indirect lighting, 225–226, 232, 316, 332
individual appliance circuits, 246
indoor transformer, 426
induced current, 426
induced secondary voltage, reversal of, 122
induced voltage, 121
inductance
  and current flow, 153–154
  defined, 182, 426
  mutual, 429
inductive AC circuits
  circuit action, 153–157
  impedance in, 161–163
  Ohm's law for, 155, 157
  power in, 157–159
  resistance in, 159–161
  uses of, 153
inductive current, 182
inductive opposition, 121
inductive reactance, 121
inductor
  defined, 426
  Ohm's law for, 154–155
  resistance in, 426
  variable, 153
industrial intercoms, 375
infrared heating devices, 208
infrared lamps, 208, 210
infrared light beam alarm systems, 359
instantaneous current, 137
instantaneous power, 137, 426
instantaneous voltage, 129, 137, 151

instant-starting fluorescent lamps, 330
instrument, defined, 427
instrument multiplier, 429
instrument transformers, 386
insulated bushing, 419
insulator(s)
  arc-over of, 418
  classes of, 35, 53
  defined, 35, 55–56
  electrical code and, 43
  impregnated, 53
  knob, 52
  plastic-insulated sheathed cables, 55
  porcelain, 51, 52
  resistance, 35, 53
  spool, 52
  for support of wires, 52–53
intercoms
  adding speakers, 371, 379
  all-call function, 375
  balanced and unbalanced lines, 371, 372
  in buildings under construction, 370, 379
  connecting, 369–371, 379
  front-door station, 370–371, 379
  home installation, 367, 379
  industrial installations, 375
  interference, avoiding, 367–368, 373, 379
  in-wall vs. on-wall, 370. 380
  layout of, 368–369, 379
  master station, 365–366, 378
  music distribution systems, 367, 374–375, 379
  National Electric Code for, 367–368, 375–376
  nonprivate installation, 366–367, 378–379
  operation of, 365
  planning details, 367, 370–371, 379
  privacy, 374
  purpose of, 365, 378
  school installations, 375–376

substations, 365–366, 378
system planning, 365–370
wiring, 368, 371
wireless systems, 365, 371–374, 379
interconnection, system, 427
interconnection tie, 427
interior lighting installation, 220
interlock, 409, 427
intermediate lamp bases, 191
intermittent duty rating, dry cell, 64
internal resistance
 defined, 63, 86, 427
 dry cell, 65
inverse square law, 213
ion, 427
IR drop, 427
iron core, 119, 120, 154
isolated magnetic pole, 4
isolating switch, 427

J
joule, 423, 427
Joule's law, 427
jumper, 427
junction, 91, 427
junction box, 418–419, 427
 connecting armored cable to, 281–282, 285–286

K
keypads, for controlling alarm systems, 363–364
kilovolt amperes. *See* kVA
kilovolts, 169
kilowatt-hour, 32
Kirchhoff's current law, 78–80, 87, 90–92, 105, 427
Kirchhoff's voltage law, 30, 61, 105, 137, 427
kitchen, lighting for, 318–320
knife switch, 28
knob insulators, 52
knockout, 427–428
krypton, 188
kVA, 428

L
lag, 166, 428
lagging phase, 166, 345, 398
lambert, 225
lamp bases, 191–194
 bayonet, 191
 bayonet candelabra, 191
 bipost, 191
 candelabra, 191
 intermediate, 191
 medium, 191
 mogul, 191
 prefocus, 191, 193
 screw-type, 191
lamp bulbs
 blackening of, 188–189, 190, 210
 light-center length, 193
 light diffusion, 193
 three-way, 316
 types of, 193–194
lamp control
 from electrolier switch, 303, 305–307
 from more than one location with three- and four-way switches, 307, 308
 multiple lamps and multiple switches installed in separate locations, 302, 303
 multiple lamps from single switch, 301, 302
 pair of lamps connected in parallel, controlled by switches in separate locations, 302, 303
 pair of lamps individually controlled by one three-way switch, 301, 302
 possible circuit arrangements, 301–303
 preplanning, 301, 332
 stairway lighting, 307
 in two separate locations, 303, 304
lamp holder, 428
lamp receptacle, 428

lamps
  aging characteristics of, 190–191
  arc, 188
  bases, 191–193
  bulbs, 193–194
  efficiency of, 190–191
  electroluminescent panels, 209
  filament resistance, 190
  flash tubes, 206–208
  fluorescent, 195, 199–200, 319, 327–331
  gas-filled, 188, 198, 194, 199
  general-service, 311
  heat, 208, 210
  helical filaments, 188, 210
  high-pressure sodium (HPS), 203
  incandescent, 186
  infrared, 208, 210
  low-pressure sodium (LPS), 203–204
  Lumiline, 311, 319
  mercury-vapor, 195, 200–201, 202
  metal halide, 202–203
  miniature, 191, 192, 193
  neon, 195
  neon glow, 195
  ozone-producing, 205–206
  possible circuitry for control of, 301–303
  RF, 330
  size, selection of, 219–220
  smashing point, 191
  socket, 428, 434
  sodium, 197, 195, 201–202, 203–204
  starting surge of current, 189–190
  styles used in various rooms by interior decorators, 322–323
  sun, 204–205, 210
  three-way, 316
  tubular, 311
  tungsten filaments, 188–189, 208, 210
  ultraviolet germicidal, 195
  vapor-discharge, 195–202
  wattage, 226–227
laundry room, lighting for, 324
laws, pertaining to electricity
  conservation of energy, 31
  electromagnetic induction, 109–110
  inverse square, 213
  Kirchhoff's current, 78–80, 87, 90–92, 105, 427
  Kirchhoff's voltage, 30, 61, 105, 137, 427
  Joule's, 427
  Lenz's, 109, 110, 428
  magnetic circuit, 21
  Ohm's, 17–19, 25–27, 30, 33–34, 71, 133–136, 154–155, 157, 166, 429
  power, in resistive AC circuits, 136–137
lay, 428
lead, 428
leading phase, 166, 398
leakage, surface, 428
leakage flux, 145, 151
left-hand rule of electricity, 14, 16, 122
leg of a circuit, 428
Lenz's law, 109
  defined, 428
  illustration of, 110
light
  artificial, 185
  defined, 185
  diffusion, 194
  distribution of, by type, 223, 224
  monochromatic sources, 203
  natural, 185
  reflection of from walls and ceiling, 219, 223
  sources of, 185–186
  wavelengths, measurement of, 185
light-center length, bulbs, 194
lighting, 185–210
  aging characteristics, 190–191
  bathroom, 323–324, 335

bedroom, 320–323, 335
branch circuit, 419
brightness, units of, 225
built-in, 311, 313
calculations. *See* lighting calculations
classification of, 223
closets, 323
colored, 317, 332
cove, 311, 314
dining room, 316–318, 332, 335
direct, 224–225, 232
electroluminescent panels, 209
entrance, 311
fixtures. *See* lighting fixtures
flash tubes, 206–208
glare, 224, 311, 332
halls, 312–314
high-pressure sodium (HPS) lamps, 203
illumination, measurement of, 186–188
indirect, 225–226, 232, 316, 332
infrared lamps, 208
interior installation, 220
kitchen, 318–320, 335
lamp bases, 191–193
lamp bulbs, 193–194
lamp wattage, 226–227
light sources, 185–186
living room, 314, 316, 332
localized, 318, 332
low-pressure sodium (LPS) lamps, 203–204
metal halide lamps, 202–203
miniature lamps, 192
outdoor, 311
outlet, 430
ozone-producing lamps, 205–206
requirements for, 311, 332
semidirect, 225, 232
semi-indirect, 225, 232
shadows, 311, 332
sodium lamps, 197, 203–204
Soffit, 319
spacing of units for, 220–223
starting surge of current, 189–190
sun lamps, 204–205
supplementary, 316, 320
troubleshooting, 32–33
tungsten filaments, 188–189
vapor-discharge lamps, 195–202
lighting calculations
inverse square law, 213
lamp wattage, 226–227
level of illumination, 226
lighting equipment, 224–226
load and length of run, 229
lumen method, 217–223, 232
panelboards, 229
perpendicular to the beam, 214
point-by-point method, 213–217, 232
service and feeders, 229–231, 232
watts per square foot, 231–232
wire capacity, 227–229
lighting control, principle of, 153
lighting equipment
for direct lighting, 224–225, 232
for indirect lighting, 225–226, 232
for semidirect lighting, 225, 232
for semi-indirect lighting, 225, 232
lighting fixtures
fixture stud, 424
mounting, 274, 275, 424
pendant-type, 314, 316, 318
shapes of, 312
lightning arrester, 119, 418
limit switch
defined, 428
thermostat, 344
line breaker, 419
line drop, 44
in parallel circuits, 81–84
in series-parallel circuits, 93–94
liquids, conduction of electricity by, 50–52
living room, lighting, 314, 316, 332

load, 28, 229
  defined, 428
  unbalanced, 100
  wire, 229
load power, 65–66, 86
load resistance, 28
load voltage, increasing, 44
localized lighting, 318, 332
locking relay, 428
lodestone, 1, 2, 33
loom. *See* flexible tubing
low-pressure sodium (LPS) lamps, 203–204
lumen, 186, 210, 218
lumen method of calculating illumination, 217–223
Lumiline lamps, 311, 319
luminosity, 428

M

magnet(s)
  attractive and repulsive forces, 3
  bar, 5, 9
  defined, 428
  earth as, 2
  experiments with, 3–7
  horseshoe, 5, 33
  molecular, 7–9, 16–17
  permanent, 5, 7–9, 30, 33, 430
  temporary, 5, 14
magnetic circuit, 20–25
  defined, 420
  law of, 21
magnetic compass, 1
magnetic energy, 111
magnetic field
  aiding, 9–10
  defined, 5, 428
  distortion of, 10
  electric current and, 13–14
  magnetic energy in, 111
  opposing, 10
  relative motion of, 107–108
  strength of, 3–4, 14
magnetic flux,
  buildup of, 121
  defined, 5, 424, 431
  and leakage reactance, 205
  total, 33
magnetic force lines, 3, 5, 14, 33
magnetic force(s), 2
  attractive vs. repulsive, 4
  unit of, 4
magnetic monopole, 4
magnetic north polarity, 2
magnetic poles, 2–3, 4, 33, 431
magnetic saturation, 9
magnetic south polarity, 2
magnetism
  coercive force, 420
  compasses, 1
  earth's source of, 3
  ferromagnetic substance, 424
  magnetic circuits, 20–25
  electrical power, 30–32
  electrical troubleshooting, 32–33
  and electricity, 1–34
  electromagnetism, 11–17
  magnetic fields, aiding and opposing, 9–10
  magnetic poles, 2–3
  magnets, experiments with, 3–7
  magnetic saturation, 9
  permanent magnets, formation of, 7–9
  retentivity, 9
magnetite, 33
magnetization curve, 22–24
magnetizing force, 22–23, 428
magnetomotive force,
  defined, 20, 428
  and electromotive force compared, 20
manually operated substation, 403, 413
mass, 428
master station, intercoms, 365–366, 378
matter, 429
maximum power transfer, 66
maximum voltage, 129, 151
measuring/monitoring devices
  ammeter, 17, 19, 33, 61–62, 86, 94, 417

frequency meter, 424
galvanometer, 108, 114, 126, 425
gaussmeter, 108
micrometer, 36
ohmmeter, 17, 33, 429
oscilloscope, 430
phase-angle meter, 397–398
potentiometer, 378, 431
power-factor meter, 400
for troubleshooting, 32–33
varmeter, 159
voltmeter, 12, 61–62, 86, 129, 131–132, 436
watt-hour meter, 32, 400
wattmeter, 33, 94, 400, 437
medium lamp bases, 191
megger (megohm meter), 53
mercury-pool arc, low-voltage mercury-vapor lamps, 195, 210
mercury-vapor arc conduction, 50
mercury-vapor lamps, 195, 200–201, 202
metal halide lamps, 202–203
metal plates, 261
meter(s)
  defined, 429
  for underground service, 240
mho, 73, 423, 429
microfarad(s), 163
micrometer, 36
mil, 36, 56
mil-foot, 37
mililambert, 225
miniature lamps, 191, 192, 193
mogul lamp bases, 191
molecular magnets, 7–9, 16–17
monochromatic light sources, 203
mounting
  boxes, 256–262
  circuit breaker, 429
mounting boxes, 255–262
  metal plates, 261
  old work, 256–258
  running boards, 261–262
  running sheathed cable, 259–260
  special mounting devices, 258–259
multimeter, 17, 18
multiple circuits, 69, 86
multiple feeder, 429
multiple joint, 429
multiplier, instrument, 429
music distribution system, 374–375
mutual inductance, 429

N

National Board of Fire Underwriters, 39
National Conference on Standard Electrical Rules, 39
National Electrical Code, 367, 375
  arc-fault circuit interrupters, 331
  changes to, 332
  described, 39
  enforcement of, 94
  flexible metal conduit, 294
  general appliance circuits, 246
  ground fault circuit interrupters, 331
  grounding, electric water heaters, 352
  intercom systems, 367–368, 375–376
  metal plates, use of, 261
  outlets, requirements for, 231
  service drop, clearances, 237
  splicing, 268
  time switches, electric water heaters, 349
  wire size, selection of, 39–43, 44, 94, 227
  wiring standards set by, 235
National Fire Protection Association, 39
natural light, 185
needle gap, 425
negative, defined, 429
negative feeder, 100, 105
negative power, 158
neon glow lamps, 195, 210

neon lamps, 195, 210
network(s)
  active, 429
  active electrical, 418
  defined, 429
  passive, 429
neutral wire, 286
nickel-cadmium cell, 69
night-lights, 311, 324
  automatic, 360
no-load current, defined, 429
nonmetallic sheathed cable, 235
  and armored cable compared, 279
nonprivate intercom installation, 366–367, 378–379
normal operation, 385–386
normally closed, 429
normally open, 429
North Pole, 2
north-seeking pole, 2
nucleus, 11

O

offset bend (conduit), 291
ohm, 122, 423
  defined, 17, 429
  to measure impedance, 161
  to measure reactance, 153
ohmmeter, 17, 33, 429
Ohm's law, 71, 133–136
  in AC circuits, 133–136
  application of to complete and partial circuits, 30
  for a capacitor, 166
  defined, 17–19, 33–34, 429
  to find resistance of a bulb, 25–27
  for an inductive circuit, 155, 157
  for an inductor, 154–155
oil circuit breaker, 386, 388
old houses
  mounting boxes in, 256–257
  new wiring in, 250–251
one watt per square foot, illumination provided by, 231
open circuit, 27–28

open-circuit alarm systems, 360
open-circuit voltage, 430
open-core transformer, 151
open-wire circuit, 430
operator (thermostat), 341
opposing magnetic fields, 10
oscilloscope, 430
out of phase voltage, 139
outdoor lighting, 314
outdoor transformer, 430
outlet(s)
  convenience, 229, 231, 311, 314, 319, 324
  defined, 430
  duplex, 231
  ground, 425, 433
  lighting, 430
  National Electric Code for, 231
  outdoor, 311
  placing, 229, 251–254, 274
  requirements for, 231
  safety, 433
outlet receptacle
  defined, 430
  installing, 271–273, 276
overload, 227
  protection, 430
ozone, 205, 430
ozone-producing lamps, 205–206

P

panelboards, 229
parallel circuits
  branch, 69, 70–71, 86
  capacitance and resistance in, 177–179
  connection of cells, 84–86
  inductance and resistance in, 175–177
  inductance, capacitance, and resistance in, 179–182
  Kirchhoff's current law for, 78–80
  line drop in, 81–84
  loads, 69, 86–87
  multiple, 69, 86
  practical problems in, 80–81

principles of, 69–72, 86–87
resistance, 73
shortcuts for, 72–73
parallel connection of cells, 84–86
paralleling generators, 392–400
parallel-resonant circuit, 179
passive network, 429
peak current, 430
peak power, 136
peak voltage, 129, 151, 430
peak-to-peak value, 430
pendant-type lighting fixtures, 314, 316, 318
  defined, 430
  rise-and-fall, 430
period, 430
permanent magnet
  defined, 4, 5, 30, 33, 430
  experiments with, 5–7
  formation of, 7–9
permeability, 430
perpendicular to the beam, 214
phase, 430–431
phase angle, 139
  defined, 431
  measurement of, 398, 399
phase-angle meter, 397–398
phasing out, 395
photoelectric bell control alarm systems, 359, 360, 378
phototransistor, 378
picture diagrams, 59
Pierce wire-holder insulator, 236, 239
pilot lights, 311
plastic-insulated sheathed cables, 55
plate hanger, 292
plug, 431
plug tip, 435
point sources, 204
point-by-point method of calculating illumination, 213–217
polarity
  defined, 431
  of the load, 137

of the source, 137
voltage, 60–62, 109, 122, 137
polarization (battery), 431
polarized devices, 352, 354, 355
pole, 2–3, 4, 33, 431
polyvinyl chloride (PVC) insulation, 55
porcelain insulators, 51, 52
porcelain-enameled space heater, 337
portable appliances, 311
portable baseboard heater, 337, 340
portable substation, 403, 413
positive, defined, 431
positive feeder, 100, 105
positive power, 158
potential difference, 431
potentiometer, 378, 431
power
  apparent, 159, 162, 417
  consumption, 158
  defined, 431
  effective, 137
  efficiency, 65, 86, 99
  in an impedance, 162–163
  in an inductive circuit, 157–159
  instantaneous, 137, 426
  load, 65–66, 86
  maximum transfer of, 66
  negative, 158
  peak, 136
  positive, 158
  reactive, 159, 162
  real, 159, 431
  in a resistive circuit, 157
  in a series-parallel circuit, 98–99
  source resistance, effect of on, 67
  thermal, 30
  true, 157, 159, 162
power-factor angle, 160–161
power-factor correction, 180–181
power-factor meter, 400
power laws, in resistive AC circuits, 136–137
prefocus lamp bases, 191, 193
primary battery, 66, 431

primary distribution mains, 403
primary distribution system, 400–401
primary windings, transformer, 114, 431
prime mover, 382
protractor, 214
proton, 431
pull box
  conduit as, 284
  defined, 419, 431
pulsating current, 431
puncture, insulation, 419, 432

Q
quick-break, 432
quick-make, 432

R
raceway, 432
rack, cable, 432
radiant energy, classification of, 185
radiant heater, 346–347, 357
radiation
  transfer of heat by, 337
  ultraviolet, 205
rate of charge flow, 18–19
rating, 432
reactance
  capacitive, 182
  defined, 153, 432
  inductive, 121
reactive power, 159, 162
reactor, 153, 166, 432
real power, 159, 431
reamer, 291, 297
receptacle(s). *See also* convenience outlet
  back-wire outlet, 271
  defined, 432
  duplex, 231
  outlet, 271–273, 430
recessed electric heater, 341, 356
recreation rooms, lighting for, 325
rectifier, 432

reduction, 94–98
  by conductances, 98
  by pairs, 97
  by product-and-sum formula, 97
  by reciprocals, 97–98
reentrant wiring, 82–83
reflection, of light from walls and ceiling, 219, 223
regular appliance circuits, 246
regulation, defined, 432
relay(s)
  AC-operated, 344–345
  defined, 432
  control, 421
  locking, 428
  thermal-type, 341
rels, 21, 432
reluctance
  defined, 20, 432
  measurement of, 21–22
  and resistance compared, 20, 153
remanence, 433
repulsive force, 4
residential underground service, 240–241
resistance, 34
  in an AC circuit, 159–161
  of a battery, 63–65, 86
  in conductors, 35, 55
  DC, 150
  defined, 17, 433
  in inductors, 426
  insulators, 35, 53
  internal, 63, 65, 86, 427
  load, 28
  Ohm's law and, 17
  parallel circuit, 73
  and reluctance compared, 20, 153
  specific, 44
  temperature coefficient of, 44–45
  winding, 150
  of a wire, 38, 44, 56
resistivity, 44
resistor,
  defined, 433

instrument multiplier, 429
resistance in, 35
retentivity, 9
return-loop wiring, 82, 83
RF lamps, 330
rheostat, 433
rho, 38
rigid steel conduit, 297, 421
ring-bus system, 391, 393
rod gap, 425
room factor, 218, 223, 227, 228
room index, 218
root-mean-square (rms) value, 422–423
rosette, 433
roughing-in, 254–255, 274
running boards, 261–262

S

saddle bend (conduit), 291
safety outlet, 433
saturation internal, 24
schematic diagrams, 59–60
school intercoms, 375–376
screw-type lamp bases, 191
secondary battery, 66, 433
secondary voltage, reversal of, 122
secondary winding, 114, 148, 150
self-induction, 110–113, 153
semiautomatic substations, 403, 410–411, 413
  automatic supervisory equipment in, 411
  dispatcher, role of, 211
  layout of, 420
  protective devices, 410–411, 413
  starting, 410
semidirect lighting, 225, 232
semidirect lighting luminaires, design data for, 416
semi-indirect lighting, 225, 232
sequence switch, 433
sequence system, 341
series, 25
series circuit, 25, 89
  capacitive reactance and resistance in, 166–171
  defined, 59, 433
  inductance, capacitance, and resistance in, 172–175
  resistance in, 69
  voltage polarities in, 60–62
series-aiding voltage drops, 61
series-opposing voltages, 66–69
series-parallel circuits, 89–106
  branch currents, 89, 105
  connection of cells, 92–93
  current flow in, 89–90
  defined, 89, 105
  horsepower, 99–100, 105
  Kirchhoff's current law, 90–92
  line drop in, 93–94
  power in, 98–99
  reduction of, 94–98
  source voltage, 105
  three-wire distribution, 100–105
  total currents, 89–106, 105
  wattmeter, use of, 94, 105
series-resonant circuit, 175, 183
service, defined, 236, 433
service cable, 429
service cap, 236
service connections, 236–240, 274
service drops, 229, 236, 237
service entrance
  armored cable for, 279–280
  conduit, entry into, 236–237, 274
  installing, 238, 239
  underground cable for, 279–280
service equipment, 423
service feeder, 229, 232
serving, of cable, 433
setscrew connectors, 297
shaded pole, 433
shading coils, 345
shadows, 311, 332
sheathed cable
  defined, 433
  running to boxes, 259–260
shell core, 433
shell-type transformer, 144, 151
short circuit, 433
shunt, 433

signaling devices, alarm systems, 363
signs, service and feeders for, 231
silver cyanide, 52
silver wire, 36
sine wave, 433
sine-wave alternating voltage, 125
single-bus system, generating stations, 385
single-pole switches, 273–274
single-ring conduit hanger, 292
single-strip flexible metal conduit, 294
sinusoidal, 433
six watts per square foot, illumination provided by, 232
sleeve wire, 434
sliding contact, 434
slip connectors, 297
slip rings, 125
smashing point, 191
smoke alarms, 360
snake
　defined, 434
　use of, 293
socket, 428, 434
sodium lamps, 195, 201–202, 210
　high-pressure (HPS), 203
　low-pressure (LPS), 203–204
Soffit lighting, 319
solderless connectors
　defined, 434
　marking, 271
　use of, 268, 270–271
solenoid, 16, 434
source voltage, 62, 105
south-seeking pole, 2
space heater
　defined, 337
　types of, 337
spare-bus, spare-circuit-breaker system, generating stations, 390
spark coils, 113, 116, 117
　defined, 434
spark gap, 121, 425
sparking voltage, 46

sparkover, 434
sparks, 46–47
specific resistance, 44
sphere gap, 425
splice, defined, 434
splicing, 268–270, 276, 285
splicing sleeve, 434
splitting the atom, 11
spool insulators, 52
spotlights, 224
square mil, 36, 56
　converting to circular mil, 37
squeeze connectors, 296–297
stairway lamp control lighting, 307
standard cell, 434
standard electrical symbols, 59–60, 247–250, 386, 434
static capacitor, 169, 170
steel-sheath space heater, 337
step-down transformer, 144, 386
step-up transformer, 143, 151, 392
storage battery, 66
　capacity of, 68
stranded wires, 43
strobe flicker, 200, 201
substation, 403–411
　automatic, 381, 407–410, 413
　defined, 381, 403
　intercom, 365–366, 378
　layout of, 403, 409
　manually operated, 403, 413
　portable, 403, 413
　semiautomatic, 403, 410–411, 413
　size of, 381
　supervisory controlled, 403, 413
　transformers, 403, 407
sun lamps, 204–205, 210
supervisory controlled substation, 403, 413
surface leakage, 428
surge
　defined, 434
　switching, 122–124
　transient, 435

switch(es)
  air, 417
  back-connected, 418
  benchboard, 418
  contactor, 421
  defined, 434
  direction of, 303
  ground, 425
  horn gap, 426
  isolating, 427
  knife, 28
  limit, 344, 428
  in living rooms, 316
  placing, 254, 276
  planning for, 316
  quick-break, 432
  quick-make, 432
  sequence, 433
  single-pole, 273–274
  three-circuit electrolier, 306–307
  three-way, 301, 304
  throw direction, 303
  two-circuit electrolier, 303, 305
  wiring, color-coding of, 307, 309–310
switching surges, 122–124
symbol(s)
  defined, 434
  standard electrical, 59–60, 247–250, 386
synchronizing, 395
synchronous generator. *See* alternator
synchroscope, 400
system interconnection, 427

T

t section, 437
tap, 435
temperature coefficient of resistance, 44–45
temporary magnet, 5, 14
terminal, 435
terminal voltage, 63
thermal power, 30
thermal-type relay, 341

thermostats, 357
  arcing and, 341
  bimetallic, 341, 356
  defined, 341
  limit switch, 344
  operation of, 341
  operator, 341
  sequence system, 341–343
  thermal-type relay, 341
threaded couplings (conduit), 291
threadless fittings (conduit), 291–292
three-circuit electrolier switch, 306–307
three-phase system, 435
three-way switches, possible circuit arrangements, 301, 304
three-wire distribution circuit, 100–105
tie-bus breaker, 388–389
tie feeder, 435
time delay
  alarm systems, 363
  defined, 435
tip, plug, 435
torque, 435
total magnetic flux, 33
total quantity of electricity, 110
transfer box, 435. *See also* pull box
transformer(s), 113–119
  in AC circuits, 142–148, 151
  AC resistance ratio of, 146
  air-blast, 417
  autotransformer, 418
  capacity, 400, 401
  closed-core, 151
  copper loss, 150
  core-type, 145, 151
  current, 386
  defined, 435
  indoor, 426
  instrument, 386
  open-core, 151
  outdoor, 430
  power rating, 147–148

464 Index

transformer(s) (*continued*)
  shell-type, 144, 151
  step-down, 144, 386
  step-up, 143, 151, 392
  substation, 403, 407
  in sun lamps, 205
  voltage, 386, 390
  windings in, 114, 147, 148, 150
transformer action, 114, 115–116
transient, 435
transmission, defined, 435
transmission line, 400
trapeze hanger, 292
troubleshooting, quick-check instruments for, 32–33
troughing, 435
true power, 157, 159, 162
tubing
  electrical metallic (EMT), 235, 284, 291–292, 297, 435
  flexible, 435
tubular lamps, 311
tungsten filaments, 188–189, 208, 210
turbines, 382, 412
  cutaway view of, 384
twin cable, 436
twisted pairs, 371
two watts per square foot, illumination provided by, 231–232
two-circuit electrolier switch, 303, 305
type AC cable, 235, 265, 279, 280
type ACT cable, 279
type ACV cable, 279
type ASE cable, 279
type BX cable, 235
type NM cable, 235, 259–260, 262, 265
type UF cable, 262, 419
type USE cable, 279

U

ultraviolet germicidal lamps, 195, 210
ultraviolet light, 48–49
ultraviolet radiation, 205
unbalanced load, 100
underground cable, 436
underground service, residential, 240–241, 242
undergrounded system, 436
Underwriter's Laboratories, Inc., 39
unidirectional current, 436
unit of magnetic force, 4
units, defined, 436

V

vacuum circuit breaker, 386, 389
value, defined, 436
vapor-discharge lamps, 195–202, 210
  ballast in, 197
  classification of, 195
  fluorescent, 195, 199–200, 210
  mercury-vapor, 200–201, 202, 210
  neon, 195, 210
  neon glow, 195, 210
  power loss, minimizing, 197, 199
  sodium, 195, 197, 210
  starters, 199-200
  types of, 195
  ultraviolet germicidal, 195, 210
vars (volt-amperes reactive), 159, 162, 164, 436
variable component, 436
variable inductor, 153
variable reactor, 153
varmeter, 159
vector, 134, 138, 436
ventilated component, 436
vibrator, 436
vibrator coils. *See* spark coils
volt(s),17, 423
  defined, 12, 436
voltage, 175
  AC, combining, 137–142. *See also* alternating current (AC)
  applied, 121, 182
  closed-circuit, 420

crest, 129, 151
DC, 132, 137
defined, 11, 34, 436
direct, 422
divider, 422
effective, 129, 151
generators, 382
induced, 121
inductive reactance and, 121
instantaneous, 129, 137, 151
Kirchhoff's law, 30, 61
load, increasing, 44
maximum, 129, 151
measurements with respect to ground, 62–63
Ohm's law and, 17
open-circuit, 430
out of phase, 139
peak, 129, 151, 430
polarity, 60–62, 109, 122, 137
secondary, reversal of, 122
self-induced, 111, 112, 154
series-opposition, 66–69
sine-wave, 138
sine-wave alternating, 125
source, 62, 105
source of, 30
sparking, 46
terminal, 63
working, 437
zero-voltage level, 437
voltage drop
 defined, 422
 Ohm's law and, 30
 safety considerations, 227
 series-aiding, 61
 symbol indicating, 63
 voltage polarities and, 60
 and wire resistance, 44
voltage magnification, 175
voltage rise, 175
voltage transformer, 386, 390
volt-amperes, 162, 165
volt-amperes reactive. *See* vars
voltmeter
 AC, 129, 131–132
 connecting, 12, 61–62, 86

 defined, 436
 measuring dry cell voltage with, 17
 testing a battery with, 63
 use of, 94

W

wall heater, 337, 338
wall switches, placing, 254
water solution, 50
watt(s), 99, 162, 165
 defined, 31, 423, 436
watt-hour, 32, 423, 436
watt-hour meter, 32, 400
wattmeter
 defined, 437
 use of, 33, 94, 400
watt-second, 32, 423, 427. *See also* joule
watts-per-square-foot basis of calculating illumination, 213, 231–232
wave, defined, 437
wavelengths of light
 eye sensitivity for, 185, 186
 measurement of, 185
weatherproof, defined, 437
white insulated wire, 286, 307
wind bracing, 437
winding resistance, 150
windings (transformer)
 primary, 114, 431
 secondary, 114, 148, 150
wiper, 437
wire(s)
 aluminum, 43–44
 American Wire Gauge, 39–43
 black insulated, 286, 307
 as conductors, 36–37, 56
 copper, 36, 40–42, 56
 current-carrying capacity, 227–229
 diameter, measuring, 36–37
 grounded, 286
 guy, 425
 insulators for support of, 52–53
 line drop and, 44

wire(s) (*continued*)
  load and length of run, 229
  National Electric Code requirements for, 39–43, 44, 94, 227
  neutral, 286
  raceway, 432
  resistance of, 38, 44–45, 56
  silver, 36
  size of, required, 230
  size of, selecting, 39, 44
  sleeve, 434
  stranded, 43
  used as conductors, 35–36, 56
  white insulated, 286, 307
wire capacity, 227–229
wire nut, 437
wireless alarm systems, 365
wireless intercom systems, 365, 371–374, 379
wiring
  basic house. *See* house wiring
  bells and chimes, 376
  checking connections in, 33
  color-coded switch, 307, 309–310
  common methods used in residential work, 235
  conduit, 292–293
  diagrams, home circuiting, 312, 317
  electric range, 320
  in flexible conduit, 293–296
  garage door openers, 326
  inspection of, 229
  installation, codes for, 227, 235, 274
  for intercom system, 368, 371
  National Electric Code standards, 235
  new homes, 235
  older homes, 235, 236
  reentrant, 82–83
  return-loop, 82, 83
wiring diagram, 286
work, defined, 437
workbenches, illumination of, 324
working voltage, 437

X

x ray
  defined, 437
  and image-retaining panels, 209
xenon, 206

Y

y joint, 437
y section, 437

Z

zero power factor, 174
zero resistance, windings, 150
zero-adjuster, 437
zero-voltage level, 437